**Lecture Notes in Statistics – Proceedings** **198**

Edited by P. Bickel, P. Diggle, S. Fienberg, U. Gather,
I. Olkin, S. Zeger



For further volumes:
http://www.springer.com/series/694

# Foreword

The workshop on "Copula Theory and Its Applications" took place at the Faculty of Mathematics, Informatics, and Mechanics at the University of Warsaw (Poland), in the period of 25th–26th September 2009. It gathered 72 participants from 18 countries, inside and outside Europe. Inspired by the nice atmosphere of an old and prestigious city in Central Europe, all the participants have been actively involved in interesting discussions and stimulating talks concerning copula theory and its applications.

The workshop was preceded by a short course on "Joint Extremes, Copulae and CDO Valuation" organized by Prof. Wolfgang Härdle from Humboldt University of Berlin (Germany), which was particularly devoted to present some challenging ideas concerning applications of copulas in finance.

As members of the Organizing Committee of the workshop, it is a great pleasure for us to present this volume collecting results and achievements discussed by the participants. It is another confirmation that this event has been fruitful for further scientific developments.

Therefore, the we would like to express our special thanks to all the participants for their delightful combination of scholarly inquiry and cheerful conviviality which confirm copula theory being such a pleasant area of research.

We also would like to acknowledge for their support the institutional organizers of the workshop: the Polish Mathematical Society; the Faculty of Mathematics, Informatics and Mechanics at the University of Warsaw; the Institute of Mathematics of Polish Academy of Sciences; the Stefan Banach International Mathematical Center. Moreover, we are honored to be supported by the Ministry of Science and Higher Education of the Republic of Poland.

The attendance of specialists from various research groups around the word, as well as the support of the institutional organizers and sponsors made the workshop a successful meeting.

| | |
|---|---|
| Warszawa, Poland | *Piotr Jaworski* |
| Linz, Austria | *Fabrizio Durante* |
| Warszawa, Poland | *Krystyna Jaworska* |
| Toruń, Poland | *Tomasz Rychlik* |
| January 2010 | |

# Preface

Copulas are mathematical objects that fully capture the dependence structure among random variables and hence, offer a great flexibility in building multivariate stochastic models. Since their introduction in the early 50s, copulas have gained a lot of popularity in several fields of applied mathematics, like finance, insurance and reliability theory. Nowadays, they represent a well-recognized tool for market and credit models, aggregation of risks, portfolio selection, etc.

Moreover, such a large interest in the applications of copulas have spurred researchers and scientists in investigating and developing new theoretical methods and tools for handling uncertainty in practical situations.

The workshop on "Copula Theory and Its Applications", which took place in Warsaw (Poland) on 25th–26th September 2009, represented a good opportunity for intensive exchange of ideas about recent developments and achievements that can contribute to the general development of the field. The talks presented at this event have focused on several interesting theoretical problems as well as empirical applications, especially to finance, insurance and reliability.

In order to make all these contributions available to a larger audience, we have planned to prepare a volume collecting selected contributions of the workshops and several original survey papers about important aspects of copula theory and its applications.

The book is divided into two main parts: Part I Surveys contains 11 manuscripts giving an up-to-date account of some aspects of copula models. Part II Contributions collects an extended version of 6 talks selected from presented at the workshop in Warsaw.

Our special thanks go to the authors for their willingness to contribute to this volume, and to our colleagues Erich Peter Klement, Radko Mesiar, José Juan Quesada Molina, Carlo Sempi, and Fabio Spizzichino, who, as members of the Scientific Committee of the workshop, contributed to the scientific success of this event.

Every paper has been submitted to, at least, one referee: we want to thank all of them for their collaboration.

The professional work of the Organizing Committee was greatly appreciated, as well as the support of the co-sponsors of this conference.

Finally, we are indebted to our publisher Springer Verlag and prepublisher Integra Software Services, in particular to Dr. Niels Peter Thomas, Alice Blanck and Sharmila Krishnamurthy for their assistance in the editorial process.

| | |
|---|---|
| Linz, Austria | *Fabrizio Durante* |
| Berlin, Germany | *Wolfgang Härdle* |
| Warszawa, Poland | *Piotr Jaworski* |
| Toruń, Poland | *Tomasz Rychlik* |
| January 2010 | |

# Contents

**Part I Surveys**

**1 Copula Theory: An Introduction** .... 3
Fabrizio Durante and Carlo Sempi
1.1 Historical Introduction .... 3
1.1.1 Outline .... 6
1.2 Preliminaries on Random Variables and Distribution Functions .... 6
1.3 Copulas: Definitions and Basic Properties .... 9
1.4 Sklar's Theorem .... 12
1.5 Copulas and Random Vectors .... 14
1.6 Families of Copulas .... 15
1.6.1 Elliptical Copulas .... 16
1.6.2 Archimedean Copulas .... 17
1.6.3 EFGM Copulas .... 19
1.7 Constructions of Copulas .... 20
1.7.1 Copulas with Given Lower Dimensional Marginals .... 20
1.7.2 Copula-to-Copula Transformations .... 21
1.7.3 Geometric Constructions of Copulas .... 22
1.8 Copula Theory: What's the Future? .... 23
References .... 24

**2 Dynamic Modeling of Dependence in Finance via Copulae Between Stochastic Processes** .... 33
Tomasz R. Bielecki, Jacek Jakubowski and Mariusz Niewęgłowski
2.1 Introduction .... 33
2.2 Lévy Copulae .... 35
2.3 Semimartingale Copulae .... 39
2.3.1 Copulae for Special Semimartingales .... 39
2.3.2 Consistent Semimartingale Copulae .... 48
2.4 Markov Copulae .... 54
2.4.1 Consistent Markov Processes .... 55

2.4.2 Markov Copulae: Generator Approach ................ 57
2.4.3 Markov Copulae: Symbolic Approach ................ 63
2.5 Applications in Finance ................................ 69
2.5.1 Pricing Rating-Triggered Step-Up Bonds via Simulation . 70
2.5.2 Model Calibration and Pricing ...................... 72
References ................................................. 75

**3 Copula Estimation** ........................................ 77
Barbara Choroś, Rustam Ibragimov and Elena Permiakova
3.1 Introduction ........................................... 77
3.2 Copula Estimation: Random Samples with Dependent Marginals . 78
3.2.1 Parametric Models: Maximum Likelihood Methods and Inference from Likelihoods for Margins ........... 78
3.2.2 Semiparametric Estimation ......................... 80
3.2.3 Nonparametric Inference and Empirical Copula Processes 81
3.3 Copula-Based Time Series and Their Estimation ............... 82
3.3.1 Copula-Based Characterizations for (Higher-Order) Markov Processes ................................ 82
3.3.2 Parametric and Semiparametric Copula Estimation Methods for Markov Processes ...................... 83
3.3.3 Nonparametric Copula Inference for Time Series ....... 84
3.3.4 Dependence Properties of Copula-Based Time Series ... 85
3.4 Further Copula Inference Methods .......................... 86
3.5 Empirical Applications .................................... 87
References ................................................. 89

**4 Pair-Copula Constructions of Multivariate Copulas** ............... 93
Claudia Czado
4.1 Introduction ........................................... 93
4.2 Pair Copula Constructions of D-Vine, Canonical and Regular Vine Distributions ........................................ 94
4.2.1 Pair-Copula Constructions of D-Vine and Canonical Vine Distributions ............................... 94
4.2.2 Regular Vines Distributions and Copulas .............. 96
4.3 Estimation Methods for Regular Vine Copulas ................. 100
4.4 Model Selection Among Vine Specifications .................. 103
4.5 Applications of Vine Distributions ......................... 105
4.6 Summary and Open Problems ................................ 106
References ................................................. 107

**5 Risk Aggregation** ........................................... 111
Paul Embrechts and Giovanni Puccetti
5.1 Motivations and Preliminaries .............................. 112
5.1.1 The Mathematical Framework ....................... 112
5.2 Bounds for Functions of Risks: The Coupling-Dual Approach .... 113
5.2.1 Application 1: Bounding Value-at-Risk ............... 115

5.2.2 Application 2: Supermodular Functions . . . . . 119
5.3 The Calculation of the Distribution of the Sum of Risks . . . . . 120
5.3.1 Open Problems . . . . . 124
References . . . . . 125

**6 Extreme-Value Copulas** . . . . . 127
Gordon Gudendorf and Johan Segers
6.1 Introduction . . . . . 127
6.2 Foundations . . . . . 128
6.3 Parametric Models . . . . . 131
6.3.1 Logistic Model or Gumbel–Hougaard Copula . . . . . 132
6.3.2 Negative Logistic Model or Galambos Copula . . . . . 132
6.3.3 Hüsler–Reiss Model . . . . . 133
6.3.4 The t-EV Copula . . . . . 134
6.4 Dependence Coefficients . . . . . 134
6.5 Estimation . . . . . 136
6.5.1 Parametric Estimation . . . . . 137
6.5.2 Nonparametric Estimation . . . . . 138
6.6 Further Reading . . . . . 140
References . . . . . 141

**7 Construction and Sampling of Nested Archimedean Copulas** . . . . . 147
Marius Hofert
7.1 Introduction . . . . . 147
7.2 Nested Archimedean Copulas . . . . . 149
7.3 A Sufficient Nesting Condition . . . . . 151
7.4 Construction of Nested Archimedean Copulas . . . . . 153
7.5 Sampling Nested Archimedean Copulas . . . . . 155
7.6 Conclusion . . . . . 159
References . . . . . 159

**8 Tail Behaviour of Copulas** . . . . . 161
Piotr Jaworski
8.1 Introduction . . . . . 161
8.2 Tail Expansions of Copulas . . . . . 163
8.2.1 Characterization and Properties of Leading Parts . . . . . 167
8.2.2 Relatively Invariant Measures on $[0,\infty)^n$ . . . . . 168
8.3 Examples of Tail Expansions . . . . . 169
8.3.1 Homogeneous Copulas . . . . . 169
8.3.2 Diagonal Copulas . . . . . 169
8.3.3 Absolutely Continuous Copulas . . . . . 171
8.3.4 Archimedean Copulas . . . . . 172
8.3.5 Multivariate Extreme Value Copulas . . . . . 177
8.4 Applications . . . . . 178
8.4.1 Tail Conditional Copulas . . . . . 178
8.4.2 Extreme Value Copulas of a Given Copula . . . . . 180

8.4.3 Regularly Varying Random Vectors with a Given Copula 181
8.4.4 Value at Risk . . . . . . . . . . 182
References . . . . . . . . . . 185

**9 Copulae in Reliability Theory (Order Statistics, Coherent Systems)** . 187
Tomasz Rychlik
9.1 Coherent Systems . . . . . . . . . . 187
9.2 Signatures . . . . . . . . . . 189
9.2.1 Components with i.i.d. Lifetimes . . . . . . . . . . 189
9.2.2 Mixed Systems . . . . . . . . . . 190
9.2.3 Components with Exchangeable Lifetimes . . . . . . . . . . 192
9.3 Bounds for Exchangeable Lifetime Components . . . . . . . . . . 194
9.3.1 Distribution Bounds . . . . . . . . . . 194
9.3.2 Expectation Bounds . . . . . . . . . . 196
9.4 Characterizations of $k$-Out-of-$n$ System Lifetime Distributions . . . 198
9.4.1 General Copula Joint Distribution . . . . . . . . . . 199
9.4.2 Absolute Continuous Copula Joint Distribution . . . . . . . . . . 200
9.4.3 Variance Bounds . . . . . . . . . . 203
9.5 Final Remarks . . . . . . . . . . 205
References . . . . . . . . . . 206

**10 Copula-Based Measures of Multivariate Association** . . . . . . . . . . 209
Friedrich Schmid, Rafael Schmidt, Thomas Blumentritt, Sandra Gaißer and Martin Ruppert
10.1 Introduction and Definitions . . . . . . . . . . 209
10.2 Aspects of Multivariate Association . . . . . . . . . . 212
10.3 Multivariate Generalizations of Spearman's Rho, Kendall's Tau, Blomqvist's Beta, and Gini's Gamma . . . . . . . . . . 215
10.3.1 Spearman's Rho . . . . . . . . . . 215
10.3.2 Kendall's Tau . . . . . . . . . . 217
10.3.3 Blomqvist's Beta . . . . . . . . . . 219
10.3.4 Gini's Gamma . . . . . . . . . . 220
10.4 Information-Based Measures of Multivariate Association . . . . . . . . . . 221
10.5 Measures of Multivariate Association Based on $L_p$-Distances . . . . 224
10.5.1 $\Phi^2$ as a $L_2$-Distance-Based Measure . . . . . . . . . . 225
10.5.2 $\sigma$ as a $L_1$-Distance-Based Measure . . . . . . . . . . 227
10.5.3 $\kappa$ as a $L_\infty$-Distance-Based Measure . . . . . . . . . . 227
10.6 Multivariate Tail Dependence . . . . . . . . . . 228
References . . . . . . . . . . 232

**11 Semi-Copulas and Interpretations of Coincidences Between Stochastic Dependence and Ageing** . . . . . . . . . . 237
Fabio Spizzichino
11.1 Introduction . . . . . . . . . . 237
11.2 Univariate Ageing and Dependence Properties of Archimedean Semi-Copulas . . . . . . . . . . 240

11.3 Dependence and Univariate Ageing in Schur-Constant Models . . . 243
11.4 Level Curves, B functions, Duality, and Interpretation of Coincidence Between Ageing and Dependence . . . . . . . . . . . . . . . . 247
11.5 Summary and Concluding Remarks . . . . . . . . . . . . . . . . . . . . . . . . . . 251
References . . . . . . . . . . . . . . . . . . . . . . . . . . . . . . . . . . . . . . . . . . . . . . . 252

**Part II Contributed Papers**

**12 A Copula-Based Model for Spatial and Temporal Dependence of Equity Markets** . . . . . . . . . . . . . . . . . . . . . . . . . . . . . . . . . . . . . . . . . . . . 257
Umberto Cherubini, Fabio Gobbi, Sabrina Mulinacci and Silvia Romagnoli
12.1 Introduction . . . . . . . . . . . . . . . . . . . . . . . . . . . . . . . . . . . . . . . . . . . . 257
12.2 A market Model in Discrete Time . . . . . . . . . . . . . . . . . . . . . . . . . . . . 258
12.3 The Martingale Property . . . . . . . . . . . . . . . . . . . . . . . . . . . . . . . . . . . 259
12.4 Applications . . . . . . . . . . . . . . . . . . . . . . . . . . . . . . . . . . . . . . . . . . . 261
12.4.1 Multivariate Digital Options . . . . . . . . . . . . . . . . . . . . . . . . . . 261
12.4.2 Basket and Spread Options . . . . . . . . . . . . . . . . . . . . . . . . . . . 263
References . . . . . . . . . . . . . . . . . . . . . . . . . . . . . . . . . . . . . . . . . . . . . . . 264

**13 Nonparametric and Semiparametric Bivariate Modeling of Petrophysical Porosity-Permeability Dependence from Well Log Data** . . . . . . . . . . . . . . . . . . . . . . . . . . . . . . . . . . . . . . . . . . . . . . . . . . . . 267
Arturo Erdely and Martin Diaz-Viera
13.1 Introduction . . . . . . . . . . . . . . . . . . . . . . . . . . . . . . . . . . . . . . . . . . . . 267
13.2 Methodology . . . . . . . . . . . . . . . . . . . . . . . . . . . . . . . . . . . . . . . . . . . 268
13.3 Data Analysis . . . . . . . . . . . . . . . . . . . . . . . . . . . . . . . . . . . . . . . . . . 271
13.4 Final Remarks . . . . . . . . . . . . . . . . . . . . . . . . . . . . . . . . . . . . . . . . . . 276
References . . . . . . . . . . . . . . . . . . . . . . . . . . . . . . . . . . . . . . . . . . . . . . . 278

**14 Testing Under the Extended Koziol-Green Model** . . . . . . . . . . . . . . . . . 279
Auguste Gaddah and Roel Braekers
14.1 Introduction . . . . . . . . . . . . . . . . . . . . . . . . . . . . . . . . . . . . . . . . . . . . 279
14.2 Asymptotic Results . . . . . . . . . . . . . . . . . . . . . . . . . . . . . . . . . . . . . . 282
14.3 Test Statistics . . . . . . . . . . . . . . . . . . . . . . . . . . . . . . . . . . . . . . . . . . . 285
14.4 Data Example: Survival with Malignant Melanoma . . . . . . . . . . . . 286
References . . . . . . . . . . . . . . . . . . . . . . . . . . . . . . . . . . . . . . . . . . . . . . . 288

**15 Parameter Estimation and Application of the Multivariate Skew *t*-Copula** . . . . . . . . . . . . . . . . . . . . . . . . . . . . . . . . . . . . . . . . . . . . . . . . . . 289
Tõnu Kollo and Gaida Pettere
15.1 Introduction . . . . . . . . . . . . . . . . . . . . . . . . . . . . . . . . . . . . . . . . . . . . 289
15.2 Preliminary Notions and Notation . . . . . . . . . . . . . . . . . . . . . . . . . . 290
15.3 Construction of a Skew *t*-Copula . . . . . . . . . . . . . . . . . . . . . . . . . . . 292
15.4 Parameter Estimation . . . . . . . . . . . . . . . . . . . . . . . . . . . . . . . . . . . . 293
15.5 Simulation . . . . . . . . . . . . . . . . . . . . . . . . . . . . . . . . . . . . . . . . . . . . . 295

15.6 Application .......... 296
References .......... 298

**16 On Analytical Similarities of Archimedean and Exchangeable Marshall-Olkin Copulas** .......... 299
Jan-Frederik Mai and Matthias Scherer
16.1 Introduction .......... 299
16.2 Complete Monotonicity and $d$-Monotonicity .......... 301
16.2.1 Definitions and Examples .......... 301
16.2.2 Probabilistic Interpretations .......... 302
16.2.3 d-Monotonicity .......... 303
16.3 Probabilistic Models and Sampling .......... 305
16.3.1 The Completely Monotone Case .......... 305
16.3.2 The Proper $d$-Monotone Case .......... 306
References .......... 308

**17 Relationships Between Archimedean Copulas and Morgenstern Utility Functions** .......... 311
Jaap Spreeuw
17.1 Introduction .......... 311
17.2 Archimedean Copulas .......... 312
17.3 Utility Functions .......... 312
17.4 Relationships Between Properties of Utility Functions and Properties of Generators .......... 315
17.5 Examples .......... 318
17.5.1 Classical Cases .......... 318
17.5.2 The HARA Family .......... 319
17.5.3 The Expo Power Utility .......... 319
17.5.4 Other Examples of Decreasing Absolute Risk Aversion (DARA) as in Pratt [9] .......... 319
17.6 Conclusion .......... 321
References .......... 321

**Index** .......... 323

# Contributors

Tomasz R. Bielecki
Department of Applied Mathematics, Illinois Institute of Technology, Chicago, IL, USA, e-mail: bielecki@iit.edu

Thomas Blumentritt
Department of Economic and Social Statistics, University of Cologne, Cologne, Germany, e-mail: blumentritt@wiso.uni-koeln.de

Roel Braekers
Interuniversity Institute for Biostatistics and Statistical Bioinformatics, Universiteit Hasselt, Diepenbeek, Belgium, e-mail: roel.braekers@uhasselt.be

Umberto Cherubini
Department of Mathematical Economics, University of Bologna, Bologna, Italy, e-mail: umberto.cherubini@unibo.it

Barbara Choroś
Institute for Statistics and Econometrics, Humboldt-Universität zu Berlin, Berlin, Germany, e-mail: barbara.choros@wiwi.hu-berlin.de

Claudia Czado
Department of Mathematics, Technische Universität München, Garching, Germany, e-mail: cczado@ma.tum.de

Martín Díaz-Viera
Programa de Investigación en Recuperación de Hidrocarburos, Instituto Mexicano del Petróleo, México D.F., México, e-mail: mdiazv@imp.mx

Fabrizio Durante
Department of Knowledge-Based Mathematical Systems, Johannes Kepler University Linz, Linz, Austria, e-mail: fabrizio.durante@jku.at

Paul Embrechts
Department of Mathematics, ETH Zurich, Zurich, Switzerland, e-mail: embrechts@math.ethz.ch

Arturo Erdely
Programa de Actuaría, División de Matemáticas e Ingeniería Facultad de Estudios Superiores Acatlán Universidad Nacional Autónoma de México e-mail: arturoerdely@yahoo.com.mx

Auguste Gaddah
Interuniversity Institute for Biostatistics and Statistical Bioinformatics, Universiteit Hasselt, Diepenbeek, Belgium, e-mail: auguste.gaddah@uhasselt.be

Sandra Gaißer
Department of Economic and Social Statistics, University of Cologne, Cologne, Germany, e-mail: sandra.gaisser@uni-koeln.de

Fabio Gobbi
Department of Mathematical Economics, University of Bologna, Bologna, Italy, e-mail: fabio.gobbi@unibo.it

Gordon Gudendorf
Institut de statistique, Université catholique de Louvain, Louvain-la-Neuve, Belgium, e-mail: gordon.gudendorf@uclouvain.be

Marius Hofert
Institute of Number Theory and Probability Theory, Ulm University, 89081 Ulm, Germany, e-mail: marius.hofert@uni-ulm.de

Rustam Ibragimov
Department of Economics, Harvard University, Cambridge, MA, USA, e-mail: ribragim@fas.harvard.edu

Jacek Jakubowski
Institute of Mathematics, University of Warsaw, Warszawa, Poland; Faculty of Mathematics and Information Science, Warsaw University of Technology, Warszawa, Poland, e-mail: jakub@mimuw.edu.pl

Piotr Jaworski
Institute of Mathematics, University of Warsaw, Warszawa, Poland, e-mail: P.Jaworski@mimuw.edu.pl

Tõnu Kollo
Institute of Mathematical Statistics, University of Tartu, Tartu, Estonia, e-mail: tonu.kollo@ut.ee

Jan-Frederik Mai
HVB-Institute for Mathematical Finance, Technische Universität München, Garching, Germany, e-mail: mai@tum.de

Sabrina Mulinacci
Department of Mathematical Economics, University of Bologna, Bologna, Italy, e-mail: sabrina.mulinacci@unibo.it

Mariusz Niewęgłowski
Faculty of Mathematics and Information Science, Warsaw University of Technology, Warszawa, Poland, e-mail: M.Nieweglowski@mini.pw.edu.pl

Elena Permiakova
N. G. Chebotarev Research Institute of Mathematics and Mechanics, Kazan State University, Kazan, Russia, e-mail: epermiakova@gmail.com

Gaida Pettere
Department of Engineering Mathematics, Riga Technical University, Riga, Latvia, e-mail: gaida@latnet.lv

Giovanni Puccetti
Department of Mathematics for Decisions, University of Firenze, Firenze, Italy, e-mail: giovanni.puccetti@dmd.unifi.it

Silvia Romagnoli
Department of Mathematical Economics, University of Bologna, Bologna, Italy, e-mail: silvia.romagnoli@unibo.it

Martin Ruppert
Graduate School of Risk Management, University of Cologne, Cologne, Germany, e-mail: ruppert@wiso.uni-koeln.de

Tomasz Rychlik
Institute of Mathematics, Polish Academy of Sciences, Toruń, Poland, e-mail: trychlik@impan.gov.pl

Matthias Scherer
HVB-Institute for Mathematical Finance, Technische Universität München, Garching, Germany, e-mail: scherer@tum.de

Friedrich Schmid
Department of Economic and Social Statistics, University of Cologne, Cologne, Germany, e-mail: schmid@wiso.uni-koeln.de

Rafael Schmidt
Risk Control, Bank for International Settlements, Basel, Switzerland, e-mail: rafael.schmidt@gmx.de

Johan Segers
Institut de statistique, Université catholique de Louvain, Louvain-la-Neuve, Belgium, e-mail: johan.segers@uclouvain.be

Carlo Sempi
Dipartimento di Matematica "Ennio De Giorgi", Università del Salento, Lecce, Italy, e-mail: carlo.sempi@unisalento.it

Fabio Spizzichino
Department of Mathematics, University La Sapienza, Rome, Italy, e-mail: Fabio.Spizzichino@uniroma1.it

Jaap Spreeuw
Cass Business School, City University London, London, UK, e-mail: J.Spreeuw@city.ac.uk

# Part I
# Surveys

# Chapter 1
# Copula Theory: An Introduction

Fabrizio Durante and Carlo Sempi

**Abstract** In this survey we review the most important properties of copulas, several families of copulas that have appeared in the literature, and which have been applied in various fields, and several methods of constructing multivariate copulas.

## 1.1 Historical Introduction

The history of copulas may be said to begin with Fréchet [70]. He studied the following problem, which is stated here in dimension 2: given the distribution functions $F_1$ and $F_2$ of two random variables $X_1$ and $X_2$ defined on the same probability space $(\Omega, \mathscr{F}, \mathbb{P})$, what can be said about the set $\Gamma(F_1, F_2)$ of the bivariate d.f.'s whose marginals are $F_1$ and $F_2$? It is immediate to note that the set $\Gamma(F_1, F_2)$, now called the *Fréchet class of* $F_1$ and $F_2$, is not empty since, if $X_1$ and $X_2$ are independent, then the distribution function $(x_1, x_2) \mapsto F(x_1, x_2) = F_1(x_1) F_2(x_2)$ always belongs to $\Gamma(F_1, F_2)$. But, it was not clear which the other elements of $\Gamma(F_1, F_2)$ were.

Preliminary studies about this problem were conducted in [65, 71, 90] (see also [31, 182] for a historical overview). But, in 1959, Sklar obtained the deepest result in this respect, by introducing the notion, and the name, of a *copula*, and proving the theorem that now bears his name [192]. In his own words [194]:

---

Fabrizio Durante
Department of Knowledge-Based Mathematical Systems, Johannes Kepler University Linz, Linz Austria
e-mail: `fabrizio.durante@jku.at`

Carlo Sempi
Dipartimento di Matematica "Ennio De Giorgi", Università del Salento, Lecce, Italy
e-mail: `carlo.sempi@unisalento.it`

P. Jaworski et al. (eds.), *Copula Theory and Its Applications*, Lecture Notes in Statistics 198,
DOI 10.1007/978-3-642-12465-5_1, 

> [...] In the meantime, Bert (Schweizer) and I had been making progress in our work on statistical metric spaces, to the extent that Menger suggested it would be worthwhile for us to communicate our results to Fréchet. We did: Fréchet was interested, and asked us to write an announcement for the Comptes Rendus [184]. This began an exchange of letters with Fréchet, in the course of which he sent me several packets of reprints, mainly dealing with the work he and his colleagues were doing on distributions with given marginals. These reprints, among the later arrivals of which I particularly single out that of Dall'Aglio [29], were important for much of our subsequent work. At the time, though, the most significant reprint for me was that of Féron [65].
>
> Féron, in studying three-dimensional distributions had introduced auxiliary functions, defined on the unit cube, that connected such distributions with their one-dimensional margins. I saw that similar functions could be defined on the unit $n$-cube for all $n \geq 2$ and would similarly serve to link $n$-dimensional distributions to their one-dimensional margins. Having worked out the basic properties of these functions, I wrote about them to Fréchet, in English. He asked me to write a note about them in French. While writing this, I decided I needed a name for these functions. Knowing the word "copula" as a grammatical term for a word or expression that links a subject and predicate, I felt that this would make an appropriate name for a function that links a multidimensional distribution to its one-dimensional margins, and used it as such. Fréchet received my note, corrected one mathematical statement, made some minor corrections to my French, and had the note published by the Statistical Institute of the University of Paris as Sklar [192].

The proof of Sklar's theorem was not given in [192], but a sketch of it was provided in [193] (see also [185]), so that for a few years practitioners in the field had to reconstruct it relying on the hand-written notes by Sklar himself; this was the case, for instance, of the second author. It should be also mentioned that some "indirect" proofs of Sklar's theorem (without mentioning copula) were later discovered by Moore and Spruill [145] and Deheuvels [37]

For about 15 years, all the results concerning copulas were obtained in the framework of the theory of Probabilistic Metric spaces [186]. The event that arose the interest of the statistical community in copulas occurred in the mid seventies, when Bert Schweizer, in his own words (see [183]),

> quite by accident, reread a paper by A. Rényi, entitled *On measures of dependence* and realized that [he] could easily construct such measures by using copulas.

See [166] for Rényi's paper. The first building blocks were the announcement by Schweizer and Wolff in the *Comptes Rendus de l'Académie des Sciences* [187] and Wolff's Ph.D. Dissertation at the University of Massachusetts at Amherst [200]. These results were presented to the statistical community in the paper [188] (compare also with [201]).

However, for several other years, Chapter 6 of the fundamental book [186] by Schweizer and Sklar, devoted to the theory of Probabilistic metric spaces and published in 1983, was the main source of basic information on copulas. Again in Schweizer's words from [183],

> After the publication of these articles and of the book ... the pace quickened as more ... students and colleagues became involved. Moreover, since interest in questions of statistical dependence was increasing, others came to the subject from different directions. In 1986 the enticingly entitled article *The joy of copulas* by C. Genest and R.C MacKay [82], attracted more attention.

In 1990, Dall'Aglio organized the first conference devoted to copulas, aptly called "Probability distributions with given marginals" [32]. This turned out to be the first in a series of conferences that greatly helped the development of the field, since each of them offered the chance of presenting one's results and to learn those of other researchers; these conferences were held in Seattle in 1993 [176], in Prague in 1996 [11], in Barcelona in 2000 [26], in Québec in 2004 [75, 76], and in Tartu in 2007 [119]; the next one is scheduled to be in São Paulo in 2010.

At end of the nineties, the notion of copulas became increasingly popular. Two books about copulas appeared and were to become the standard references for the following decade. In 1997 Joe published his book on multivariate models [104], with a great part devoted to copulas and families of copulas. In 1999 Nelsen published the first edition of his introduction to copulas [150] (reprinted with some new results in [151]).

But, the main reason of this increased interest has to be found in the discovery of the notion of copulas by researchers in several applied field, like finance. Here we should like briefly to describe this explosion by quoting Embrechts's comments [57]:

> As we have seen so far, the notion of copula is both natural as well as easy for looking at multivariate d.f.'s. But why do we witness such an incredible growth in papers published starting the end of the nineties (recall, the concept goes back to the fifties and even earlier, but not under that name). Here I can give three reasons: finance, finance, finance. In the eighties and nineties we experienced an explosive development of quantitative risk management methodology within finance and insurance, a lot of which was driven by either new regulatory guidelines or the development of new products; see for instance Chapter 1 in [138] for the full story on the former. Two papers more than any others "put the fire to the fuse": the [...] 1998 RiskLab report [58] and at around the same time, the Li credit portfolio model [121].

The advent of copulas in finance [79] originated a wealth of investigations about copulas and, especially, applications of copulas. See, for example, the books [19, 130, 138, 181]. At the same time, different fields like hydrology [77, 177] discovered the importance of this concept for constructing more flexible multivariate models. Nowadays, it is near to impossible to give a complete account of all the applications of copulas to the many fields where they have be used. As Schweizer wrote [183]:

> The "era of i.i.d." is over: and when dependence is taken seriously, copulas naturally come into play. It remains for the statistical community at large to recognize this fact. And when every statistics text contains a section or chapter on copulas, the subject will have come of age.

However, a word of caution is in order here. Several criticisms have been recently raised about copulas and their applications, and several people started to speak about "copula craze" [57]. See, for example, the very interesting discussion related to the paper by Mikosch [140, 141] (see also [56, 85, 94, 105, 126, 161, 189]).

From our point of view, these criticisms were a quite natural reaction to such a wide diffusion of applications of copulas, not always in a well motivated way. It should be said that several people have wrongly interpreted copulas as the solution to "all problems of stochastic dependence". This is definitely not the case! Copulas are an indispensable tool for understanding several problems about stochastic dependence, but they are not the "panacea" for all stochastic models.

Despite this broad range of interest about copulas, we still believe that this concept is still in "its infancy" [151] and several other investigations may (and should) be conducted in order to stress whether copulas, or related copula-based concepts, can be really considered as a "strong" mathematical concept worth of use in several applications (see Sect. 1.8).

### *1.1.1 Outline*

This paper is so organized. Section 1.2 presents some basic notions of probability theory that will be used in the sequel. A basic introduction to copulas is given in Sect. 1.3, while the importance of copulas for stochastic models is illustrated in Sect. 1.4 and 1.5. Families of copulas and construction methods are illustrated in Sect. 1.6. Finally (Section 1.8), we present a discussion about possible open problems in the field.

A final remark should be given here. This survey is intended to be a basic introduction to multivariate copulas, focusing on some theoretical aspects that we consider essential for understanding any application. We do not present any result about the statistical procedure for fitting copulas to data. The reader will find such an information in the rest of this book (see [20, 180]). In writing it, we have tried to provide a list of references as complete as possible. Obviously, it may happen that several important papers have been not cited: we apologize in advance for this.

Other surveys concerning copulas (from various perspectives) can be found as well in the literature; just to make few same examples, we refer the interested reader to [31, 57, 74, 77, 95, 118, 159, 182, 190, 194, 196].

## 1.2 Preliminaries on Random Variables and Distribution Functions

In this section, we recall the bare minimum that is necessary in order to understand the meaning and the use of copula. All this material can be found in standard books on probability theory, like [13, 106, 199].

To begin with, we need to establish basic notation. Let $d \in \mathbb{N}$. In the sequel, $\mathbf{x}$ denotes a vector $(x_1, x_2, \dots, x_d)$ in $\mathbb{R}^d$ (or $\overline{\mathbb{R}}^d = [-\infty, +\infty]^d$). If not otherwise stated, all expressions such as $\min\{\mathbf{x}, \mathbf{y}\}$ or $\mathbf{x} \le \mathbf{y}$ are intended to be componentwise operations. The symbol $\mathbb{I}$ will denote the unit interval $[0,1]$. We shall always use the expressions, "increasing" and "decreasing" in the weak sense; thus, a real function $\varphi$ defined on a subset $(a,b)$ of the real line $\mathbb{R}$, will be said to be increasing (respectively, decreasing) if, for all $x$ and $y$ in $(a,b)$ with $x < y$, one has $\varphi(x) \le \varphi(y)$ (respectively, $\varphi(x) \ge \varphi(y)$). Moreover, a real-valued function $\varphi$ is said to be *posi-*

*tive* (respectively, *strictly positive*) if $\varphi(x) \geq 0$ (respectively, $\varphi(x) > 0$) for every $x$ belonging to the domain of $\varphi$.

A *probability space* is a triplet $(\Omega, \mathscr{F}, \mathbb{P})$, where $\Omega$ is a nonempty set, $\mathscr{F}$ is a $\sigma$-*algebra* of subsets of $\Omega$ and $\mathbb{P}$ is a *probability measure* on $\mathscr{F}$. A $d$-dimensional *random vector* is a measurable mapping $\mathbf{X} : \Omega \to \overline{\mathbb{R}}^d$; in this case, the word measurable means that the counter image $\mathbf{X}^{-1}(B)$ of every Borel set $B$ in $\mathscr{B}(\overline{\mathbb{R}}^d)$ belongs to $\mathscr{F}$. It can be proved that a random vector $\mathbf{X}$ can be represented in the form $\mathbf{X} = (X_1, X_2, \ldots, X_d)$ where, for every $j \in \{1, 2\ldots, d\}$, $X_j$ is a 1-dimensional random vector, also called *random variable*. Usually, the abbreviation r.v. will denote a random vector (possibly, univariate).

When a r.v. $\mathbf{X} = (X_1, X_2, \ldots, X_d)$ is given, two problems are interesting:

- to study the probabilistic behaviour of each one of its components;
- to investigate the relationship among them.

It will be seen how copulas allow to answer the second one of these problems in an admirable and thorough way.

It is a general fact that in probability theory, theorems are proved in the probability space $(\Omega, \mathscr{F}, \mathbb{P})$, while computations are usually carried out in the measurable space $(\overline{\mathbb{R}}^d, \mathscr{B}(\overline{\mathbb{R}}^d))$ endowed with the law of the random vector $\mathbf{X}$. The study of the law $\mathbb{P}_{\mathbf{X}}$ is made easier by the knowledge of the *distribution function*, as defined here.

**Definition 1.2.1.** Given a random vector $\mathbf{X} = (X_1, X_2, \ldots, X_d)$ on the probability space $(\Omega, \mathscr{F}, \mathbb{P})$, its *distribution function* $F_{\mathbf{X}} : \overline{\mathbb{R}}^d \to \mathbb{I}$ is defined by

$$F_{\mathbf{X}}(x_1, x_2, \ldots, x_d) = \mathbb{P}\left(\bigcap_{i=1}^{d} \{X_i \leq x_i\}\right) \tag{1.1}$$

if all the $x_i$'s are in $\mathbb{R}$, while:

(DF1) $F_{\mathbf{X}}(x_1, x_2, \ldots, x_d) = 0$, if at least one of the arguments equals $-\infty$,
(DF2) $F_{\mathbf{X}}(+\infty, +\infty, \ldots, +\infty) = 1$.

Very often, the abbreviation d.f. will be used instead of "distribution function". Sometimes, we also use the term *joint d.f.* for denoting the d.f. of a random vector having, at least, two components.

In order to describe the properties of a d.f., we need to introduce some preliminary notations.

**Definition 1.2.2.** Given two points $\mathbf{a}, \mathbf{b} \in \overline{\mathbb{R}}^d$ with $\mathbf{a} \leq \mathbf{b}$, a *d-box* $[\mathbf{a}, \mathbf{b}]$ (also called *orthotope*) is the cartesian product

$$[\mathbf{a}, \mathbf{b}] = \times_{i=1}^{d} [a_i, b_i], \tag{1.2}$$

Let $A$ be a convex set in $\overline{\mathbb{R}}^d$. For a bounded function $H : A \to \overline{\mathbb{R}}$, the *H-volume* $V_H$ of the $d$-box $[\mathbf{a}, \mathbf{b}] \subseteq A$ is defined by

$$V_H([\mathbf{a},\mathbf{b}]) := \sum_{\mathbf{v}\in \times_{i=1}^d \{a_i,b_i\}} (-1)^{N(\mathbf{v})} H(\mathbf{v}), \tag{1.3}$$

where $N(\mathbf{v}) = \text{card}\{j \mid v_i = a_i\}$, provided that $a_i < b_i$ for every $i \in \{1,2,\dots,d\}$. Otherwise, we set $V_H([\mathbf{a},\mathbf{b}]) = 0$.

**Theorem 1.2.1.** *The d.f. $F_{\mathbf{X}}$ of the r.v. $\mathbf{X} = (X_1, X_2, \dots, X_d)$ has the following properties:*

(DF3) $F_{\mathbf{X}}$ *is* isotonic, *i.e.* $F_{\mathbf{X}}(\mathbf{x}) \le F_{\mathbf{X}}(\mathbf{y})$ *for all* $\mathbf{x},\mathbf{y} \in \mathbb{R}^d$, $\mathbf{x} \le \mathbf{y}$;
(DF4) *for all* $(x_1,\dots,x_{i-1},x_{i+1},\dots,x_d) \in \mathbb{R}^{d-1}$, *the function*

$$\mathbb{R} \ni t \mapsto F_{\mathbf{X}}(x_1,\dots,x_{i-1},t,x_{i+1},\dots,x_d)$$

*is right-continuous;*
(DF5) *the $F_{\mathbf{X}}$-volume $V_{F_{\mathbf{X}}}$ of every d-box $[\mathbf{a},\mathbf{b}]$ is positive, i.e.,* $V_{F_{\mathbf{X}}}([\mathbf{a},\mathbf{b}]) \ge 0$.

Condition (DF5) in the previous theorem is called the *d-increasing property* (or *quasi-monotone property*) of a $d$-dimensional d.f.; it simply means that the probability that the r.v. $\mathbf{X}$ takes values in the box $[\mathbf{a},\mathbf{b}]$ is positive. Analogously, a function $F\colon \overline{\mathbb{R}}^d \to \mathbb{R}$ satisfies (DF5) if, and only if, it has positive finite differences of order $d$, namely if

$$\Delta^d_{a_d,b_d} \cdots \Delta^1_{a_1,b_1} F \ge 0$$

for all $a_i \le b_i$ for every $i \in \{1,2,\dots,d\}$, where $\Delta^i_{a_i,b_i}$ is the finite difference operator given by

$$\Delta^i_{a_i,b_i} F = F(t_1,\dots,t_{i-1},b_i,t_{i+1},\dots,t_d) - F(t_1,\dots,t_{i-1},a_i,t_{i+1},\dots,t_d).$$

Note that, one can prove (see, e.g., [13]) that for every $F\colon \mathbb{R}^d \to \mathbb{I}$ satisfying conditions (DF1)–(DF5), there is a probability space $(\Omega,\mathscr{F},\mathbb{P})$ and a random vector $\mathbf{X}$ on it such that $F$ is the d.f. of $\mathbf{X}$. We will use the symbol $\mathbf{X} \sim F$ in order to denote the fact that $F$ is the d.f. of $\mathbf{X}$ (or, equivalently, $\mathbf{X}$ is distributed according to $F$).

A fundamental notion will be that of *marginal distribution* of a $d$–d.f. $F$.

**Definition 1.2.3.** Let $d \ge 2$ and let $F$ be a $d$-dimensional d.f.. Let $\sigma = (j_1,\dots,j_m)$ a subvector of $(1,2,\dots,d)$, $1 \le m \le d-1$. We call *$\sigma$-marginal* of $F$ the d.f. $F_\sigma\colon \overline{\mathbb{R}}^m \to \mathbb{I}$ defined by setting $d-m$ arguments of $F$ equal to $+\infty$, namely, for every $x_1,\dots,x_m \in \overline{\mathbb{R}}$,

$$F_\sigma(x_1,\dots,x_m) = F(y_1,\dots,y_d), \tag{1.4}$$

where, for every $j \in \{1,2,\dots,d\}$, $y_j = x_j$ if $j \in \{j_1,\dots,j_m\}$, and $y_j = +\infty$ otherwise.

In particular, when $\sigma = \{j\}$, $F_{(j)}$ is usually called 1-*dimensional marginal* and it is denoted by $F_j$.

Obviously, if $F$ is the d.f. of the r.v. $\mathbf{X} = (X_1,X_2,\dots,X_d)$, then the $\sigma$-marginal of $F$ is simply the d.f. of the subvector $(X_{j_1},\dots,X_{j_m})$.

As is well known, if the r.v.'s $X_1, X_2, \dots, X_d$ are independent and if $F_{X_i}$ denotes the d.f. of $X_i$ $(i = 1,2,\dots,d)$, then the $d$-dimensional d.f. of the random vector $\mathbf{X} = (X_1,X_2,\dots,X_d)$ is the product of the marginals, i.e. for all $x_1,x_2,\dots,x_d \in \overline{\mathbb{R}}$,

$$F_{\mathbf{X}}(x_1,x_2,\dots,x_d) = \prod_{i=1}^{d} F_{X_i}(x_i).$$

A random vector $\mathbf{X}$ (equivalently, its d.f. $F_{\mathbf{X}}$) is said to be *absolutely continuous* if there exists a positive measurable function $f_{\mathbf{X}} : \mathbb{R}^d \to \mathbb{R}_+$ (called the *density* of $F$) such that

$$\int_{\mathbb{R}^d} f_{\mathbf{X}} \, d\lambda_d = 1,$$

where $\lambda_d$ is the Lebesgue measure on $\mathbb{R}^d$, and the d.f. $F_{\mathbf{X}}$ can be represented in the form

$$F_{\mathbf{X}}(\mathbf{x}) = \int_{-\infty}^{x_1} dt_1 \int_{-\infty}^{x_2} dt_2 \dots \int_{-\infty}^{x_d} f_X(t_1,t_2,\dots,t_d)\, dt_d. \tag{1.5}$$

For a univariate absolutely continuous d.f. $F$, we shall use the notion $F \sim \mathscr{U}([a,b])$ (read: $F$ is uniformly distributed on $[a,b]$) to denote the fact that $F$ is absolutely continuous with density $f(t) = (b-a)^{-1}\mathbf{1}_{(a,b)}(t)$, where $\mathbf{1}_{(a,b)}$ denoted the indicator function of $(a,b)$.

*Remark 1.2.1.* Let $\mathbf{X}$ be a r.v. such that $\mathbb{P}(\mathbf{X} \in A) = 1$ for some $A \subseteq \mathbb{R}^d$. Then, the d.f. $F_{\mathbf{X}}$ is uniquely determined by the values that it assumes on $A$ and, as a consequence, one usually refrains from specifying the value of $F_{\mathbf{X}}$ outside $A$. For such a situation, we may say shortly that $F_{\mathbf{X}}$ is a d.f. on $A$.

*Remark 1.2.2.* In several contexts (especially, reliability theory), it is more useful to consider the *survival function* $\overline{F}_{\mathbf{X}}$ associated with a given random vector $\mathbf{X} = (X_1,X_2,\dots,X_d)$, where, for every $i \in \{1,2,\dots,d\}$, $X_i \geq 0$ almost surely (i.e., $X_i$ can be interpreted as a "lifetime"), and given by

$$\overline{F}_{\mathbf{X}}(x_1,x_2,\dots,x_d) = \mathbb{P}\left(\bigcap_{i=1}^{d} \{X_i > x_i\}\right). \tag{1.6}$$

The univariate survival marginal of $\overline{F}_{\mathbf{X}}$ are then defined in an analogous way by setting some of the arguments of $\overline{F}$ equal to 0. Note that survival functions are decreasing in each argument.

## 1.3 Copulas: Definitions and Basic Properties

We start with the definition of a copula.

**Definition 1.3.1.** For every $d \geq 2$, a *d-dimensional copula* (shortly, $d$-copula) is a $d$-variate d.f. on $\mathbb{I}^d$ whose univariate marginals are uniformly distributed on $\mathbb{I}$.

Thus, each $d$-copula may be associated with a r.v. $\mathbf{U} = (U_1, U_2, \dots, U_d)$ such that $U_i \sim \mathscr{U}(\mathbb{I})$ for every $i \in \{1, 2, \dots, d\}$ and $\mathbf{U} \sim C$. Conversely, any r.v. whose components are uniformly distributed on $\mathbb{I}$ is distributed according to some copula.

The class of all $d$-copulas will be denoted by $\mathscr{C}_d$.

Since copulas are multivariate d.f.'s, as a consequence of the results stated in Sect. 1.2, they can be characterized in the following equivalent way.

**Theorem 1.3.1.** *A function $C \colon \mathbb{I}^d \to \mathbb{I}$ is a copula if, and only if, the following properties hold:*

(C1) *for every $j \in \{1, 2, \dots, d\}$, $C(\mathbf{u}) = u_j$ when all the components of $\mathbf{u}$ are equal to 1 with the exception of the $j$-th one that is equal to $u_j \in \mathbb{I}$;*
(C2) *$C$ is isotonic, i.e. $C(\mathbf{u}) \le C(\mathbf{v})$ for all $\mathbf{u}, \mathbf{v} \in \mathbb{I}^d$, $\mathbf{u} \le \mathbf{v}$;*
(C3) *$C$ is $d$-increasing.*

As an easy consequence, we can prove also that $C(\mathbf{u}) = 0$ for every $\mathbf{u} \in \mathbb{I}^d$ having at least one of its components equal to 0. Another interesting property of a $d$-copula $C$ is that it is a *Lipschitz function*, namely, for all $\mathbf{u}, \mathbf{v} \in \mathbb{I}^d$, one has

$$|C(\mathbf{u}) - C(\mathbf{v})| \le \sum_{i=1}^{d} |u_i - v_i|. \tag{1.7}$$

By using Ascoli-Arzelá Theorem, one can show that $\mathscr{C}_d$ is a compact set in the set of all continuous functions from $\mathbb{I}^d$ into $\mathbb{I}$ equipped with the product topology, which corresponds to the topology of pointwise convergence. Moreover, in $\mathscr{C}_d$ pointwise and uniform convergence are equivalent (see also [33]).

Basic examples of copulas are:

- the *independence copula* $\Pi_d(\mathbf{u}) = u_1 u_2 \cdots u_d$ associated with a random vector $\mathbf{U} = (U_1, U_2, \dots, U_d)$ whose components are independent and uniformly distributed on $\mathbb{I}$;
- the *comonotonicity copula* $M_d(\mathbf{u}) = \min\{u_1, u_2, \dots, u_d\}$ associated with a vector $\mathbf{U} = (U_1, U_2, \dots, U_d)$ of r.v.'s uniformly distributed on $\mathbb{I}$ and such that $U_1 = U_2 = \cdots = U_d$ almost surely;
- the *countermonotonicity copula* $W_2(u_1, u_2) = \max\{u_1 + u_2 - 1, 0\}$ associated with a vector $\mathbf{U} = (U_1, U_2)$ of r.v.'s uniformly distributed on $\mathbb{I}$ and such that $U_1 = 1 - U_2$ almost surely.

By using Theorem 1.3.1 it is easy to show that the set $\mathscr{C}_d$ is convex. Convex combinations of copulas have the following probabilistic interpretation (compare with [143]).

*Example 1.3.1 (Convex combinations of copulas).* Let $\mathbf{U}^1$ and $\mathbf{U}^2$ be two $d$-dimensional r.v.'s on the probability space $(\Omega, \mathscr{F}, \mathbb{P})$ distributed according to the copulas $C^1$ and $C^2$, respectively. Let $Z$ be a Bernoulli r.v. such that $\mathbb{P}(Z = 1) = \alpha$ and $\mathbb{P}(Z = 2) = 1 - \alpha$ for some $\alpha \in \mathbb{I}$. Suppose that $\mathbf{U}^1$, $\mathbf{U}^2$ and $Z$ are independent. Now, consider the $d$-dimensional r.v. $\mathbf{U}^*$

$$\mathbf{U}^* = \sigma_1(Z)\mathbf{U}^1 + \sigma_2(Z)\mathbf{U}^2$$

where, for $i \in \{1,2\}$, $\sigma_i(x) = 1$, if $x = i$, $\sigma_i(x) = 0$, otherwise. Then, it could be proved that $\mathbf{U}^*$ is distributed according to the copula $\alpha C^1 + (1-\alpha)C^2$.

*Example 1.3.2 (Fréchet–Mardia copulas).* Let $C_d^{\mathbf{FM}}$ be the $d$-copula given by

$$C_d^{\mathbf{FM}}(\mathbf{u}) = \alpha\Pi_d(\mathbf{u}) + (1-\alpha)M_d(\mathbf{u}) \tag{1.8}$$

for every $\lambda \in \mathbb{I}$. These copulas can be considered as a multivariate version of the bivariate families by Fréchet and Mardia (see [72, 132]). They are obtained as convex sum of the copulas $\Pi_d$ and $M_d$ .

*Example 1.3.3.* Let $\mathbf{X} = (X_1, X_2, \dots, X_d)$ be a r.v. whose components are independent and identically distributed according to $F_{X_i}(t) = t^\alpha$ for every $t \in \mathbb{I}$ and for some $\alpha \in \mathbb{I}$. Let $Z$ be a random variable, independent of $\mathbf{X}$, whose d.f. is given by $F_Z(t) = t^{1-\alpha}$ on $\mathbb{I}$. Intuitivelly, $Z$ might be interpreted as a *shock* that will change the dependence structure of $\mathbf{X}$. We define another r.v. $\mathbf{Y}$ such that, for every $i \in \{1,2,\dots,d\}$, $Y_i = \max\{X_i, Z\}$. Now, it can be easily proved that the joint d.f. of $Y$ is, in fact, a copula given by

$$C_d^{\mathbf{CA}}\mathbf{u} = (\Pi_d(\mathbf{u}))^\alpha (M_d(\mathbf{u}))^{1-\alpha}, \tag{1.9}$$

which belongs to the *Cuadras-Augé* family of copulas (see [24, 25]). The idea of considering families of d.f.'s associated with some *shock model* had its origin in the seminal paper by Marshall and Olkin (see [134]), where the multivariate exponential distribution was considered (see also [122, 123, 148])). Further generalizations of these methods have recently provided several constructions of copulas, as can be seen from [44, 50, 127–129].

The following result gives upper and lower bounds in $\mathscr{C}_d$ (see, for example, [151, Theorem 2.10.12]).

**Theorem 1.3.2 (Fréchet–Hoeffding bounds).** *For every $C_d \in \mathscr{C}_d$ and for every $\mathbf{u} \in \mathbb{I}^d$,*

$$W_d(\mathbf{u}) = \max\left\{\sum_{i=1}^d u_i - d + 1, 0\right\} \le C(\mathbf{u}) \le M_d(\mathbf{u}). \tag{1.10}$$

*Moreover, the following bounds are sharp, in the sense that the pointwise infimum and supremum of all the elements of $\mathscr{C}_d$ coincide, respectively, with $W_d$ and $M_d$, i.e. for all $\mathbf{u} \in \mathbb{I}^d$:*

$$\inf_{C\in\mathscr{C}_d} C(\mathbf{u}) = W_d(\mathbf{u}), \qquad \sup_{C\in\mathscr{C}_d} C(\mathbf{u}) = M_d(\mathbf{u}).$$

Notice that, while $W_2$ is a copula, $W_d$ is not a copula for $d \ge 3$.

The Fréchet–Hoeffding bounds appeared for the first time in the present form in an article by Fréchet (see [70]). An earlier version had already be given by Hoeffding ([96]), but with reference to the square $[-1/2, 1/2]^2$. A very general formulation of it, including the proof of the sharpness of these bounds, has been given by Rüschendorf [173].

Several investigations have been conducted about the bounds for multivariate d.f.'s (see [15, 154, 169]), especially when some additional information is given, like lower dimensional marginals [38, 46, 104], and measures of association [152, 156]. In this context, the concept of *quasi-copula* plays a special rôle [2, 27, 84, 153, 157, 172].

## 1.4 Sklar's Theorem

Sklar's theorem is the building block of the theory of copulas; without it, the concept of copula would be one in a rich set of joint distribution functions.

**Theorem 1.4.1.** *Let $F$ be a $d$-dimensional d.f. with univariate margins $F_1$, $F_2$,..., $F_d$. Let $A_j$ denote the range of $F_j$, $A_j := F_j(\overline{\mathbb{R}})$ $(j = 1,2,\dots,d)$. Then there exists a copula $C$ such that for all $(x_1,x_2,\dots,x_d) \in \overline{\mathbb{R}}^d$,*

$$F(x_1,x_2,\dots,x_d) = C(F_1(x_1),F_2(x_2),\dots,F_d(x_d)). \tag{1.11}$$

*Such a $C$ is uniquely determined on $A_1 \times A_2 \times \cdots \times A_d$ and, hence, it is unique when $F_1$, $F_2$,..., $F_d$ are all continuous.*

Sklar's theorem has been announced in [192], however its first proof for the bivariate case appeared in [185]. Curiously, it should be noted that in [192], the author "Abe Sklar" is named as "M. Sklar" (we conjecture that this "M." should be intended as "Monsieur").

Another (bivariate) proof can be also found in [16], based on the so-called "checkerboard copulas". A multivariate proof, based on the *distributional transform*, has been recently presented in [175] (compare also with [145, Lemma 3.2]). Another possible proof can be also derived (for positive random variables) from [14]. Sklar's Theorem on more abstract spaces has been given in [179].

Theorem 1.4.1 also admits the following converse implication, usually very important when one wants to construct statistical models by considering, separately, the univariate behaviour of the components of a random vector and their dependence properties as captured by some copula.

**Theorem 1.4.2.** *If $F_1$, $F_2$,..., $F_d$ are univariate d.f.'s, and if $C$ is any $d$-copula, then the function $F : \overline{\mathbb{R}}^d \to \mathbb{I}$ defined by* (1.11) *is a $d$-dimensional distribution function with margins $F_1$, $F_2$,..., $F_d$.*

By summarizing, from any $d$-variate d.f. $F$ one can derive a copula $C$ via (1.11). Specifically, when $F_i$ is continuous for every $i \in \{1,2,\dots,d\}$, $C$ can be obtained by means of the formula

$$C(u_1,u_2,\dots,u_d) = F(F_1^{-1}(u_1),F_2^{-1}(u_2),\dots,F_d^{-1}(u_d)), \tag{1.12}$$

where $F_i^{-1}$ denoted the pseudo-inverse of $F_i$ given by $F_i^{-1}(s) = \inf\{t \mid F_i(t) \geq s\}$. Thus, copulas are essentially a way for transforming the r.v. $(X_1, X_2 \dots, X_d)$ into another r.v. $(U_1, U_2, \dots, U_d) = (F_1(X_1), F_2(X_2), \dots, F_d(X_d))$ having the margins uniform on $\mathbb{I}$ and preserving the dependence among the components.

On the other hand, any copula can be combined with different univariate d.f.'s in order to obtain a $d$-variate d.f. by using (1.11). In particular, copulas can serve for modelling situations where a different distribution is needed for each marginal, providing a valid alternative to several classical multivariate d.f.'s such Gaussian, Pareto, Gamma, etc. (compare with [74]). This fact represents one of the main advantage of the copula's idea, as underlined by Mikosch [140]:

> There is no simple alternative to the Gaussian distribution in the non-Gaussian world. In particular, one needs multivariate models for portfolios with different marginal distributions (including different tail behavior) and a dependence structure which is determined not only by covariances. Many of the well known multivariate distributions are not flexible enough to allow for different tail behavior in different components. Therefore copulas seem to be the right tools in order to overcome the mentioned difficulties: they generate all multivariate distributions with flexible marginals.

*Remark 1.4.1.* The copula representation is usually very convenient, as stressed for example by Kimeldorf and Sampson [109], who referred to it as *uniform representation*. However, following [57], it should be stressed that

> there is absolutely no real, compelling mathematical reason for transforming the marginal d.f.'s of $F$ to uniform d.f.'s on $\mathbb{I}$, though it may be useful from a statistical point of view. In his 1940 paper [98], Hoeffding used the interval $\left[-\frac{1}{2}, \frac{1}{2}\right]$. In multivariate Extreme Value Theory, it is standard to transform to unit Fréchet marginal d.f.'s. In this context, Resnick [167, page 265] writes "How one standardizes is somewhat arbitrary and depends on taste. Different specifications have led to (superficially) different representations in the literature".

As investigated recently in [115], other ways for transforming r.v.'s to having some standard margins may be more convenient in some cases.

*Remark 1.4.2.* Theorem 1.4.1 should be used with some caution when the margins have jumps. In fact, even if there exists a copula representation for not-continuous joint d.f.'s, it is no longer unique. In such cases, modelling and interpreting dependence through copulas is subject to caution. The interested readers should refer to the seminal paper by Marshall [133] and to the in-depth discussion by Genest and Nešlehová [83].

Finally, notice that Theorems 1.4.1 and 1.4.2 can be formulated in an analogous way in terms of survival functions instead of d.f.'s. Specifically, given a r.v. $\mathbf{X} = (X_1, X_2, \dots X_d)$ with joint survival function $\overline{F}$ and univariate survival marginals $\overline{F}_i$ $(i = 1, 2, \dots, d)$, for all $(x_1, x_2, \dots, x_n) \in \overline{\mathbb{R}}^d$ it holds that:

$$\overline{F}(x_1, x_2, \dots, x_d) = \widehat{C}\left(\overline{F}_1(x_1), \overline{F}_2(x_2), \dots, \overline{F}_d(x_d)\right). \tag{1.13}$$

for some copula $\widehat{C}$, usually called *survival copula* of $\mathbf{X}$ (to be intended as a copula associated with the survival function of $\mathbf{X}$).

In particular, let $C$ be the copula of $\mathbf{X}$ and let $\mathbf{U} = (U_1, U_2, \dots, U_d)$ be a vector such that $\mathbf{U} \sim C$. Then, one has

$$\widehat{C}(\mathbf{u}) = \overline{C}(1-u_1, 1-u_2, \dots, 1-u_d),$$

where $\overline{C}(\mathbf{u}) = \mathbb{P}(U_1 > u_1, U_2 > u_2, \dots, U_d > u_d)$ is the survival function associated with $C$, explicitly given by

$$\overline{C}(\mathbf{u}) = 1 + \sum_{k=1}^{d} (-1)^k \sum_{1 \le i_1 < i_2 < \dots < i_k \le n} C_{i_1 i_2 \cdots i_k}(u_{i_1}, u_{i_2}, \dots, u_{i_k}), \tag{1.14}$$

with $C_{i_1 i_2 \cdots, i_k}$ denoting the marginal of $C$ related to $(i_1, i_2, \cdots, i_k)$.

## 1.5 Copulas and Random Vectors

Here, we should like to discuss some probabilistic properties of copulas that can be inferred from Sklar's Theorem. These results appeared already in [193] (for the proofs, see [138, 151]).

First, we state the invariance of the copula of the r.v. $\mathbf{X}$ with respect to any increasing rescaling of the components of $\mathbf{X}$.

**Proposition 1.5.1.** *Let $\mathbf{X} = (X_1, \dots, X_d)$ be a r.v. with continuous joint d.f. $F$ and copula $C$. Let $T_1, \dots, T_d$ be strictly increasing functions from $\mathbb{R}$ to $\mathbb{R}$. Then $C$ is also the copula of the r.v. $(T_1(X_1), \dots, T_d(X_d))$.*

Thus, copulas that describe the dependence of the components of a random vector are invariant under increasing transformations of each coordinate. Therefore, as said in [188],

> the study of rank statistics – insofar as it is the study of properties invariant under such transformations – may be characterized as the study of copulas and copula-invariant properties.

The next results characterize some special structures of r.v.'s in terms of the basic copulas $\Pi_d$, $M_d$, and $W_2$.

**Proposition 1.5.2.** *Let $(X_1, X_2, \dots, X_d)$ be a r.v. with continuous joint d.f. $F$. Then the copula of $(X_1, \dots, X_d)$ is $\Pi_d$ if, and only if, $X_1, \dots, X_d$ are independent.*

**Proposition 1.5.3.** *Let $(X_1, X_2, \dots, X_d)$ be a r.v. with continuous joint d.f. $F$. Then the copula of $(X_1, \dots, X_d)$ is $M_d$ if, and only if, there exists a r.v. $Z$ and increasing functions $T_1, \dots, T_d$ such that $\mathbf{X} = (T_1(Z), \dots, T_d(Z))$ almost surely.*

**Proposition 1.5.4.** *Let $(X_1, X_2)$ be a r.v. with continuous joint d.f. $F$. Then $(X_1, X_2)$ has copula $W_2$ if, and only if, for some strictly decreasing function $T$, $X_2 = T(X_1)$ almost surely.*

Obviously Proposition 1.5.4 cannot be extended to dimension $d$ with $d \geq 3$; in fact, it is impossible for each one of the $d$ random variables $X_1, X_2, \ldots, X_d$ to be a decreasing function of the remaining ones.

A r.v. $\mathbf{X}$ as in Proposition 1.5.3 is said to have a *comonotone* dependence. This concept has several important consequences in applications, as described in [39, 40]. The related concept described in Proposition 1.5.4 is known as *countermonotonicity*.

A recent generalization of comonotonicity has been provided in [162].

## 1.6 Families of Copulas

As underlined, copulas play an important role in the construction of multivariate d.f.'s and, as a consequence, having at one's disposal a variety of copulas can be very useful for building stochastic models having different properties that are sometimes indispensable in practice (e.g., heavy tails, asymmetries, etc.). Therefore, several investigations have been carried out concerning the construction of different families of copulas and their properties. Here, we present just a few of them, by focusing on the families that seem to be more popular in the literature. Different families (or construction methods) are discussed in the rest of this book (see [28, 89, 97]).

Before presenting these examples, we would like briefly to discuss the use of these families for copula-fitting procedures. For several researchers and practitioners it seems enough to try to fit *any* stochastic model with the most convenient family and check that the fitting procedure is not so "bad". However, it should be stressed that any fitting procedure may be misleading if one were to describe any situation with families of copulas satisfying some unnecessary assumptions (e.g., exchangeability, light tails, etc.). In our opinion, before applying any statistical tool, one should not forget to analyse the main characteristics of the model under consideration. At the same time, one should have clearly in mind the final output of the investigation. Usually, in fact, fitting a copula (or a joint d.f.) to some data is just a tool for deriving some quantities of interest for the problem at hand (e.g., VaR of a portfolio, return period of an extreme event, etc.). For such problems, a dramatic underestimation of the risk can be obtained when one tries to fit with copulas that do not exhibit any peculiar behaviour in the tails. This was exactly one of the main criticisms to the use of Li's model for credit risk using Gaussian copulas (see, for example, [121, 198]).

Inspired by some considerations of Joe [104], we discuss here some general properties that a "good" family of multivariate copulas $\{C_\theta\}$, where $\theta$ is a parameter belonging to a (usually, compact) subset $\Theta \subseteq \mathbb{R}^p$ $(p \geq 1)$, should have for being considered "interesting" in statistical applications.

- *Interpretability*. The members of the family should have some *probabilistic interpretation* suggesting "natural" situations where this family could be considered. For instance, as we have seen in Example 1.3.3, Cuadras-Augé copulas have a quite direct interpretation in terms of shock models, making this class suitable

for modelling situations where a common shock has consequences on a system composed by several components (credit portfolio, lifetime system, etc.).
- *Flexible and wide range of dependence*. The members of the family should describe different types of dependence, including the independence copula $\Pi_d$ and one of the Fréchet-Hoeffding bounds (possibly, as a limiting case with respect to the parameter). Having members with a variety of tail dependencies and asymmetries is a desirable property as well.
- *Easy-to-handle*. The members of the family should be expressed in a *closed form* or, at least, should be easily simulated by means of some known algorithm. Notice, in fact, that several goodness-of-fit procedures are based on the fact that the fitted family can be readily sampled (see, for example, [12, 86]).

Notice that several families of copulas recently introduced seem not to be "good" in the above sense. This should be not surprising. On one hand, the above properties are intentionally be given in a very restrictive way, on the other hand, one should consider that several new constructions have been introduced and studied more for their mathematical elegance and interest than for a practical implementation. It is our opinion that in the future more attention should be devoted to this latter aspect: there are no clever applications without a beautiful theory!

### 1.6.1 Elliptical Copulas

A random vector $\mathbf{X} = (X_1, X_2 \dots, X_d)$ is said to have an *elliptical distribution* with mean vector $\mu \in \mathbb{R}^d$, covariance matrix $\Sigma = (\sigma_{ij})$ and generator $g\colon [0,+\infty[ \to [0,+\infty[$, and one writes $\mathbf{X} \sim \mathscr{E}(\mu, \Sigma, g)$, if it can be expressed in the form

$$\mathbf{X} = \mu + R\mathbf{A}\mathbf{U}, \tag{1.15}$$

where $\mathbf{A}\mathbf{A}^{\mathbf{T}} = \Sigma$ is the Cholesky decomposition of $\Sigma$, $\mathbf{U}$ is a $d$-dimensional random vector uniformly distributed on the sphere $\mathbb{S}^{d-1} = \left\{\mathbf{u} \in \mathbb{R}^d : u_1^2 + \dots + u_d^2 = 1\right\}$, and $R$ is a positive random variable independent of $\mathbf{U}$, with density given, for every $r > 0$, by

$$f_g(r) = \frac{2\pi^{d/2}}{\Gamma(d/2)} r^{d-1} g(r^2).$$

The density function (if it exists) of an elliptical distribution is given, for every $\mathbf{x} \in \mathbb{R}^d$, by

$$h_g(\mathbf{x}) = |\Sigma|^{-1/2} g\left((\mathbf{x}-\mu)^T \Sigma^{-1} (\mathbf{x}-\mu)\right). \tag{1.16}$$

For more details, see [62, 63, 68, 78, 138].

For instance, when $g(t) = (2\pi)^{-d/2} \exp(-t/2)$, then $\mathbf{X}$ has a multivariate Gaussian distribution. Similarly, $g(t) = c(1+t/\nu)^{-(d+\nu)/2}$, for a suitable constant $c$, generates the multivariate $t$-Student distribution with $\nu$ degrees of freedom.

One of the characteristics of an elliptical distribution is that the scaled components $X_1/\sqrt{\sigma_{11}}, \dots, X_d/\sqrt{\sigma_{dd}}$ are identically distributed according to a d.f. $F_g$. This fact represents a limitation to the use of such distributions for modelling stochastic systems when the components are not similar. In order to avoid this, it is useful to calculate the copula of a multivariate elliptical distributions and use it, together with some univariate marginal d.f.'s, for obtaining more flexible models. These distributions, constructed by means of Theorem 1.4.2, are usually called *meta-elliptical distribution* [63]. For these reasons, we give the following

**Definition 1.6.1.** Let $\mathbf{X}$ be an elliptical random vector, $\mathbf{X} \sim \mathscr{E}_d(\mu, \Sigma, g)$. Suppose that, for every $i \in \{1, 2, \dots, d\}$, $(X_i/\sqrt{\sigma_{ii}}) \sim F_g$. We call *elliptical copula* the distribution function of the random vector

$$\left(F_g\left(\frac{X_1}{\sqrt{\sigma_{11}}}\right), F_g\left(\frac{X_2}{\sqrt{\sigma_{22}}}\right), \dots, F_g\left(\frac{X_d}{\sqrt{\sigma_{dd}}}\right)\right).$$

An elliptical copula is typically not available in closed form. In order just to provide an example, the bivariate Gaussian copula is given by

$$C_\theta(u_1, u_2) = \int_{-\infty}^{\Phi^{-1}(u_1)} \int_{-\infty}^{\Phi^{-1}(u_2)} \frac{1}{2\pi\sqrt{1-\theta^2}} \left(-\frac{s^2 - 2\theta st + s^2}{2(1-\theta^2)}\right) \mathrm{d}s\mathrm{d}t,$$

where $\theta \in [-1, 1]$, and $\Phi^{-1}$ denotes the inverse of the univariate Gaussian distribution.

## 1.6.2 Archimedean Copulas

Here we present the basic properties and examples of the Archimedean class of copulas. Basically, we follow the approach in [137] (see also [136]).

First, we introduce some notations. We call *Archimedean generator* any decreasing and continuous function $\psi\colon [0, \infty[ \to \mathbb{I}$ that satisfies the conditions $\psi(0) = 1$, $\lim_{t\to\infty} \psi(t) = 0$ and which is strictly decreasing on $[0, \inf\{t \mid \psi(t) = 0\}[$. By convention, $\psi(+\infty) = 0$ and $\psi^{-1}(0) = \inf\{t \geq 0 \mid \psi(t) = 0\}$, where $\psi^{-1}$ denotes the pseudo-inverse of $\psi$.

**Definition 1.6.2.** A $d$-dimensional copula $C$ is called *Archimedean* if it admits the representation

$$C(\mathbf{u}) = \psi\left(\psi^{-1}(u_1) + \psi^{-1}(u_2) + \cdots + \psi^{-1}(u_d)\right) \tag{1.17}$$

for all $\mathbf{u} \in \mathbb{I}^d$ and for some Archimedean generator $\psi$.

Functions of type (1.17) have been widely considered and studied in the literature, especially in connexion with the investigations about semigroups of the unit interval. The reader interested about all these aspects should look at the following books [1, 112, 186] and the references therein.

The following result characterizes Archimedean copulas in terms of the properties of their generators; it can be found in [137].

**Theorem 1.6.1.** *Let $\psi$ be an Archimedean generator. Let $C_\psi$ be the function given by* (1.17)*. Then $C_\psi$ is a d-dimensional copula if, and only if, the restriction of $\psi$ to* $]0,\infty[$ *is d*-monotone, *i.e. it satisfies:*

(a) *$\psi$ is differentiable up to the order $d-2$ in $]0,+\infty[$ and the derivatives satisfy*

$$(-1)^k \psi^{(k)}(t) \geq 0 \qquad \text{for } k \in \{0,1,\dots,d-2\}$$

*for every $t > 0$,*
(b) *$(-1)^{d-2}\psi^{(d-2)}$ is decreasing and convex in $]0,+\infty[$.*

*Example 1.6.1 (Gumbel–Hougaard copula).* The standard expression for members of this family of $d$-copulas is

$$C_\theta^{\mathbf{GH}}(\mathbf{u}) = \exp\left(-\left(\sum_{i=1}^{d}(-\log(u_i))^\theta\right)^{1/\theta}\right) \tag{1.18}$$

where $\theta \geq 1$. For $\theta = 1$ we obtain the independence copula as a special case, and the limit of $C_\theta^{\mathbf{GH}}$ for $\theta \to +\infty$ is the comonotonicity copula. The Archimedean generator of this family is given by $\psi(t) = \exp\left(-t^{1/\theta}\right)$. Each member of this class is absolutely continuous.

Copulas of this type can be derived from the work by Gumbel [92] and have been further considered by Hougaard [99]. For this reason, this family is named as *Gumbel–Hougaard family* [100].

*Example 1.6.2 (Mardia–Takahasi–Clayton copula).* The standard expression for members of this family of $d$-copulas is

$$C_\theta^{\mathbf{MTC}}(u,v) = \max\left\{\left(\sum_{i=1}^{d} u_i^{-\theta} - (d-1)\right)^{-1/\theta}, 0\right\} \tag{1.19}$$

where $\theta \geq \frac{-1}{d-1}$, $\theta \neq 0$. The limiting case $\theta = 0$ corresponds to the independence copula.

The Archimedean generator of this family is given by

$$\psi_\theta(t) = (\max\{1+\theta t, 0\})^{-1/\theta}.$$

It was proved in [137] that, for every $d$-dimensional Archimedean copula $C$ and for every $\mathbf{u} \in \mathbb{I}^d$, $C_{\theta_L}^{\mathbf{MTC}}(\mathbf{u}) \leq C(\mathbf{u})$ for $\theta_L = -\frac{1}{d-1}$.

Copulas of this type can be derived from two types of multivariate distributions: the Pareto distribution by Mardia [131] and the Burr distribution by Takahasi [195]. They were mentioned as bivariate copulas in [109] and as multivariate copulas in [22], subsequently extended in [81]. A special member of this family can also be derived from the bivariate logistic model of first type by Gumbel [93]. As showed by Oakes [158], bivariate copulas of type (1.19) are associated with the Clayton's model [21]. For this reason, many people referred to this family as the Clayton's family of copulas. Here, we prefer to name it as *Mardia–Takahasi–Clayton* family of copulas.

*Example 1.6.3 (Frank copula).* The standard expression for members of this family of $d$-copulas is

$$C_{\theta}^{\mathbf{Fr}}(\mathbf{u}) = -\frac{1}{\theta}\log\left(1 + \frac{\prod_{i=1}^{d}\left(e^{-\theta u_i} - 1\right)}{\left(e^{-\theta} - 1\right)^{d-1}}\right), \tag{1.20}$$

where $\theta > 0$. The limiting case $\theta = 0$ corresponds to $\Pi_d$. For the case $d = 2$, the parameter $\theta$ can be extended also to the case $\theta < 0$.

Copulas of this type have been introduced by Frank [69] in relation with a problem about associative functions on $\mathbb{I}$. They are absolutely continuous.

The Archimedean generator is given by $\psi_{\theta}(t) = -\frac{1}{\theta}\log\left(1 - (1 - e^{-\theta})e^{-t}\right)$.

### 1.6.3 EFGM Copulas

The so-called *Eyraud–Farlie–Gumbel–Morgenstern* (shortly, EFGM) distributions have been considered by Morgenstern [146] and Gumbel [90, 91], further developed by Farlie [64]. However, the idea of considering such distributions originated in an earlier and, for many years, forgotten work by Eyraud [61]. Here, we present the EFGM family of copulas that can be derived from the papers just mentioned.

Let $d \geq 2$. Let $\mathscr{S}$ be the class of all subsets of $\{1, 2, \ldots, d\}$ having at least 2 elements. Trivially, $\mathscr{S}$ contains $2^d - d - 1$ elements. To each $S \in \mathscr{S}$, we associate a real number $\alpha_S$, with the convention that, when $S = \{i_1, i_2, \ldots, i_k\}$, $\alpha_S = \alpha_{i_1 i_2 \ldots i_k}$.

An EFGM copula can be expressed in the following form:

$$C_d^{\mathbf{EFGM}}(\mathbf{u}) = \prod_{i=1}^{d} u_i \left(1 + \sum_{S \in \mathscr{S}} \alpha_S \prod_{j \in S}(1 - u_j)\right), \tag{1.21}$$

for suitable values of the $\alpha_S$'s.

For the bivariate and trivariate cases, respectively, EFGM copulas have the following expressions:

$$C_2^{\mathbf{EFGM}}(u_1, u_2) = u_1 u_2 \left(1 + \alpha_{12}(1 - u_1)(1 - u_2)\right), \tag{1.22}$$

and

$$\begin{aligned} C_3^{\mathbf{EFGM}}(u_1, u_2, u_3) = u_1 u_2 u_3 [&1 + \alpha_{12}(1 - u_1)(1 - u_2) + \alpha_{13}(1 - u_1)(1 - u_3) \\ &+ \alpha_{23}(1 - u_2)(1 - u_3) + \alpha_{123}(1 - u_1)(1 - u_2)(1 - u_3)]. \end{aligned} \tag{1.23}$$

It is not so difficult to show that any FGM copula is absolutely continuous with density given by

$$c_d^{\mathbf{EFGM}}(\mathbf{u}) = 1 + \sum_{S \in \mathscr{S}} \alpha_S \prod_{j \in S} (1 - 2u_j). \tag{1.24}$$

As a consequence, the parameters $\alpha_S$'s have to satisfy the following inequality

$$1 + \sum_{S \in \mathscr{S}} \alpha_S \prod_{j \in S} \xi_j \geq 0$$

for any $\xi_j \in \{-1, 1\}$. In particular, it holds that $|\alpha_S| \leq 1$.

On account of the fact that EFGM copulas do not allow the modelling of large dependence among the random variables involved, several extensions have been proposed in the literature, starting with the works by Farlie [64] and Saramanov [178]. A complete survey about these generalized EFGM models of dependence is given in [41], where a list of several other references can be also found. More recent investigations are also provided in [4–6, 66, 170]. Another possible approach for extending EFGM copulas is based on the construction of copulas that are quadratic in one variable [164, 171].

## 1.7 Constructions of Copulas

Several constructions of copulas have been developed during the years from a variety of perspectives. At an abstract level, all these methods start with some known copulas and/or some auxiliary functions (sometimes, possible sections of copulas) and generate in an automatic way "new" copulas. Essentially, three kinds of such constructions can be distinguished.

### *1.7.1 Copulas with Given Lower Dimensional Marginals*

These constructions are strictly related to the original Fréchet problem of considering distribution functions with fixed marginals (eventually overlapping). In this context the most interesting results have been obtained by means of the "conditioning method", as used by Dall'Aglio [29, 30] and Rüschendorf [174]. Joe presented several results of this type in [104, Chap. 3]. A powerful recent method based on these ideas is the so-called *pair-copula construction* (for more details, see [28] and the references therein). For related recent studies, see [47, 117]. Other constructions can be found in [163] (*direct compatibility*), [97] (*nested constructions*), and [17, 67, 116, 124].

### 1.7.2 Copula-to-Copula Transformations

Constructions of this second kind aim at transforming $d$-copulas into other $d$-copulas having possibly some additional features (for example, having a larger number of parameters). Specifically, the following cases have been extensively studied.

#### 1.7.2.1 Ordinal Sums

The ordinal sum construction for (bivariate) copulas is fully described in [151, 186]. This construction was introduced in an algebraic framework, namely the theory of semigroups. Then, it was translated into the language of *triangular norms* (briefly, $t$-norms), which are binary operations on $\mathbb{I}$ that are associative, commutative, monotonic and with neutral element 1 (see [1, 112, 186]), and, finally, it was applied as well to bivariate copulas (which, in fact, can be seen also as special binary operations on $\mathbb{I}$). An extension of the ordinal sum construction to $\mathscr{C}_d$ has been recently discussed in [102, 103, 139]. This method is essentially based on a kind of "patchwork procedure", consisting of redefining the value that a copula assume on a $d$-box $B$ of $\mathbb{I}^d$ by plugging-in a suitable rescaling of another copula. For other recent investigations of this type, see [34, 51, 191]. Following [139], we give the following result.

**Theorem 1.7.1.** *Let $\mathscr{J}$ be a finite or countable subset of $\mathbb{N}$ and let $(]a_k,b_k[)_{k\in\mathscr{J}}$ be a family of sub-intervals of $\mathbb{I}$ indexed by $\mathscr{J}$ and let $(C_k)_{k\in\mathscr{J}}$ be a family of copulas in $\mathscr{C}_d$ also indexed by $\mathscr{J}$. It is required that any two of the intervals $]a_k,b_k[$ $(k\in\mathscr{J})$ have at most an endpoint in common. Then the* ordinal sum *$C$ of $(C_k)_{k\in\mathscr{J}}$ with respect to family of intervals $(]a_k,b_k[)_{k\in\mathscr{J}}$ is the d-copula defined, for all $\mathbf{u}\in\mathbb{I}^d$ by*

$$C(\mathbf{u}) := \begin{cases} a_k + (b_k - a_k)\,C_k\left(\frac{\min\{u_1,b_k\}-a_k}{b_k-a_k},\dots,\frac{\min\{u_d,b_k\}-a_k}{b_k-a_k}\right), \\ \qquad \textit{if } \min\{u_1,u_2,\dots,u_d\}\in\,]a_k,b_k[\ \textit{for some } k\in\mathscr{J}, \\ \min\{u_1,u_2,\dots,u_d\}, \qquad \textit{elsewhere}. \end{cases} \tag{1.25}$$

*For such a $C$ one writes $C = (\langle a_k,b_k,C_k\rangle)_{k\in\mathscr{J}} = \bigoplus_{\{(a_k,b_k):k\in\mathscr{J}\}} C_k$.*

#### 1.7.2.2 Distortions

Given a copula $C$ and an increasing bijection $\psi:\mathbb{I}\to\mathbb{I}$, the *distortion* of $C$ is defined as the function $C_\psi:\mathbb{I}^d\to\mathbb{I}$,

$$C_\psi(\mathbf{u}) = \psi(C(\psi^{-1}(u_1),\psi^{-1}(u_2),\dots,\psi^{-1}(u_d))). \tag{1.26}$$

Such a transformation has originated from the study of distorted probability distribution functions (especially, power distortions), and has been considered by several authors like [3, 18, 43, 53, 55, 74, 87, 113, 114, 147, 151]. In reliability theory, this kind of transformation is used in order to introduce the so-called bivariate ageing function that are used for the definition of bivariate notion of ageing (see [8, 9, 54, 149]). In the context of synthetic Collateralized Debt Obligations (CDOs), distortions of copulas have been recently used in order to produce a heavy tailed portfolio loss distribution [23].

#### 1.7.2.3 Pointwise Composition of Copulas

Given two copulas $A$ and $B$ in $\mathscr{C}_d$, we define as composition of $A$ and $B$ through some suitable functions $H\colon \mathbb{I}^2 \to \mathbb{I}$, $f_i\colon \mathbb{I} \to \mathbb{I}$ and $g_i\colon \mathbb{I} \to \mathbb{I}$ $(i = 1,2,\dots,d)$, any mapping $C_{A,B}\colon \mathbb{I}^d \to \mathbb{I}$ given by

$$C_{A,B}(\mathbf{u}) = H\left(A(f_1(u_1),\dots,f_d(u_d)),B(g_1(u_1),\dots,g_d(u_d))\right). \tag{1.27}$$

To the best of our knowledgement, the idea of such constructions arose in the Khoudraji's work [108], and, then, studied and further generalized in [24, 42, 80, 125]. Although, formally, distortions of copulas can be included in this class (for a suitable choice of $H$), we prefer to distinguish these two constructions, following their historical development.

#### 1.7.2.4 Shuffles of Copulas

These constructions are based on the transformation of a copula $C$ into another one by means of a suitable rearrangement of the original mass distribution of $C$. The idea goes back to the notion of shuffles of Min, as introduced in [144], and is related to some modifications of the copula $M_d$ (see also [142]). A recent generalization is discussed in [52].

### *1.7.3 Geometric Constructions of Copulas*

The third kind of constructions refers to methods for originating copulas starting with some information about their structure (for example, support, diagonals, sections). For a good overview to these constructions, the reader should refer to [151, Chap. 3]. More recent investigations are listed below.

- copulas with given support: see [73];
- copulas with given horizontal and/or vertical sections: see [49, 111, 168, 171, 197];
- copulas with given diagonal sections: see [35, 36, 45, 48, 60, 102, 103, 135, 155];
- copulas with given affine sections [110, 165].

## 1.8 Copula Theory: What's the Future?

Above we have described, sometimes in a sketchy manner, the state of some, if not most, of the known results about copula theory; thus, the title of the present section poses a very natural question. It is hard to foresee the future, but there certainly are a few directions that we feel the investigations about copulas and their applications are likely to take. Running the risk of being completely, or even partially, proved wrong, we venture to put forward the following suggestions for likely directions of future investigations:

*New constructions of copulas.* The search for families of copulas having properties desirable for specific applications in various fields ought to continue to be important. Having at one's disposal several families of copulas (spanning different behaviour) is essential in order to create a wider spectrum of possible scenarios for the stochastic model at hand. This is of special interest to assist in decision making the risk managers and, under Basel Accords, it is mandatory for large banks to determine their risky positions. In particular, we think that special emphasis will be devoted to the search for copulas exhibiting different *asymmetries*, (non-exchangeable copulas, copulas with different tail behaviour, etc.).

*The compatibility problem.* Given that one just has some vague idea about the dependence of a r.v. $\mathbf{X}$ (for example, one knows the lower dimensional marginals of $\mathbf{X}$ or some dependence measures among its components), the question is whether one can describe the set of all possible copulas of $\mathbf{X}$, *compatible* with the given information. As said, this problem has its roots at early works on the Fréchet classes, but its popularity have recently increased due to its connection with several problems arising in risk aggregation (see, e.g., [59]). a copula from them without any loss of information?

*Copulas and stochastic processes.* Starting with the seminal paper by Darsow, Nguyen and Olsen [33] linking copulas and Markov processes, it is still a matter of discussion whether copulas can be really useful for describing space-time dependence structure (see also [120]). In this respect, the recent concept of Lévy-copula (see [7, 10, 101, 107]) seems quite powerful and promising for modelling the large class of Lévy processes.

Other investigations related related to copulas and time-changing dependence structure can be found in [88, 160].

**Acknowledgements** First of all, we would like to express our gratitude to several Colleagues with whom through the years we had the pleasure to discuss ideas on copulas and their applications. Some of them have also received, and commented on, a first version of the present report. We thank them for their comments and suggestions: we feel that these have improved our presentation.

## References


1. Alsina, C., Frank, M.J., Schweizer, B.: Associative Functions. Triangular Norms and Copulas. World Scientific Publishing Co. Ptv. Ltd., Hackensack, NJ (2006)
2. Alsina, C., Nelsen, R.B., Schweizer, B.: On the characterization of a class of binary operations on distribution functions. Stat. Probab. Lett. **17**(2), 85–89 (1993)
3. Alvoni, E., Papini, P.L., Spizzichino, F.: On a class of transformations of copulas and quasi-copulas. Fuzzy Sets Syst. **160**(3), 334–343 (2009)
4. Amblard, C., Girard, S.: Une famille semi-paramétrique de copules symétriques bivariées. C. R. Acad. Sci. Paris Sér. I Math. **333**(2), 129–132 (2001)
5. Amblard, C., Girard, S.: Symmetry and dependence properties within a semiparametric family of bivariate copulas. J. Nonparametr. Stat. **14**(6), 715–727 (2002)
6. Amblard, C., Girard, S.: A new symmetric extension of FGM copulas. Metrika **70**(1), 1–17 (2009)
7. Barndorff-Nielsen, O.E., Lindner, A.M.: Lévy copulas: dynamics and transforms of Upsilon type. Scand. J. Stat. **34**(2), 298–316 (2007)
8. Bassan, B., Spizzichino, F.: Bivariate survival models with Clayton aging functions. Insur. Math. Econ. **37**(1), 6–12 (2005)
9. Bassan, B., Spizzichino, F.: Relations among univariate aging, bivariate aging and dependence for exchangeable lifetimes. J. Multivar. Anal. **93**(2), 313–339 (2005)
10. Bäuerle, N., Blatter, A., Müller, A.: Dependence properties and comparison results for Lévy processes. Math. Methods Oper. Res. **67**(1), 161–186 (2008)
11. Beneş, V., Štěpán, J. (eds.): Distributions with Given Marginals and Moment Problems. Kluwer Academic Publishers, Dordrecht (1997)
12. Berg, G.: Copula goodness-of-fit testing: an overview and power comparison. Eur. J. Finance **15**(7–8), 675–701 (2009)
13. Billingsley, P.: Probability and Measure, 3rd edn. Wiley Series in Probability and Mathematical Statistics. Wiley Inc., New York, NY (1995). A Wiley-Interscience Publication
14. Burchard, A., Hajaiej, H.: Rearrangement inequalities for functionals with monotone integrands. J. Funct. Anal. **233**(2), 561–582 (2006)
15. Carley, H.: Maximum and minimum extensions of finite subcopulas. Commun. Stat. Theory Methods **31**(12), 2151–2166 (2002)
16. Carley, H., Taylor, M.D.: A new proof of Sklar's theorem. In: Cuadras, C.M., Fortiana, J., Rodriguez-Lallena, J.A. (eds.) Distributions with Given Marginals and Statistical Modelling, pp. 29–34. Kluwer Academic Publishers, Dordrecht (2002)
17. Chakak, A., Koehler, K.J.: A strategy for constructing multivariate distributions. Commun. Stat. Simul. Comput. **24**(3), 537–550 (1995)
18. Charpentier, A.: Dynamic dependence ordering for Archimedean copulas and distorted copulas. Kybernetika (Prague) **44**(6), 777–794 (2008)
19. Cherubini, U., Luciano, E., Vecchiato, W.: Copula Methods in Finance. Wiley Finance Series. Wiley, Chichester (2004)
20. Choroś, B., Ibragimov, R., Permiakova, E.: Copula estimation. In: Jaworski, P., Durante, F., Härdle, W., Rychlik, T. (eds.) Copula Theory and Its Applications, Proceedings of the Workshop, Warsaw, 25–26 Sept 2009. Springer, Dordrecht (2010)
21. Clayton, D.G.: A model for association in bivariate life tables and its application in epidemiological studies of familial dependency in chronic disease incidence. Biometrika **65**, 141–151 (1978)
22. Cook, R.D., Johnson, M.E.: A family of distributions for modelling nonelliptically symmetric multivariate data. J. Roy. Stat. Soc. Ser. B **43**(2), 210–218 (1981)
23. Crane, G., van der Hoek, J.: Using distortions of copulas to price synthetic CDOs. Insur. Math. Econ. **42**(3), 903–908 (2008)
24. Cuadras, C.M.: Constructing copula functions with weighted geometric means. J. Stat. Plan. Infer. **139**(11), 3766–3772 (2009)

25. Cuadras, C.M., Augé, J.: A continuous general multivariate distribution and its properties. Commun. Stat. A Theory Methods **10**(4), 339–353 (1981)
26. Cuadras, C.M., Fortiana, J., Rodriguez-Lallena, J.A. (eds.): Distributions with Given Marginals and Statistical Modelling, Papers from the meeting, Barcelona, 17–20 Jul 2000. Kluwer Academic Publishers, Dordrecht (2002)
27. Cuculescu, I., Theodorescu, R.: Copulas: diagonals, tracks. Rev. Roumaine Math. Pures Appl. **46**(6), 731–742 (2002) (2001)
28. Czado, C.: Pair-copula constructions. In: Jaworski, P., Durante, F., Härdle, W., Rychlik, T. (eds.) Copula Theory and Its Applications, Proceedings of the Workshop, Warsaw, 25–26 Sept 2009. Springer, Dordrecht (2010)
29. Dall'Aglio, G.: Sulla compatibilità delle funzioni di ripartizione doppia. Rend. Mat. e Appl. **18**(5), 385–413 (1959)
30. Dall'Aglio, G.: Fréchet classes and compatibility of distribution functions. In: Symposia Mathematica, vol. IX (Convegno di Calcolo delle Probabilità, INDAM, Rome, 1971), pp. 131–150. Academic Press, London (1972)
31. Dall'Aglio, G.: Fréchet classes: the beginnings. In: Dall'Aglio, G., Kotz, S., Salinetti, G. (eds.) Advances in Probability Distributions with Given Marginals (Rome, 1990). Mathematics and Its Applications, vol. 67, pp. 1–12. Kluwer Academic Publishers, Dordrecht (1991)
32. Dall'Aglio, G., Kotz, S., Salinetti, G. (eds.): Advances in Probability Distributions with Given Marginals. Mathematics and Its Applications, vol. 67. Kluwer Academic Publishers, Dordrecht (1991)
33. Darsow, W.F., Nguyen, B., Olsen, E.T.: Copulas and Markov processes. Ill. J. Math. **36**(4), 600–642 (1992)
34. De Baets, B., De Meyer, H.: Orthogonal grid constructions of copulas. IEEE Trans. Fuzzy Syst. **15**(6), 1053–1062 (2007)
35. De Baets, B., De Meyer, H., Mesiar, R.: Asymmetric semilinear copulas. Kybernetika (Prague) **43**(2), 221–233 (2007)
36. De Baets, B., De Meyer, H., Úbeda-Flores, M.: Opposite diagonal sections of quasi-copulas and copulas. Int. J. Uncertain. Fuzziness Knowl-Based Syst. **17**(4), 481–490 (2009)
37. Deheuvels, P.: Caractérisation complète des lois extrêmes multivariées et de la convergence des types extrêmes. Publ. Inst. Stat. Univ. Paris **23**(3–4), 1–36 (1978)
38. Deheuvels, P.: Indépendance multivariée partielle et inégalités de Fréchet. In: Studies in Probability and Related Topics, pp. 145–155. Nagard, Rome (1983)
39. Dhaene, J., Denuit, M., Goovaerts, M.J., Kaas, R., Vyncke, D.: The concept of comonotonicity in actuarial science and finance: applications. Insur. Math. Econ. **31**(2), 133–161 (2002)
40. Dhaene, J., Denuit, M., Goovaerts, M.J., Kaas, R., Vyncke, D.: The concept of comonotonicity in actuarial science and finance: theory. Insur. Math. Econ. **31**(1), 3–33 (2002). 5th IME Conference (University Park, PA, 2001)
41. Drouet-Mari, D., Kotz, S.: Correlation and dependence. Imperial College Press, London (2001)
42. Durante, F.: Construction of non-exchangeable bivariate distribution functions. Stat. Papers **50**(2), 383–391 (2009)
43. Durante, F., Foschi, R., Sarkoci, P.: Distorted copulas: constructions and tail dependence. Commun. Stat. Theory Methods (2010). In press
44. Durante, F., Hofert, M., Scherer, M.: Multivariate hierarchical copulas with shocks. Methodol. Comput. Appl. Probab. (2009). In press
45. Durante, F., Jaworski, P.: Absolutely continuous copulas with given diagonal sections. Commun. Stat. Theory Methods **37**(18), 2924–2942 (2008)
46. Durante, F., Klement, E., Quesada-Molina, J.: Bounds for trivariate copulas with given bivariate marginals. J. Inequal. Appl. **2008**, 1–9 (2008). Article ID 161537
47. Durante, F., Klement, E., Quesada-Molina, J., Sarkoci, P.: Remarks on two product-like constructions for copulas. Kybernetika (Prague) **43**(2), 235–244 (2007)

48. Durante, F., Kolesárová, A., Mesiar, R., Sempi, C.: Copulas with given diagonal sections: novel constructions and applications. Int. J. Uncertain. Fuzziness Knowl-Based Syst. **15**(4), 397–410 (2007)
49. Durante, F., Kolesárová, A., Mesiar, R., Sempi, C.: Copulas with given values on a horizontal and a vertical section. Kybernetika (Prague) **43**(2), 209–220 (2007)
50. Durante, F., Quesada-Molina, J.J., Úbeda-Flores, M.: On a family of multivariate copulas for aggregation processes. Inform. Sci. **177**(24), 5715–5724 (2007)
51. Durante, F., Saminger-Platz, S., Sarkoci, P.: Rectangular patchwork for bivariate copulas and tail dependence. Commun. Stat. Theory Methods **38**(15), 2515–2527 (2009)
52. Durante, F., Sarkoci, P., Sempi, C.: Shuffles of copulas. J. Math. Anal. Appl. **352**(2), 914–921 (2009)
53. Durante, F., Sempi, C.: Copula and semicopula transforms. Int. J. Math. Math. Sci. **2005**(4), 645–655 (2005)
54. Durante, F., Spizzichino, F.: Semi-copulas, capacities and families of level curves. Fuzzy Sets Syst. **161**(2), 269–276 (2009)
55. Durrleman, V., Nikeghbali, A., Roncalli, T.: A simple transformation of copulas (2000). Available at SSRN: http://ssrn.com/abstract=1032543
56. Embrechts, P.: Discussion of: "Copulas: tales and facts" by T. Mikosch [Extremes **9** (2006), no. 1, 3–20]. Extremes **9**(1), 45–47 (2006)
57. Embrechts, P.: Copulas: a personal view. J. Risk Ins. **76**(3), 639–650 (2009)
58. Embrechts, P., McNeil, A.J., Straumann, D.: Correlation and dependence in risk management: properties and pitfalls. In: Dempster, M. (ed.) Risk Management: Value at Risk and Beyond, pp. 176–223. Cambridge University Press, Cambridge (2002)
59. Embrechts, P., Puccetti, G.: Risk aggregation. In: Jaworski, P., Durante, F., Härdle, W., Rychlik, T. (eds.) Copula Theory and Its Applications, Proceedings of the Workshop, Warsaw, 25–26 Sept 2009. Springer, Dordrecht (2010)
60. Erdely, A., González-Barrios, J.M.: On the construction of families of absolutely continuous copulas with given restrictions. Commun. Stat. Theory Methods **35**(4–6), 649–659 (2006)
61. Eyraud, H.: Les principes de la mesure des correlations. Ann. Univ. Lyon III. Ser. Sect. A **1**, 30–47 (1936)
62. Fang, H.B., Fang, K.T., Kotz, S.: The meta-elliptical distributions with given marginals. J. Multivar. Anal. **82**(1), 1–16 (2002)
63. Fang, K.T., Kotz, S., Ng, K.W.: Symmetric multivariate and related distributions. Monographs on Statistics and Applied Probability, vol. 36. Chapman and Hall Ltd., London (1990)
64. Farlie, D.J.G.: The performance of some correlation coefficients for a general bivariate distribution. Biometrika **47**, 307–323 (1960)
65. Féron, R.: Sur les tableaux de corrélation dont les marges sont données. Cas de l'espace a trois dimensions. Publ. Inst. Stat. Univ. Paris **5**, 3–12 (1956)
66. Fischer, M., Klein, I.: Constructing generalized FGM copulas by means of certain univariate distributions. Metrika **65**(2), 243–260 (2007)
67. Fischer, M., Köck, C., Schlüter, S., Weigert, F.: An empirical analysis of multivariate copula models. Quant. Finance **9**(7), 839–854 (2009)
68. Frahm, G., Junker, M., Szimayer, A.: Elliptical copulas: applicability and limitations. Stat. Probab. Lett. **63**(3), 275–286 (2003)
69. Frank, M.J.: On the simultaneous associativity of $F(x, y)$ and $x+y-F(x, y)$. Aequationes Math. **19**(2–3), 194–226 (1979)
70. Fréchet, M.: Sur les tableaux de corrélation dont les marges sont données. Ann. Univ. Lyon. Sect. A. **14**(3), 53–77 (1951)
71. Fréchet, M.: Sur les tableaux de corrélation dont les marges sont données. C. R. Acad. Sci. Paris **242**, 2426–2428 (1956)
72. Fréchet, M.: Remarques au sujet de la note précédente. C. R. Acad. Sci. Paris **246**, 2719–2720 (1958)
73. Fredricks, G.A., Nelsen, R.B., Rodríguez-Lallena, J.A.: Copulas with fractal supports. Insur. Math. Econ. **37**(1), 42–48 (2005)

74. Frees, E.W., Valdez, E.A.: Understanding relationships using copulas. N. Am. Actuar. J. **2**(1), 1–25 (1998)
75. Genest, C.: Preface [International Conference on Dependence Modelling: Statistical Theory and Applications in Finance and Insurance (DeMoSTAFI)]. Canad. J. Stat. **33**(3), 313–314 (2005). Held in Québec City, QC, 20–22 May 2004
76. Genest, C.: Preface [Special issue: Papers presented at the DeMoSTAFI Conference]. Insur. Math. Econ. **37**(1), 1–2 (2005). Held in Québec, QC, 20–22 May 2004
77. Genest, C., Favre, A.C.: Everything you always wanted to know about copula modeling but were afraid to ask. J. Hydrol. Eng. **12**(4), 347–368 (2007)
78. Genest, C., Favre, A.C., Béliveau, J., Jacques, C.: Metaelliptical copulas and their use in frequency analysis of multivariate hydrological data. Water Resour. Res. **43**, W09,401 (2007). Doi: 10.1029/2006WR005275
79. Genest, C., Gendron, M., Bourdeau-Brien, M.: The advent of copulas in finance. Eur. J. Finance **15**(7–8), 609–618 (2009)
80. Genest, C., Ghoudi, K., Rivest, L.P.: "Understanding relationships using copulas," by Edward Frees and Emiliano Valdez, January 1998. N. Am. Actuar. J. **2**(3), 143–149 (1998)
81. Genest, C., MacKay, R.J.: Copules archimédiennes et familles de lois bidimensionnelles dont les marges sont données. Canad. J. Stat. **14**(2), 145–159 (1986)
82. Genest, C., MacKay, R.J.: The joy of copulas: bivariate distributions with uniform marginals. Am. Stat. **40**(4), 280–283 (1986)
83. Genest, C., Nešlehová, J.: A primer on copulas for count data. Astin Bull. **37**(2), 475–515 (2007)
84. Genest, C., Quesada-Molina, J.J., Rodríguez-Lallena, J.A., Sempi, C.: A characterization of quasi-copulas. J. Multivar. Anal. **69**(2), 193–205 (1999)
85. Genest, C., Rémillard, B.: Discussion of: "Copulas: tales and facts" by T. Mikosch [Extremes **9** (2006), no. 1, 3–20]. Extremes **9**(1), 27–36 (2006)
86. Genest, C., Rémillard, B., Beaudoin, D.: Goodness-of-fit tests for copulas: a review and a power study. Insur. Math. Econ. **44**(2), 199–213 (2009)
87. Genest, C., Rivest, L.P.: On the multivariate probability integral transformation. Stat. Probab. Lett. **53**(4), 391–399 (2001)
88. Giacomini, E., Härdle, W., Spokoiny, V.: Inhomogeneous dependence modeling with timevarying copulae. J. Bus. Econ. Stat. **27**(2), 224–234 (2009)
89. Gudendorf, G., Segers, J.: Extreme value theory and copulae. In: Jaworski, P., Durante, F., Härdle, W., Rychlik, T. (eds.) Copula Theory and Its Applications, Proceedings of the Workshop, Warsaw, 25–26 Sept 2009. Springer, Dordrecht (2010)
90. Gumbel, E.J.: Distributions à plusieurs variables dont les marges sont données. C. R. Acad. Sci. Paris **246**, 2717–2719 (1958)
91. Gumbel, E.J.: Bivariate exponential distributions. J. Am. Stat. Assoc. **55**, 698–707 (1960)
92. Gumbel, E.J.: Distributions des valeurs extrêmes en plusieurs dimensions. Publ. Inst. Stat. Univ. Paris **9**, 171–173 (1960)
93. Gumbel, E.J.: Bivariate logistic distributions. J. Am. Stat. Assoc. **56**, 335–349 (1961)
94. de Haan, L.: Discussion of: "Copulas: tales and facts" by T. Mikosch [Extremes **9** (2006), no. 1, 3–20]. Extremes **9**(1), 21–22 (2006)
95. Härdle, W., Okhrin, O.: De copulis non est disputandum. Copulae: an overview. AStA Adv. Stat. Anal. **94**(1), 1–31 (2010).
96. Hoeffding, W.: Maßstabinvariante Korrelationstheorie. Schriften des Mathematischen Instituts und des Instituts für Angewandte Mathematik der Universität Berlin **5**(3), 179–233 (1940). (Reprinted as "Scale-invariant correlation theory" in Fisher, N.I., Sen, P.K. (eds.) The Collected Works of Wassily Hoeffding, pp. 57–107. Springer, New York, NY, 1994)
97. Hofert, M.: Construction and sampling of nested Archimedean copulas. In: Jaworski, P., Durante, F., Härdle, W., Rychlik, T. (eds.) Copula Theory and Its Applications, Proceedings of the Workshop, Warsaw, 25–26 Sept 2009. Springer, Dordrecht (2010)
98. Höffding, W.: Maszstabinvariante Korrelationstheorie. Schr. Math. Inst. u. Inst. Angew. Math. Univ. Berlin **5**, 181–233 (1940)

99. Hougaard, P.: A class of multivariate failure time distributions. Biometrika **73**(3), 671–678 (1986)
100. Hutchinson, T.P., Lai, C.D.: Continuous Bivariate Distributions, Emphasising Applications. Rumsby Scientific Publishing, Adelaide (1990)
101. Ibragimov, R.: Copula-based characterizations for higher order Markov processes. Economet. Theor. **25**(3), 819–846 (2009)
102. Jaworski, P.: On copulas and their diagonals. Inform. Sci. **179**(17), 2863–2871 (2009)
103. Jaworski, P., Rychlik, T.: On distributions of order statistics for absolutely continuous copulas with applications to reliability. Kybernetika (Prague) **44**(6), 757–776 (2008)
104. Joe, H.: Multivariate Models and Dependence Concepts. Monographs on Statistics and Applied Probability, vol. 73. Chapman & Hall, London (1997)
105. Joe, H.: Discussion of: "Copulas: tales and facts" by T. Mikosch [Extremes **9** (2006), no. 1, 3–20]. Extremes **9**(1), 37–41 (2006)
106. Kallenberg, O.: Foundations of Modern Probability, 2nd edn. Probability and Its Applications (New York). Springer, New York, NY (2002)
107. Kallsen, J., Tankov, P.: Characterization of dependence of multidimensional Lévy processes using Lévy copulas. J. Multivar. Anal. **97**(7), 1551–1572 (2006)
108. Khoudraji, A.: Contributions à l'étude des copules et à la modélisation des valeurs extremes bivariées. Ph.D. thesis, Université de Laval, Québec, Canada (1995)
109. Kimeldorf, G., Sampson, A.: Uniform representations of bivariate distributions. Commun. Stat. **4**(7), 617–627 (1975)
110. Klement, E.P., Kolesárová, A.: Intervals of 1-Lipschitz aggregation operators, quasi-copulas, and copulas with given affine section. Monatsh. Math. **152**(2), 151–167 (2007)
111. Klement, E.P., Kolesárová, A., Mesiar, R., Sempi, C.: Copulas constructed from horizontal sections. Commun. Stat. Theory Methods **36**(13–16), 2901–2911 (2007)
112. Klement, E.P., Mesiar, R., Pap, E.: Triangular Norms. Trends in Logic—Studia Logica Library, vol. 8. Kluwer Academic Publishers, Dordrecht (2000)
113. Klement, E.P., Mesiar, R., Pap, E.: Archimax copulas and invariance under transformations. C. R. Math. Acad. Sci. Paris **340**(10), 755–758 (2005)
114. Klement, E.P., Mesiar, R., Pap, E.: Transformations of copulas. Kybernetika (Prague) **41**(4), 425–434 (2005)
115. Klüppelberg, C., Resnick, S.I.: The Pareto copula, aggregation of risks, and the emperor's socks. J. Appl. Probab. **45**(1), 67–84 (2008)
116. Koehler, K.J., Symanowski, J.T.: Constructing multivariate distributions with specific marginal distributions. J. Multivar. Anal. **55**(2), 261–282 (1995)
117. Kolesárová, A., Mesiar, R., Sempi, C.: Measure-preserving transformations, copulæ and compatibility. Mediterr. J. Math. **5**(3), 325–339 (2008)
118. Kolev, N., dos Anjos, U., Mendes, B.: Copulas: a review and recent developments. Stoch. Models **22**(4), 617–660 (2006)
119. Kollo, T.: Preface. J. Stat. Plan. Infer. **139**(11), 3740 (2009)
120. Lagerås, A.N.: Copulas for Markovian dependence. Bernoulli **16**(2), 331–342 (2010)
121. Li, D.: On default correlation: a copula function approach. J. Fixed Income **9**, 43–54 (2001)
122. Li, H.: Duality of the multivariate distributions of Marshall-Olkin type and tail dependence. Commun. Stat. Theory Methods **37**(11–12), 1721–1733 (2008)
123. Li, H.: Tail dependence comparison of survival Marshall-Olkin copulas. Methodol. Comput. Appl. Probab. **10**(1), 39–54 (2008)
124. Li, H., Scarsini, M., Shaked, M.: Linkages: a tool for the construction of multivariate distributions with given nonoverlapping multivariate marginals. J. Multivar. Anal. **56**(1), 20–41 (1996)
125. Liebscher, E.: Construction of asymmetric multivariate copulas. J. Multivar. Anal. **99**(10), 2234–2250 (2008)
126. Lindner, A.: Discussion of: "Copulas: tales and facts" by T. Mikosch [Extremes **9** (2006), no. 1, 3–20]. Extremes **9**(1), 43–44 (2006)

127. Mai, J.F., Scherer, M.: Efficiently sampling exchangeable Cuadras-Augé copulas in high dimensions. Inform. Sci. **179**(17), 2872–2877 (2009)
128. Mai, J.F., Scherer, M.: Lévy-Frailty copulas. J. Multivar. Anal. **100**(7), 1567–1585 (2009)
129. Mai, J.F., Scherer, M.: Reparameterizing Marshall-Olkin copulas with applications to high-dimensional sampling. J. Stat. Comput. Simul. (2009) In press
130. Malevergne, Y., Sornette, D.: Extreme Financial Risks. Springer, Berlin (2006)
131. Mardia, K.V.: Multivariate Pareto distributions. Ann. Math. Stat. **33**, 1008–1015 (1962)
132. Mardia, K.V.: Families of Bivariate Distributions. Hafner Publishing Co., Darien, Conn. (1970). Griffin's Statistical Monographs and Courses, No. 27
133. Marshall, A.W.: Copulas, marginals, and joint distributions. In: Distributions with Fixed Marginals and Related Topics (Seattle, WA, 1993). IMS Lecture Notes Monograph Series, vol. 28, pp. 213–222. Institute of Mathematical Statistics, Hayward, CA (1996)
134. Marshall, A.W., Olkin, I.: A multivariate exponential distribution. J. Am. Stat. Assoc. **62**, 30–44 (1967)
135. Mayor, G., Mesiar, R., Torrens, J.: On quasi-homogeneous copulas. Kybernetika (Prague) **44**(6), 745–756 (2008)
136. McNeil, A.J., Nešlehová, J.: From Archimedean to Liouville copulas. J. Multivariate Anal. **101**(8), 1772–1790 (2010)
137. McNeil, A.J., Nešlehová, J.: Multivariate Archimedean copulas, d-monotone functions and $\lambda_1$-norm symmetric distributions. Ann. Stat. **37**(5B), 3059–3097 (2009)
138. McNeil, A.J., Frey, R., Embrechts, P.: Quantitative risk management. Concepts, techniques and tools. Princeton Series in Finance. Princeton University Press, Princeton, NJ (2005)
139. Mesiar, R., Sempi, C.: Ordinal sums and idempotents of copulas. Aequationes Math. **79** (1–2), 39–52 (2010)
140. Mikosch, T.: Copulas: tales and facts. Extremes **9**(1), 3–20 (2006)
141. Mikosch, T.: "Copulas: tales and facts" [Extremes **9** (2006), no. 1, 3–20]—rejoinder. Extremes **9**(1), 55–62 (2006)
142. Mikusiński, P., Taylor, M.D.: Some approximations of $n$-copulas. Metrika (2009). In press
143. Mikusiński, P., Sherwood, H., Taylor, M.D.: Probabilistic interpretations of copulas and their convex sums. In: Advances in Probability Distributions with Given Marginals (Rome, 1990). Mathematics and Its Applications, vol. 67, pp. 95–112. Kluwer Academic Publishers, Dordrecht (1991)
144. Mikusiński, P., Sherwood, H., Taylor, M.D.: Shuffles of Min. Stochastica **13**(1), 61–74 (1992)
145. Moore, D.S., Spruill, M.C.: Unified large-sample theory of general chi-squared statistics for tests of fit. Ann. Stat. **3**, 599–616 (1975)
146. Morgenstern, D.: Einfache Beispiele zweidimensionaler Verteilungen. Mitteilungsbl. Math. Stat. **8**, 234–235 (1956)
147. Morillas, P.M.: A method to obtain new copulas from a given one. Metrika **61**(2), 169–184 (2005)
148. Nadarajah, S.: Marshall and Olkin's distributions. Acta Appl. Math. **103**(1), 87–100 (2008)
149. Nappo, G., Spizzichino, F.: Kendall distributions and level sets in bivariate exchangeable survival models. Inform. Sci. **179**(17), 2878–2890 (2009)
150. Nelsen, R.B.: An Introduction to Copulas. Lecture Notes in Statistics, vol. 139. Springer, New York, NY (1999)
151. Nelsen, R.B.: An Introduction to Copulas, 2nd edn. Springer Series in Statistics. Springer, New York, NY (2006)
152. Nelsen, R.B., Quesada-Molina, J.J., Rodríguez-Lallena, J.A., Úbeda-Flores, M.: Bounds on bivariate distribution functions with given margins and measures of association. Commun. Stat. Theory Methods **30**(6), 1155–1162 (2001)
153. Nelsen, R.B., Quesada-Molina, J.J., Rodríguez-Lallena, J.A., Úbeda-Flores, M.: Some new properties of quasi-copulas. In: Cuadras, C., Fortiana, J., Rodrí guez Lallena, J. (eds.) Distributions with Given Marginals and Statistical Modelling, pp. 187–194. Kluwer, Dordrecht (2003)

154. Nelsen, R.B., Quesada-Molina, J.J., Rodríguez-Lallena, J.A., Úbeda-Flores, M.: Best possible bounds on sets of bivariate distribution functions. J. Multivar. Anal. **90**(2), 348–358 (2004)
155. Nelsen, R.B., Quesada-Molina, J.J., Rodríguez-Lallena, J.A., Úbeda-Flores, M.: On the construction of copulas and quasi-copulas with given diagonal sections. Insur. Math. Econ. **42**(2), 473–483 (2008)
156. Nelsen, R.B., Úbeda-Flores, M.: A comparison of bounds on sets of joint distribution functions derived from various measures of association. Commun. Stat. Theory Methods **33**(10), 2299–2305 (2004)
157. Nelsen, R.B., Úbeda-Flores, M.: The lattice-theoretic structure of sets of bivariate copulas and quasi-copulas. C. R. Math. Acad. Sci. Paris **341**(9), 583–586 (2005)
158. Oakes, D.: A model for association in bivariate survival data. J. Roy. Stat. Soc. Ser. B **44**(3), 414–422 (1982)
159. Owzar, K., Sen, P.K.: Copulas: concepts and novel applications. Metron **61**(3), 323–353 (2004) (2003)
160. Patton, A.J.: Copula-based models for financial time series. In: Andersen, T.G., Davis, R.A., Kreiss, J.P., Mikosch, T. (eds.) Handbook of Financial Time Series, pp. 767–785. Springer, New York, NY (2009)
161. Peng, L.: Discussion of: "Copulas: tales and facts" by T. Mikosch [Extremes **9** (2006), no. 1, 3–20]. Extremes **9**(1), 49–50 (2006)
162. Puccetti, G., Scarsini, M.: Multivariate comonotonicity. J. Multivar. Anal. **101**(1), 291–304 (2010)
163. Quesada-Molina, J.J., Rodríguez-Lallena, J.A.: Some advances in the study of the compatibility of three bivariate copulas. J. Ital. Stat. Soc. **3**(3), 397–417 (1994)
164. Quesada-Molina, J.J., Rodríguez-Lallena, J.A.: Bivariate copulas with quadratic sections. J. Nonparametr. Stat. **5**(4), 323–337 (1995)
165. Quesada-Molina, J.J., Saminger-Platz, S., Sempi, C.: Quasi-copulas with a given subdiagonal section. Nonlinear Anal. **69**(12), 4654–4673 (2008)
166. Rényi, A.: On measures of dependence. Acta Math. Acad. Sci. Hungar. **10**, 441–451 (1959)
167. Resnick, S.I.: Extreme Values, Regular Variation, and Point Processes. Applied Probability. A Series of the Applied Probability Trust, vol. 4. Springer, New York, NY (1987)
168. Rodríguez-Lallena, J.A.: A class of copulas with piecewise linear horizontal sections. J. Stat. Plan. Infer. **139**(11), 3908–3920 (2009)
169. Rodríguez-Lallena, J.A., Úbeda-Flores, M.: Best-possible bounds on sets of multivariate distribution functions. Commun. Stat. Theory Methods **33**(4), 805–820 (2004)
170. Rodríguez-Lallena, J.A., Úbeda-Flores, M.: A new class of bivariate copulas. Stat. Probab. Lett. **66**(3), 315–325 (2004)
171. Rodríguez-Lallena, J.A., Úbeda-Flores, M.: Multivariate copulas with quadratic sections in one variable. Metrika (2009). In press
172. Rodríguez-Lallena, J.A., Úbeda-Flores, M.: Some new characterizations and properties of quasi-copulas. Fuzzy Sets Syst. **160**(6), 717–725 (2009)
173. Rüschendorf, L.: Sharpness of Fréchet-bounds. Z. Wahrsch. Verw. Gebiete **57**(2), 293–302 (1981)
174. Rüschendorf, L.: Construction of multivariate distributions with given marginals. Ann. Inst. Stat. Math. **37**(2), 225–233 (1985)
175. Rüschendorf, L.: On the distributional transform, Sklar's Theorem, and the empirical copula process. J. Stat. Plan. Infer. **139**(11), 3921–3927 (2009)
176. Rüschendorf, L., Schweizer, B., Taylor, M. (eds.): Distributions with Fixed Marginals and Related Topics. Institute of Mathematical Statistics Lecture Notes—Monograph Series, vol. 28. Institute of Mathematical Statistics, Hayward, CA (1996)
177. Salvadori, G., De Michele, C., Kottegoda, N.T., Rosso, R.: Extremes in Nature. An Approach Using Copulas. Water Science and Technology Library, vol. 56. Springer, Dordrecht (NL) (2007)

178. Sarmanov, O.V.: Generalized normal correlation and two-dimensional Fréchet classes. Dokl. Akad. Nauk SSSR **168**, 32–35 (1966)
179. Scarsini, M.: Copulae of probability measures on product spaces. J. Multivar. Anal. **31**(2), 201–219 (1989)
180. Schmid, F.: Copula-based measures of multivariate association. In: Jaworski, P., Durante, F., Härdle, W., Rychlik, T. (eds.) Copula Theory and Its Applications, Proceedings of the Workshop, Warsaw, 25–26 Sept 2009. Springer, Dordrecht (2010)
181. Schönbucker, P.: Credit Derivatives Pricing Models: Models, Pricing, Implementation. Wiley Finance Series. Wiley, Chichester (2003)
182. Schweizer, B.: Thirty years of copulas. In: Dall'Aglio, G., Kotz, S., Salinetti, G. (eds.) Advances in Probability Distributions with Given Marginals (Rome, 1990). Mathematics and Its Applications, vol. 67, pp. 13–50. Kluwer Academic Publishers, Dordrecht (1991)
183. Schweizer, B.: Introduction to copulas. J. Hydrol. Eng. **12**(4), 346–346 (2007)
184. Schweizer, B., Sklar, A.: Espaces métriques aléatoires. C. R. Acad. Sci. Paris **247**, 2092–2094 (1958)
185. Schweizer, B., Sklar, A.: Operations on distribution functions not derivable from operations on random variables. Studia Math. **52**, 43–52 (1974)
186. Schweizer, B., Sklar, A.: Probabilistic Metric Spaces. North-Holland Series in Probability and Applied Mathematics. North-Holland Publishing Co., New York, NY (1983)
187. Schweizer, B., Wolff, E.F.: Sur une mesure de dépendance pour les variables aléatoires. C. R. Acad. Sci. Paris Sér. A **283**, 659–661 (1976)
188. Schweizer, B., Wolff, E.F.: On nonparametric measures of dependence for random variables. Ann. Stat. **9**(4), 879–885 (1981)
189. Segers, J.: Efficient estimation of copula parameter. Discussion of: "Copulas: tales and facts" [Extremes **9** (2006), no. 1, 3–20] by T. Mikosch. Extremes **9**(1), 51–53 (2006)
190. Sempi, C.: Copulæ and their uses. In: Doksum, K., Lindquist, B. (eds.) Mathematical and Statistical Methods in Reliability, pp. 73–86. World Scientific, Singapore (2003)
191. Siburg, K.F., Stoimenov, P.A.: Gluing copulas. Commun. Stat. Theory Methods **37**(19), 3124–3134 (2008)
192. Sklar, A.: Fonctions de répartition à *n* dimensions et leurs marges. Publ. Inst. Stat. Univ. Paris **8**, 229–231 (1959)
193. Sklar, A.: Random variables, joint distribution functions, and copulas. Kybernetika (Prague) **9**, 449–460 (1973)
194. Sklar, A.: Random variables, distribution functions, and copulas—a personal look backward and forward. In: Distributions with Fixed Marginals and Related Topics (Seattle, WA, 1993). IMS Lecture Notes Monograph Series, vol. 28, pp. 1–14. Institute of Mathematical Statistics, Hayward, CA (1996)
195. Takahasi, K.: Note on the multivariate Burr's distribution. Ann. Inst. Stat. Math. **17**, 257–260 (1965)
196. Trivedi, P.K., Zimmer, D.M.: Copula modeling: an introduction for practitioners. Now Publishers, Hanover, Mass. (2007)
197. Úbeda-Flores, M.: Multivariate copulas with cubic sections in one variable. J. Nonparametr. Stat. **20**(1), 91–98 (2008)
198. Whitehouse, M.: How a formula ignited market that burned some big investors. Wall St. J. (2005). Published on 12 Sept 2005
199. Williams, D.: Probability with martingales. Cambridge Mathematical Textbooks. Cambridge University Press, Cambridge (1991)
200. Wolff, E.F.: Measures of dependence derived from copulas. Ph.D. thesis, University of Massachusetts, Amherst (1977)
201. Wolff, E.F.: *n*-dimensional measures of dependence. Stochastica **4**(3), 175–188 (1980)

# Chapter 2
# Dynamic Modeling of Dependence in Finance via Copulae Between Stochastic Processes

Tomasz R. Bielecki, Jacek Jakubowski and Mariusz Niewęgłowski

**Abstract** Modeling of stochastic dependence is crucial to pricing and hedging of basket derivatives, as well as to pricing and hedging of some other financial products, such as rating-triggered corporate step-up bonds. The classical approach to modeling of dependence in finance via static copulae (and Sklar's theorem) is inadequate for consistent valuation and hedging in time. In this survey we present recent developments in the area of modeling of dependence between stochastic processes with given marginal laws. Some of these results have already been successfully applied in finance in connection with the portfolio credit risk.

## 2.1 Introduction

Dynamic modeling of dependence between financial risks is crucial to achieving consistent calibration through time to market data, as well as to dynamic hedging of these risks.

The classical approaches to modeling dependence in finance were typically rooted in the static copula theory (see e.g. [7]). A standard example is the Gaussian copula model introduced by David Li [25], which was widely used by practitioners.

---

Tomasz R. Bielecki
Department of Applied Mathematics, Illinois Institute of Technology, Chicago, IL, USA
e-mail: bielecki@iit.edu

Jacek Jakubowski
Institute of Mathematics, University of Warsaw, Warszawa, Poland; Faculty of Mathematics and Information Science, Warsaw University of Technology, Warszawa, Poland
e-mail: jakub@mimuw.edu.pl

Mariusz Niewęgłowski
Faculty of Mathematics and Information Science, Warsaw University of Technology, Warszawa, Poland
e-mail: M.Nieweglowski@mini.pw.edu.pl

P. Jaworski et al. (eds.), *Copula Theory and Its Applications*, Lecture Notes in Statistics 198,
DOI 10.1007/978-3-642-12465-5_2, 

But the static copulae models were not capable of effectively dealing with the dynamic aspects of dependence between financial risks. Some people went to the extreme of blaming the recent financial crisis on the use of static models (cf. e.g. [28]), accusations that definitely went too far (for a critique of these accusations see e.g. [14] or [33]).

Recently, an effort has been started to model dependence between financial risks in a dynamic way. The so called Lévy copulae were studied in Kallsen and Tankov [21]. Markov copulae were introduced in Bielecki, Jakubowski, Vidozzi and Vidozzi [4], and subsequently studied in Bielecki, Vidozzi and Vidozzi [6] and Bielecki, Jakubowski and Niewęgłowski [3]. Motivated by the results in [4] the so called semimartingale copulae were formally defined and studied in L. Vidozzi [32]. Stochastic dependence between the components of a multivariate Markov process in terms of its infinitesimal operator was investigated in A. Vidozzi [31]. A related, but different line of research devoted to modeling dynamic dependence between stopping times and applications to credit risk was originated in El Karoui, Jeanblanc and Jiao [12], where the conditional density approach is used.

In this chapter we shall describe an approach to dynamic modeling of dependence in finance, based on modeling of dependence between stochastic processes, and using the results from Bielecki et al. [4, 6], Vidozzi [32] and Bielecki et al. [3], Cont and Tankov [8] and Kallsen and Tankov [21]. In particular, we shall describe various ways of modeling dependence between stochastic processes so that the laws of individual components of a multivariate process agree with some prescribed laws. Therefore, with an abuse of terminology, we shall refer to relevant constructions as *semimartingale copulae* and *Markov copulae*. It needs to be stressed, though, that the term "copula" is used here for convenience and for its historical connotation only.

The objective of the methodology outlined in this article is different from that in Lageras [22], in which results of Darsow et al. [11] are extended. Those two papers aim at relating the classical concept of copula and the concept of Markov property. In this context they investigate dependence along the time line in the case of a one-dimensional Markov process, and characterize the Markov property in terms of copulae. Next, Ibragimov [17] generalized results of Darsow et al. [11] to higher order Markov processes. The problem that we present here is also different from [12], since their interest is not in building models with prescribed marginal laws.

Sections 2.3 and 2.4.2 are, for the most part, taken from [31, 32], respectively. The proofs are skipped though in this survey article. We refer the interested reader to [4, 31, 32] for a comprehensive treatment of the relevant topics. Analogous remarks apply to Sect. 2.4.3 based on [3], and Sect. 2.5 based on [5].

The paper is organized as follows. Section 2.2 describes Lévy copulae. Semimartingale copulae are defined and investigated in Sect. 2.3. We consider semimartingales that are uniquely characterized, in the sense of their probability laws, by their characteristics. We construct a process $X$ whose $i$-th univariate law, i.e., the law of the $i$-th component $X^i$, is the same as the law of a given process $Y^i$, $i = 1,\dots,n$. Section 2.4 is devoted to Markov copulae. We present two different approaches: generator based and symbolic. The first is based on infinitesimal generators, the second applies pseudo-differential operators. The last section presents an application to finance of copulae so defined.

## 2.2 Lévy Copulae

As is well known, the law of a multivariate Lévy process is entirely determined by any of its one-dimensional distributions. Thus, creating dependence between univariate components of a multivariate Lévy process essentially amounts to creating dependence between finite dimensional random variables. The problem is that if one wants to do this things in terms of the Lévy characteristics of the process, then one needs, among other, to create dependence between the marginal Lévy measures, which, in general, are not finite measures. This leads to certain technical difficulties that however were successfully dealt with in the papers by Tankov [30] and Kallsen and Tankov [21].

Tankov [30] introduced Lévy copulae to characterize dependence between components of a multidimensional Lévy process. His construction is for Lévy processes with positive jumps in every component. Later, Kallsen and Tankov [21] generalized this concept to arbitrary Lévy processes. A Lévy copula is a counterpart of the notion of copula for multivariate distributions. Copulae give a characterization of possible dependence structures of a random vector, given the margins, and allow one to construct a multidimensional distribution with specified dependence from a collection of one-dimensional distributions. Similarly, the aim of Lévy copulae is to provide a way to construct multivariate Lévy processes with given marginals.

Since the dependence structure of the Brownian motion part of a Lévy process $X$ is characterized entirely by its covariance matrix, and since the Brownian motion part of $X$ is independent of the jump part, it remains to describe the dependence structure of the purely discontinuous part of $X$, and this is done by means of Lévy copulae.

Now, we present formal definitions.

**Definition 2.2.1.** Let $\overline{\mathbb{R}} := (-\infty, +\infty]$. A function $F : \overline{\mathbb{R}}^d \to \mathbb{R}$ is $d$-increasing if $F(u_1, \ldots, u_d) \neq \infty$ for $(u_1, \ldots, u_d) \neq (\infty, \ldots, \infty)$ and

$$\sum_{c \in \{a_1, b_1\} \times \ldots \times \{a_d, b_d\}} (-1)^{N(c)} F(c) \geq 0$$

for any $-\infty < a_i \leq b_i \leq \infty$ and $N(c) := \#\{k : c_k = a_k\}$.

For a $d$-increasing function we can define margins in a similar way as for a probability distribution function. To do this, we set

$$\operatorname{sgn} x = \begin{cases} 1, & x \geq 0, \\ -1, & x < 0. \end{cases}$$

**Definition 2.2.2.** Let $F$ be a $d$-increasing function. For any nonempty index set $I \subset \{1,\dots,d\}$ the $I$-margin of $F$ is the function $F^I : \overline{\mathbb{R}}^I \to \overline{\mathbb{R}}$ defined by

$$F^I((u_i)_{i\in I}) := \lim_{k\to\infty} \sum_{(u_j)_{j\in I^c}\in\{-k,\infty\}^{I^c}} F(u_1,\dots,u_d) \prod_{j\in I^c} \operatorname{sgn} u_j,$$

where $I^c := \{1,\dots,d\} \setminus I$.

In particular, for $I = \{i\}$, $F^{\{i\}}$, the $i$-th margin of $F$, is given by

$$F^{\{i\}}(x) := \lim_{c\to-\infty} \left(F(+\infty,\dots,+\infty,x,+\infty,\dots,+\infty) - F(c,\dots,c,x,c,\dots,c)\right).$$

**Definition 2.2.3.** A function $F : \overline{\mathbb{R}}^d \to \overline{\mathbb{R}}$ is called a *Lévy copula* if

1. $F(u_1,\dots,u_d) = 0$ if $u_i = 0$ for at least one $i \in \{1,\dots,d\}$,
2. $F$ is $d$-increasing,
3. $F^{\{i\}}(u) = u$ for any $i \in \{1,\dots,d\}$, $u \in \mathbb{R}$.

It is worth noting that Lévy copulae have properties similar to ordinary copulae; in particular, they are Lipschitz continuous. In order to use Lévy copulae to investigate dependence between components of a general multivariate Lévy process we have to define a tail integral.

**Definition 2.2.4.** Let $X$ be an $\mathbb{R}^d$-valued Lévy process with Lévy measure $\nu$. The tail integral of $X$ is the function $U : (\mathbb{R}\setminus\{0\})^d \to \mathbb{R}$ defined by

$$U(x_1,\dots,x_d) := \prod_{i=1}^d \operatorname{sgn}(x_i)\nu\left(\prod_{j=1}^d \mathfrak{I}(x_j)\right),$$

where, for $x \in \mathbb{R}$, we denote

$$\mathfrak{I}(x) := \begin{cases} (x,\infty), & x \geq 0, \\ (-\infty,x], & x < 0. \end{cases}$$

The tail integral does not determine the Lévy measure uniquely, in general, since it does not give any information about mass on coordinate axes.

This motivates introducing the *$I$-marginal tail integral* $U^I$, for nonempty set $I \subset \{1,\dots,d\}$. It is the tail integral of the Lévy process $(X^i)_{i\in I}$, or equivalently the tail integral of the $I$-marginal of the Lévy measure $\nu$, that is, the measure $\nu^I$ defined by

$$\nu^I(A) := \nu(x \in \mathbb{R}^d : (x_i)_{i\in I} \in A\setminus\{0\}) \quad \text{for } A \in \mathscr{B}(\mathbb{R}^{|I|}).$$

It turns out that to determine the Lévy measure uniquely we have to know all marginal tail integrals, i.e., we have to know $U^I$ for all $I \subset \{1,\dots,d\}$ (see [21, Lemma 3.5]). In fact, there is one-to-one correspondence between the Lévy measure and the set of all marginal tail integrals.

The most important feature of Lévy copulae is that they allow separating the margins and the dependence structure of Lévy measures. This is clear from the following counterpart of Sklar's theorem, proved by Kallsen and Tankov [21].

**Theorem 2.2.1.** *1) Let $X = (X_1,\dots,X_d)$ be an $\mathbb{R}^d$-valued Lévy process. Then there exists a Lévy copula $F$ such that the tail integrals of $X$ satisfy*

$$U^I\left((x_i)_{i\in I}\right) = F^I\left((U_i(x_i))_{i\in I}\right) \tag{2.1}$$

*for any nonempty $I \subset \{1,\dots,d\}$ and any $(x_i)_{i\in I} \in (\mathbb{R}\setminus\{0\})^I$. The Lévy copula $F$ is unique on $\prod_{i=1}^d \overline{\operatorname{Ran} U_i}$.*
*2) Let $F$ be a $d$-dimensional Lévy copula and $U_i$, $i = 1,\dots,d$, be the tail integrals of real valued Lévy processes. Then there exists an $\mathbb{R}^d$-valued Lévy process whose components have tail integrals $U_1,\dots,U_d$ and whose marginal tail integral satisfies condition* (2.1) *for any nonempty $I \subset \{1,\dots,d\}$ and any $x \in (\mathbb{R}\setminus\{0\})^I$. The Lévy measure $\nu$ of $X$ is uniquely determined by $F$ and $U_i$, $i = 1,\dots,d$.*

We now proceed with a few examples of Lévy copulae.

*Example 2.2.1.* In Kallsen and Tankov [21] it is shown that a pure jump Lévy process has independent coordinates if and only if its Lévy copula is given by the following formula:

$$F_\perp(x_1,\dots,x_d) := \sum_{i=1}^d x_i \prod_{j\neq i} 1_{\{\infty\}}(x_j).$$

*Example 2.2.2.* Kallsen and Tankov [21] introduce an *Archimedean Lévy copula*, analogously to an ordinary Archimedean copula, by setting

$$F(x_1,\dots,x_d) := \varphi\left(\prod_{i=1}^d \tilde{\varphi}(u_i)\right),$$

where $\varphi : [-1,1] \to [-\infty,\infty]$ is a strictly increasing continuous function with $\varphi(-1) = -\infty$, $\varphi(0) = 0$, $\varphi(1) = \infty$, and having derivatives up to order $d$ on the intervals $(-1,0)$ and $(0,1)$, satisfying

$$\frac{\partial^d \varphi(e^x)}{\partial x^d} \geq 0, \quad \frac{\partial^d \varphi(-e^x)}{\partial x^d} \leq 0, \quad x \in (-\infty,0),$$

and where $\tilde{\varphi}$ is defined by

$$\tilde{\varphi}(u) := 2^{d-2}(\varphi(u) - \varphi(-u)).$$

*Example 2.2.3.* Bauerle et al. [2] observed that in the case $d > 2$ the family of Archimedean Lévy copulae fails to generate positively dependent Lévy processes.[1] In or-

[1] Three concepts of dependence are introduced in [2]: (positive) association, positive orthant dependence (POD), and positive supermodular dependence (PSMD). According to Corollary 3.10 in [2], all three concepts are equivalent in the case of multivariate Lévy processes. Thus, in the context of Lévy processes we give the same name to all three concepts: *positive dependence*.

der to overcome this problem they proposed to generalize this family. In the first step they noted that for $\phi : (0,\infty) \to (0,\infty)$ a strictly decreasing function with alternating signs of derivatives up to order $d$, and with $\lim_{t\downarrow 0}\phi(t) = \infty$ and $\lim_{t\downarrow\infty}\phi(t) = 0$ the function defined by

$$F_\phi(u_1,\dots,u_d) := \phi\left(\sum_{i=1}^{d}\phi^{-1}(u_i)\right), \quad u_1,\dots,u_d > 0, \tag{2.2}$$

is a Lévy copula on $(0,\infty)^d$. Then, the main idea of [2] was to spread these positive Lévy copulae on all orthants with additional weighting functions. This construction is a generalization of an ordinary Archimedean copula that uses the additive generator rather than the multiplicative one. To make the above idea precise, given functions $F_{\phi_\mathbf{i}}$ defined by (2.2) for $\mathbf{i} \in I := \{-1,1\}^d$, let

$$F(x_1,\dots,x_d) := \begin{cases} \sum_{\mathbf{i}\in I}\left(\eta(\mathbf{i})F_{\phi_\mathbf{i}}(|u_1|,\dots,|u_d|)1_{\{u\in O_\mathbf{i}\}}\prod_{i=1}^{d}\operatorname{sgn}(u_i)\right) & \text{if } |u_j|>0, \\ & j=1,\dots,d, \\ 0 & \text{otherwise,} \end{cases}$$

where $O_\mathbf{i}$ denotes the orthant with signs in $\mathbf{i}$, i.e.,

$$O_\mathbf{i} := \left\{x\in\mathbb{R}^d : sgn(x_j) = i_j, j = 1,\dots,d\right\},$$

and $\eta : I \to [0,1]$ is a weight function having the property

$$\sum_{\mathbf{i}:i_k=-1}\eta(\mathbf{i}) = \sum_{\mathbf{i}:i_k=1}\eta(\mathbf{i}) = 1.$$

The above function $F$ defines a Lévy copula on $\mathbb{R}^d$ which generates positively dependent Lévy processes if and only if $\eta(1,1,...,1) = \eta(-1,-1,...,-1) = 1$.

*Example 2.2.4.* One can obtain a Clayton type Lévy copula by choosing $\varphi(u) = u^{-1/\theta}$ and $\eta(-1,-1) = \eta(1,1) = 1$. Then

$$F_\theta(u_1,u_2) = \left(u_1^{-\theta}+u_2^{-\theta}\right)^{-1/\theta}1_{\{u\in\mathbb{R}^2_{++}\}} + \left((-u_1)^{-\theta}+(-u_2)^{-\theta}\right)^{-1/\theta}1_{\{u\in\mathbb{R}^2_{--}\}}$$

with $\mathbb{R}^2_{++} := \mathbb{R}_+\times\mathbb{R}_+$ and $\mathbb{R}^2_{--} := \mathbb{R}_-\times\mathbb{R}_-$, is a Lévy copula for positively dependent Lévy processes.

*Remark 2.2.1.* In various applications we often need an appropriate algorithm for Monte Carlo simulation of *dependent* Lévy processes. Tankov [29] discusses the issue of generation of sample paths of Lévy processes with given Lévy copulae by using series representations of Lévy processes.

## 2.3 Semimartingale Copulae

In this section, which is based on [4, 32], we study certain aspects of stochastic dependence between some classes of finite dimensional semimartingale processes in terms of their infinitesimal characteristics. We shall only consider the semimartingales that are uniquely characterized, in the sense of their probability laws, by their characteristics.

Let $(\Omega, \mathscr{F}, \mathbb{P})$ be some underlying probability space with $\Omega = \Omega_1 \times \Omega_2 \times \cdots \times \Omega_n$, and let $X = (X^1, X^2, \ldots, X^n)$ be an $\mathbb{R}^n$-valued semimartingale with respect to some filtration, defined on this probability space. Let also $Y^1, Y^2, \ldots, Y^n$ be a collection of semimartingales on $(\Omega_i, \mathscr{F}_i, \mathbb{P}_i)$, with respect to some filtrations. All filtrations in what follows are assumed to satisfy the usual conditions.

For the most part of this section, for simplicity of presentation, we shall only consider the bivariate case, that is, $n = 2$, although the results presented can be generalized to higher dimensions in a straightforward manner.

Our study is motivated by the question which arises naturally in various applications, such as valuation and hedging of financial derivatives written on baskets of underlying securities: What conditions on the local characteristics of a process $X$ are sufficient for the law of $X^i$ to be the same as the law of $Y^i$, $i = 1, 2, \ldots, n$, where the $Y^i$ are given processes. So our aim is to construct a process $X$ so that its $i$-th univariate law, i.e., the law of the $i$-th component $X^i$, is the same as the law of a given process $Y^i$, $i = 1, 2, \ldots, n$.

In this context, the question is reminiscent of the concept of copula functions, and the celebrated Sklar theorem (see [26]). Unfortunately, the complex structure of the cylindrical sigma algebras on canonical spaces does not allow a direct extension of Sklar's results to random variables on function spaces. However, infinitesimal characteristics of a stochastic process are often available, so we study dependence between processes in terms of those infinitesimal characteristics. Consequently, for historical reasons, we somewhat abuse terminology when using the term "copula": various "copulae" that we define below are not really copula functions. Nevertheless, we find this terminology useful and convenient.

Our approach was in part inspired by Tankov [30] and Kallsen and Tankov [21] (cf. Sect. 2.2). Although very appealing, their approach cannot be extended to construct more general processes, as its validity relies on the fact that the jump characteristic of a Lévy process is a measure on a finite dimensional space.

The key role in this section will be played by the canonical characteristics of a semimartingale, that is, the characteristics expressed as functions of the trajectory of the process.

### 2.3.1 *Copulae for Special Semimartingales*

Assume that we are given two real valued semimartingales $Y^1, Y^2$, whose finite dimensional distributions are uniquely determined by the corresponding infinitesimal characteristics. The processes are possibly defined on different (canonical) probability spaces, say $(\Omega_i, \mathscr{F}^{Y^i}, \mathbb{P}^i)$, $i = 1,2$, endowed with the canonical filtrations $\mathbb{F}^{Y^i}$. We would like to construct a probability space $(\Omega, \mathscr{F}^X, \mathbb{P})$, with $\Omega = \Omega_1 \times \Omega_2$, such that the finite dimensional distributions of the components of the canonical process $X = (X^1, X^2)$ on that space are identical to those of $Y^1, Y^2$. In what follows we use the notation:

$$\mathscr{F}_t^X = \bigcap_{s>t} \sigma(X_r, r \le s), \quad \mathscr{F}^X := \mathscr{F}_\infty^X := \bigvee_{t>0} \mathscr{F}_t^X, \quad \mathbb{F}^X := \{\mathscr{F}_t^X\}_{t \ge 0},$$

for a semimartingale $X$.

Consider a bivariate semimartingale, $X = (X^1, X^2)$, defined on a stochastic basis $(\Omega, \mathscr{F}, \mathbb{F}, \mathbb{P})$, where the components $X^1, X^2$ are real valued semimartingales. We assume that the finite dimensional distributions of the vector process $X$ are uniquely determined by its $\mathbb{F}^X$ characteristic triple.[2]

We first examine the problem of finding the $\mathbb{F}^{X^i}$ characteristic triple of $X^i, i = 1,2$ (i.e., the canonical characteristic triple of the coordinate processes) knowing the $\mathbb{F}^X$ characteristic triple of the components $X^i, i = 1,2$. We provide a characterization of the canonical characteristics of $X^i$ in terms of projections of their $\mathbb{F}^X$ characteristic triple. Next we illustrate the theory on some examples, for which we compute explicitly the $\mathbb{F}^{X^i}$ characteristics of the coordinate processes. Finally, we explore how to extend these to determine the distribution of $X = (X^1, X^2)$ in some cases of interest.

**Characteristics of the coordinate processes.** Assume that $X$ is a semimartingale $X = (X^1, X^2)$ taking values in $\mathbb{R}^2$ and defined on the stochastic basis $(\Omega, \mathscr{F}, \mathbb{F}, \mathbb{P})$. Moreover we are given the $\mathbb{F}^X$-characteristics of $X$, say, $(B, C, \nu)$, where $B = (B^i)$ and $C = [C^{ij}]$ with $i, j = 1,2$ are predictable processes taking values in $\mathbb{R}^2$ and $\mathbb{R}^{2\times 2}$ respectively, and $\nu$ is a predictable random measure on $\mathscr{B}(\mathbb{R}^2) \otimes \mathscr{B}(\mathbb{R}_+)$ (the dual predictable projection of the integer valued, optional random measure $\mu$ counting the jumps of $X$).

We introduce the following notation:

- $\mu^i$ is the integer valued, optional random measure on $\mathscr{B}(\mathbb{R}) \otimes \mathscr{B}(\mathbb{R}_+)$ counting the jumps of the process $X^i$;
- $\nu^i$ is the compensator of $\mu^i$ in the filtration $\mathbb{F}^X$;
- $\widetilde{\nu}^i$ is the compensator of $\mu^i$ with respect to $\mathbb{F}^{X^i}$;
- ${}^{o_i}(Z)$ (or ${}^{p_i}(Z)$) is the optional (resp. predictable) projection of the process $Z$ on $\mathbb{F}^{X^i}$;
- $(Z)^{p_i}$ (or $(\mu)^{p_i}$) is the dual predictable projection of the process $Z$ (resp. of the random measure $\mu$) on $\mathbb{F}^{X^i}$.

[2] The precise meaning of this statement is the following: If $(\Omega, \mathscr{F}, \mathbb{F}, \mathbb{P})$ and $(\Omega, \mathscr{F}, \mathbb{F}, \mathbb{Q})$ are two probability spaces, where $(\Omega, \mathscr{F}, \mathbb{F})$ is a canonical space endowed with canonical filtration such that the canonical process $X$ is a semimartingale on both stochastic bases with the same characteristics, then $\mathbb{P} = \mathbb{Q}$. This implies that two semimartingales, defined on $(\Omega, \mathscr{F}, \mathbb{F})$, that have the same characteristics, also have the same law. This uniqueness property can be verified in terms of uniqueness of the so called *martingale problem* (cf. [32] for details).

Clearly $\mu^2(dx,dt) = \mu(\mathbb{R},dx,dt)$, $\mu^1(dx,dt) = \mu(dx,\mathbb{R},dt)$ and likewise for $\nu^1, \nu^2$. We make the following standing assumptions:
A1. For $i = 1,2$ and for all $\mathbb{F}^X$ local martingales under consideration, there exists a fundamental sequence of $\mathbb{F}^{X^i}$ stopping times.
A2. The process $X$ is a special semimartingale.

Assumption A1 ensures that we do not destroy the local martingale property by taking projections (if a local martingale is a genuine martingale this assumption is trivially satisfied: consider the sequence $T_n = n$). Assumption A2 ensures that $X$ has a unique canonical semimartingale decomposition. The following results yield the $\mathbb{F}^{X^i}$ characteristics of the processes $X^i$.

**Proposition 2.3.1 (see [32]).** *Let $X^i = X_0^i + M^i + B^i$ denote the canonical decomposition of the semimartingale $X^i$ in the filtration $\mathbb{F}^X$ and let $B^{i+}$ and $B^{i-}$ denote the processes in the Jordan decomposition of $B^i$. Then $X^i$ admits the following canonical decomposition in the filtration $\mathbb{F}^{X^i}$ :*

$$X^i = X_0^i + \widetilde{M}^i + \widetilde{B}^i,$$

*where*

$$\widetilde{M}^i = {}^{o_i}(M^i) + L^{i+} - L^{i-},$$

*with $L^{i+}$ and $L^{i-}$ the local martingale parts of the Doob Meyer decomposition of ${}^{o_i}(B^{i+})$ and ${}^{o_i}(B^{i-})$ respectively, and*

$$\widetilde{B}^i = ({}^{o_i}(B^{i+}))^{p_i} - ({}^{o_i}(B^{i-}))^{p_i}. \tag{2.3}$$

The above proposition yields the first two $\mathbb{F}^{X^i}$ characteristics of the process $X^i$, i.e., $\widetilde{B}^i$ is given by (2.3), and $\widetilde{C}^{ii} = \langle (\widetilde{M}^i)^c, (\widetilde{M}^i)^c \rangle$. Using similar arguments, we compute the jump characteristic of $X^i$ in the filtration $\mathbb{F}^{X^i}$. To this end, we shall need the following results:

**Proposition 2.3.2 (see [32]).** *Fix $i = 1,2$, $s \geq 0$, $A \in \mathscr{B}(\mathbb{R})$ and $B \in \mathscr{F}_s^{X^i}$. Then the process*

$$L_t^i = \mathbb{1}_B \left( \mu^i((s,t],A) - ({}^{o_i}(\nu^i((s,t],A)))^{p_i} \right), \quad t \geq s,$$

*is an $\mathbb{F}^{X^i}$ local martingale.*

From now on we assume that the jumps of the processes $X^1$ and $X^2$ take values in a finite set,[3] say $E = \{x_1, \ldots, x_M\} \subset \mathbb{R}$. For every $x \in E$ and every interval $(s,t]$ we put

$$\widetilde{\nu}^i((s,t],x) := ({}^{o_i}(\nu^i((s,t],x)))^{p_i}.$$

[3] The set $E$ can be interpreted as the mark space.

One can uniquely extend $\widetilde{\nu}^i$ to a measure on $\mathscr{B}(\mathbb{R}_+)\otimes 2^E$. The next proposition shows that this unique extension, denoted by $\widetilde{\nu}^i(dt,dx)$, is indeed the $\mathbb{F}^{X^i}$ compensator of $\mu_i(dt,dx)$.

**Proposition 2.3.3.** *The measure $\widetilde{\nu}^i(dt,dx)$ is the $\mathbb{F}^{X^i}$ dual predictable projection of the counting measure $\mu^i(dt,dx)$.*

If the compensator is absolutely continuous, then we can compute the projections $({}^{o_i}(\nu^i([s,t],A_n)))^{p_i}$ in a simple way:

**Lemma 2.3.1 (see [32]).** *Assume that $\nu^i((s,t],A)$ is (locally) integrable for every set $A$ in $\mathscr{B}(\mathbb{R})$. In addition, assume that $\nu^i((s,t],A)$ is absolutely continuous, i.e.,*

$$\nu^i((s,t],A)(\omega) = \int_s^t \int_A K^i(u,\omega,dx)\,du$$

*for some $\mathscr{F}^X \otimes \mathscr{B}(\mathbb{R})$ measurable kernel $K^i$. Then, for any $s<t<\infty$,*

$$({}^{o_i}(\nu^i((s,t],A)))^{p_i} = \int_s^t {}^{o_i}(K^i(u,A))\,du.$$

We can obtain a similar result for the process $\widetilde{B}^i$, that is, for the finite variation part of the $\mathbb{F}^{X^i}$ semimartingale decomposition of $X^i$.

**Lemma 2.3.2 (see [32]).** *Assume that, for $i=1,2$, $B^i$ is (locally) integrable and absolutely continuous, i.e.,*

$$B^i_t = \int_0^t b^i_s\,ds$$

*for some progressively measurable process $b^i$. Then, for any $t<\infty$,*

$$\widetilde{B}^i_t = ({}^{o_i}(B^{i+}))^{p_i}_t - ({}^{o_i}(B^{i-}))^{p_i}_t = \int_0^t {}^{o_i}(b^i_s)\,ds.$$

Now, we compute explicitly the $\mathbb{F}^{X^i}$ characteristics of the vector semimartingale $X=(X^1,X^2)$ in some special cases.

*Example 2.3.1.* Consider the stochastic basis $(\Omega,\mathscr{F},\mathbb{F},\mathbb{P})$, where $\mathbb{F}:=\mathbb{F}^W\vee\mathbb{F}^{X^2}$, $W$ is an SBM and $X^2$ is a Markov chain that takes values in $\{e_1,\dots,e_N\}$, where $(e_i)_{i=1}^N$ is the standard basis in $\mathbb{R}^N$. We assume that $X^2$ admits a constant generator matrix $A$. Consider the vector process $X=(X^1,X^2)$, where

$$dX^1_t = \langle b,X^2_t\rangle\,dt + \sigma dW_t, \quad X^1_0 = x^1 \in \mathbb{R},$$

with $\sigma\in\mathbb{R}^+$ and $b\in\mathbb{R}^N$. Note that $\mathbb{F}^X=\mathbb{F}$ and the first $\mathbb{F}^X$ characteristic of $X^1$ is given by the process $B^1_t = \int_0^t \langle b,X^2_s\rangle\,ds$. In view of Lemma 2.3.2, the first $\mathbb{F}^{X^1}$ characteristic of $X^1$ is given by the process $\widetilde{B}^i_t = \int_0^t \langle b,p_s\rangle\,ds$, where $p_s := {}^{o_1}(X^2_s)$.

It is a well known result in filtering theory (the so called Wonham filter, see Elliot [13]) that the process $p$ satisfies the following vector SDE:

$$p_t = p_0 + \int_0^t A^T p_s\, ds + \frac{1}{\sigma^2}\int_0^t \mathrm{diag}(p_s)(b - \langle b, p_s\rangle \mathbb{1})(dX_s^1 - \langle b, p_s\rangle\, ds),$$

where $p_0$ is the initial distribution of the chain $X^2$, and $\mathbb{1} := (1,\dots,1)^\top$.

*Example 2.3.2.* Consider a stochastic basis $(\Omega, \mathscr{F}^N, \mathbb{F}^N, \mathbb{P})$, where $\Omega = \Omega_1 \times \Omega_2$ is the canonical space of a bivariate one-point process, and a canonical process, say $N = (N^1, N^2)$, on this space. Observe that $N$ can be identified with a pair of positive random variables $T_1 : \Omega_1 \to \mathbb{R}_+$ and $T_2 : \Omega_2 \to \mathbb{R}_+$ given by $T_1 := \inf\{t > 0 : \Delta N_t^1 \neq 0\}$ and $T_2 := \inf\{t > 0 : \Delta N_t^2 \neq 0\}$. We assume that, under $\mathbb{P}$, the joint probability of $(T_1, T_2)$ admits a density function $f(u,v)$. We first compute the $\mathbb{F}^N$ compensator of $N^i$. By a straightforward application of the Fubini theorem, the $\mathbb{F}^N$ jump characteristic of $N^1$ is

$$\nu^1(ds,dx) = \delta_1(dx)\left(\frac{\int_s^\infty f(s,v)\,dv}{\int_s^\infty\int_s^\infty f(u,v)\,du\,dv}\mathbb{1}_{\{s\le T_1\wedge T_2\}} + \frac{f(s,T_2)}{\int_s^\infty f(u,T_2)\,du}\mathbb{1}_{\{T_2<s\le T_1\}}\right)ds.$$

To compute the canonical jump characteristic of the coordinate processes $N^1$, say $\widetilde{\nu}^1([0,t),\{1\})$, we use Propositions 2.3.2 and Lemma 2.3.1:

$$\begin{aligned}\widetilde{\nu}^1(ds,\{1\}) &= \mathbb{E}\left(\frac{\int_s^\infty f(s,v)\,dv}{\int_s^\infty\int_s^\infty f(u,v)\,du\,dv}\mathbb{1}_{\{s\le T_1\wedge T_2\}}\Big|\mathscr{F}_s^{N^1}\right)ds\\ &\quad + \mathbb{E}\left(\frac{f(s,T_2)}{\int_s^\infty f(u,T_2)\,du}\mathbb{1}_{\{T_2<s\le T_1\}}\Big|\mathscr{F}_s^{N^1}\right)ds.\end{aligned}$$

Since the process $\mathbb{1}_{\{T_1\ge t\}}$ is predictable, it is adapted to $(\mathscr{F}_{t-}^{N^1}, t\ge 0)$. Therefore, for any $s<\infty$, we have

$$\begin{aligned}\widetilde{\nu}^1(ds,\{1\}) &= \frac{\int_s^\infty f(s,v)\,dv}{\int_s^\infty\int_s^\infty f(u,v)\,du\,dv}\mathbb{P}(\{T_2\ge s\}\big|\{T_1\ge s\})\mathbb{1}_{\{T_1\ge s\}}\,ds\\ &\quad+\mathbb{E}\left(\frac{f(s,T_2)}{\int_s^\infty f(u,T_2)\,du}\mathbb{1}_{\{T_2<s\le T_1\}}\big|\{T_1\ge s\}\right)\mathbb{1}_{\{T_1\ge s\}}\,ds\\ &= \frac{\int_s^\infty f(s,v)\,dv}{\int_s^\infty\int_s^\infty f(u,v)\,du\,dv}\,\frac{\int_s^\infty\int_s^\infty f(u,v)\,du\,dv}{\int_s^\infty\int_0^\infty f(u,v)\,du\,dv}\mathbb{1}_{\{T_1\ge s\}}\,ds\\ &\quad+\int_0^s \frac{f(s,v)}{\int_s^\infty f(u,v)\,du}\,\frac{\int_s^\infty f(u,v)\,du}{\int_s^\infty\int_0^\infty f(u,v)\,du\,dv}\,dv\mathbb{1}_{\{T_1\ge s\}}\,ds\\ &= \frac{\int_0^\infty f(s,v)\,dv}{\int_s^\infty\int_0^\infty f(u,v)\,du\,dv}\mathbb{1}_{\{T_1\ge s\}}\,ds.\end{aligned}$$

*Remark 2.3.1.* Let $X = (X^1, X^2)$ be a two dimensional semimartingale defined on the stochastic basis $(\Omega, \mathscr{F}^X, \mathbb{F}^X, \mathbb{P})$ and such that its $\mathbb{F}^X$ characteristic triple uniquely determines its finite dimensional distributions. We know, at least in some special cases, how to compute the $\mathbb{F}^{X^i}$ characteristics of the components $X^i, i = 1,2$, from the corresponding $\mathbb{F}^X$ characteristic triple. We still have to establish whether the $\mathbb{F}^{X^i}$ characteristic triple uniquely determines the finite dimensional distributions of the components $X^i$, say $(\widetilde{B}^i, \widetilde{C}^i, \widetilde{\nu}^i)$. We can give a positive answer under the assumption that there exists a unique probability measure $\mathbb{P}^i$ on $\mathscr{F}^{X^i}$ such that $X^i$ is a semimartingale with $\mathbb{F}^{X^i}$ characteristic triple $(\widetilde{B}^i, \widetilde{C}^i, \widetilde{\nu}^i)$. Then necessarily the restriction $\mathbb{P}|_{\mathscr{F}^{X^i}}$ must coincide with $\mathbb{P}^i$. This implies that, at least in this case, the $\mathbb{F}^{X^i}$ characteristics indeed determine the finite dimensional distributions of $X^i$.

**Semimartingale copulae.** We are now ready to proceed with the presentation of semimartingale copulae. In fact, the discussion earlier in this section, and, in particular, the discussion in Remark 2.3.1 indicate a recipe for constructing bivariate semimartingales with given margins. We construct $(\Omega, \mathscr{F}^X, \mathbb{F}^X, \mathbb{P})$ in such a way that the $\mathbb{F}^{X^i}$ characteristics of the coordinate processes $X^i$, $i = 1,2$, are identical (as functions of trajectories) to the $\mathbb{F}^{Y^i}$ characteristics of $Y^i$, $i = 1,2$. This implies that for $i = 1,2$, $X^i$ and $Y^i$ are equal in law.

Indeed, $X^i$ and $Y^i$ live in the same canonical space $\Omega_i$ (this means that the canonical $\sigma$-algebras $\mathscr{F}^{Y^i}$ and $\mathscr{F}^{X^i}$ contain the same events), and the $\mathbb{F}^{X^i}$ characteristics of $X^i$ and the $\mathbb{F}^{Y^i}$ characteristics of $Y^i$ coincide (as functions of trajectories), so uniqueness implies that $X^i$ and $Y^i$ have the same finite dimensional distributions.

We now proceed to define the concept of semimartingale copula for two dimensional semimartingales (this definition can be readily extended to higher dimensional processes).

Let $Y^1, Y^2$ be two $\mathbb{R}$-valued semimartingales defined on possibly different (canonical) filtered probability spaces $(\Omega_i, \mathscr{F}^{Y^i}, \mathbb{F}^{Y^i}, \mathbb{P}^i)$. Let $(\widehat{B}^i, \widehat{C}^i, \widehat{\nu}^i)$ denote the characteristics of $Y^i$, $i = 1,2$, and assume that the finite dimensional distributions of $Y^i$ are uniquely determined by its characteristic triple. Let $X$ denote the vector valued, canonical process on the filtered canonical stochastic basis $(\Omega, \mathscr{F}^X, \mathbb{F}^X)$, where $\Omega = \Omega_1 \times \Omega_2$.

**Definition 2.3.1.** We say that a triple $(B, C, \nu)$ defined on the basis $(\Omega, \mathscr{F}^X, \mathbb{F}^X)$ is a semimartingale copula for $Y^i, i = 1,2$, if the following conditions hold:

i) there is a unique probability measure $\mathbb{P}$ on $\mathscr{F}^X$ such that the canonical process on the stochastic basis $(\Omega, \mathscr{F}^X, \mathbb{F}^X)$ is a semimartingale with characteristic triple $(B, C, \nu)$;

ii) under $\mathbb{P}$, the $\mathbb{F}^{X^i}$ characteristics of $X^i$, say $(\widetilde{B}^i, \widetilde{C}^i, \widetilde{\nu}^i)$, are equal (as functions of trajectories) to $(\widehat{B}^i, \widehat{C}^i, \widehat{\nu}^i)$ for $i = 1,2$.

Now we introduce a suitable measure of dependence between components of the process $X$. Let $\mathscr{J} = 2^{\{1,\ldots,d\}}$, $\mathscr{J}_i = \{S \in \mathscr{J} : S \text{ contains at least } i \text{ elements}\}$ and let $\mathrm{card}(S)$ denote the cardinality of the set $S$.

**Definition 2.3.2.** Let $X = (X^1, X^2, \ldots, X^d)$ be an $E = \times_{i=1}^d \mathbb{E}_i \subset \mathbb{R}^d$ valued locally square integrable semimartingale. Let $T < \infty$. The *d-volume* is defined as

$$\mathrm{Dvol}(X_T) = \mathbb{E}\left(\int_0^T \frac{1}{2}\sum_{i,j=1,i\neq j}^{m} |d\langle X^i,X^j\rangle_s^c| + \sum_{S\in\mathscr{I}_2} \mathrm{card}(S)\nu^S(T)\right), \tag{2.4}$$

where $\langle X^i,X^j\rangle^c$ is the process compensating $[X^i,X^j]^c$ in the filtration $\mathbb{F}^X$, $\nu_t^S$ is the $\mathbb{F}^X$-dual predictable projection of the process

$$J_t^S := \int_0^t \int_{E_S\times\mathbf{0}_{S^c}} \mu(dx,ds),$$

$\mu(dx,ds)$ is an integer valued random measure counting the jumps of the process $X$, and $E_S\times\mathbf{0}_{S^c}$ is a set in $\mathbb{R}^d$ defined as $\mathsf{X}_{i\in S}\mathbb{E}_i \times \mathsf{X}_{i\in S^c}\{0\}$.

*Remark 2.3.2.* The d-volume is a measure of "dynamic" dependence between components of a multivariate semimartingale $X$. The dependence between processes is also related to the dependence structure intrinsic to the initial state $X_0$, but this "static" dependence is not accounted for in the d-volume.

To motivate this choice, we give the following simple example.

*Example 2.3.3.* Let $X = (X^1,X^2)$ be a nondegenerate jump diffusion with

$$\begin{aligned} dX^1 &= \mu_1(X^1)dt + \sigma_{11}(X^1)dW^1(t) + \sigma_{12}(X^1)dW^2(t) + dN^1(t) + dN^2(t),\\ dX^2 &= \mu_2(X^2)dt + \sigma_{21}(X^2)dW^1(t) + \sigma_{22}(X^2)dW^2(t) + dN^2(t) + dN^3(t), \end{aligned}$$

where $W^1,W^2$ are independent Brownian motions and $N^1,N^2,N^3$ are independent Poisson processes with intensities $\lambda_1,\lambda_2,\lambda_3$ respectively. In this case the covariance between the continuous components of $X^1$ and $X^2$ is measured by

$$dc_{ij} = \left(\sigma_{12}(X^1)\sigma_{21}(X^2) + \sigma_{12}(X^1)\sigma_{22}(X^2)\right)dt,$$

while the tendency of the processes to jump together is measured by

$$\nu(dt) = \lambda_2\,dt.$$

**Proposition 2.3.4.** *Let $X = (X^1,X^2,\ldots,X^m)$ be an $E$-valued semimartingale. Then*

$$0 \le \mathrm{Dvol}(X_T) \le \mathbb{E}\left(\frac{1}{2}\sum_{i,j=1,i\neq j}^{m} (\langle X^i,X^i\rangle_T^c)^{\frac{1}{2}}(\langle X^j,X^j\rangle_T^c)^{\frac{1}{2}} + \sum_{S\in\mathscr{I}_1}\mathrm{card(S)}\nu^S(T)\right).$$

*Proof.* This is an immediate consequence of the Kunita-Watanabe inequality (see Protter [27, Chap. II, Theorem 25]) and the observation that $\sum_{S\in\mathscr{I}_2}\mathrm{card}(S)\nu_T^S \le \sum_{S\in\mathscr{I}_1}\mathrm{card}(S)\nu_T^S$. □

*Example 2.3.4.* We construct now a semimartingale copula for a vector one-point process. Suppose we are given two one-point processes $Y^i$ defined on the basis $(\Omega_i,\mathscr{F}^{Y^i},\mathbb{F}^{Y^i},\mathbb{P}^i)$, where $\Omega_i$ is the canonical space of one-point processes on $\mathbb{R}$. Let $\widetilde{T}_1$ and $\widetilde{T}_2$ denote the jump times of $Y^1$ and $Y^2$, respectively, and assume that, under $\mathbb{P}^i$, $\widetilde{T}^i$ is exponentially distributed with parameter $\lambda_i$, $i = 1,2$. Let $F_i$ denote the corresponding distribution function. It can be easily verified that $Y^i$ admits jump characteristic $\widetilde{\nu}^i(dt,dx) = \delta_1(dx)\lambda_i \mathbb{1}_{\{\widetilde{T}_i \geq t\}}\, dt$. Let $N = (N^1, N^2)$ denote the canonical point process on the stochastic basis $(\Omega, \mathscr{F}^N, \mathbb{F}^N)$, where $\Omega = \Omega_1 \times \Omega_2$. Next define two positive random variables $T_1$, $T_2$ as follows: $T_1 = \inf\{t \geq 0 : \Delta N_t^1 = 1\}$, $T_2 = \inf\{t \geq 0 : \Delta N_t^2 = 1\}$. Let $C(\cdot,\cdot)$ be an arbitrary two dimensional absolutely continuous copula function, and $c(\cdot,\cdot)$ the density of the distribution function $C(F_1(\cdot), F_2(\cdot))$. In addition, define the following random measure on $\mathscr{B}(\mathbb{R}_+) \otimes 2^E$ (here $E = \{0,1\}^2$):

$$
\begin{aligned}
\nu(ds,\{(0,0)\}) &= 0, \\
\nu(ds,\{(1,0)\}) &= \left( \frac{\int_s^\infty c(s,v)dv}{\int_s^\infty \int_s^\infty c(u,v)dudv} I_{\{s \leq T_1 \wedge T_2\}} + \frac{c(s,T_2)}{\int_s^\infty c(u,T_2)du} I_{\{T_2 < s \leq T_1\}} \right) ds, \\
\nu(ds,\{(0,1)\}) &= \left( \frac{\int_s^\infty c(u,s)du}{\int_s^\infty \int_s^\infty c(u,v)dudv} I_{\{s \leq T_1 \wedge T_2\}} + \frac{c(T_1,s)}{\int_s^\infty c(T_1,v)dv} I_{\{T_1 < s \leq T_2\}} \right) ds, \\
\nu(ds,\{(1,1)\}) &= 0.
\end{aligned}
$$

First we prove that the measure $\nu(dt,dx)$ is $\mathbb{F}^N$ predictable. To this end consider the simple random function $W = \mathbb{1}_A(\mathbb{1}_{C_0}\mathbb{1}_{\{t \leq T_1 \wedge T_2\}} + \mathbb{1}_{C_1}\mathbb{1}_{\{T_1 \wedge T_2 < t \leq T_1 \vee T_2\}} + \mathbb{1}_{C_2}\mathbb{1}_{\{t > T_1 \vee T_2\}})$, where $A$ is a set in $E$, $C_0 \in \mathscr{F}_0^N, C_1 \in \mathscr{F}_{T_1 \wedge T_2}^N$ and $C_2 \in \mathscr{F}_{T_1 \vee T_2}^N$. Since $W * \nu$ is of the form $D_0\mathbb{1}_{\{t \leq T_1 \wedge T_2\}} + D_1\mathbb{1}_{\{T_1 \wedge T_2 < t \leq T_1 \vee T_2\}} + D_2\mathbb{1}_{\{t > T_1 \vee T_2\}}$, where $D_0 \in \mathscr{F}_0^N$, $D_1 \in \mathscr{F}_{T_1 \wedge T_2}^N$ and $D_2 \in \mathscr{F}_{T_1 \vee T_2}^N$ we see that $W * \nu$ is an $\mathbb{F}^N$ predictable process (Lemma III 1.29 in [20]). Hence $\nu$ is an $\mathbb{F}^N$ predictable random measure by a monotone class argument. By direct verification we see that the probability measure $\mathbb{P}$ defined by the distribution function

$$
\mathbb{P}(T_1 \leq t_1, T_2 \leq t_2) = C(F_1(t_1), F_2(t_2))
$$

is a solution to the martingale problem for $\nu$. In fact, if $\mu(dt,dx)$ denote the optional counting measure associated to $N$, then $\mu((0,t],(1,0)) = N_t^1$ and $\mu((0,t],(0,1)) = N_t^2$, and the martingale property of $\mu((0,t],A) - \nu((0,t],A)$, for all $A \in 2^E$, follows from a straightforward application of the Fubini theorem. Next, we deduce that, for all predictable simple random functions $W$, the process $W * (\mu - \nu)$ is a martingale, implying that $\mathbb{P}$ is a solution to the martingale problem for $\nu$. It follows from known results (e.g. [24, Chap. 4, Theorem 5]) that $\mathbb{P}$ is the unique probability measure on $\mathscr{F}^N$ such that the canonical process $N$ is a bivariate one-point process with compensator $\nu$.

Moreover, by arguments analogous to those used in Example 2.3.2, the $\mathbb{F}^{N^1}$ dual predictable projection of $N^1$ is given by

$$
\begin{aligned}
\widetilde{\nu}^1(ds,dx) &= \delta_1(dx) \frac{\int_0^\infty c(s,s_2)ds_2}{\int_s^\infty \int_0^\infty c(s_1,s_2)ds_2\, ds_2} \mathbb{1}_{\{s \leq T_1\}}\, ds \\
&= \delta_1(dx) \frac{\lambda_1 \exp(-\lambda_1 s)}{\exp(-\lambda_1 s)} \mathbb{1}_{\{s \leq T_1\}}\, ds = \delta_1(dx)\lambda_1 \mathbb{1}_{\{s \leq T_1\}}\, ds.
\end{aligned}
$$

We conclude that the random measure $\nu(dt,dx)$ is a semimartingale copula for $Y^1, Y^2$.

The above example relied on the fact that we are able to compute the projections of the $\mathbb{F}^N$ jump characteristic of $N^i, i = 1,2$. In the general case, if we had to compute projections in order to construct a semimartingale copula, the practical usefulness of the theory presented so far would be rather restricted. So we try to construct nontrivial semimartingale copulae without computing projections. We start from a construction of a vector one-point process $N = (N^1, N^2)$ different from that given in Example 2.3.4.

*Example 2.3.5.* Let the setting be as in Example 2.3.4. Again, we would like to construct a probability measure on $(\Omega, \mathscr{F}^N, \mathbb{F}^N)$ such that the distributions of the jump time of the components $N^i$, $i = 1,2$, are exponential with intensity $\lambda_i$, i.e., $N^i$ are equal in law to $Y^i$ for $i = 1,2$. To do this, let $\lambda_{12}$ be a positive real number satisfying the condition $\lambda_{12} \le \lambda_1 \wedge \lambda_2$ and define the following random measure on $\mathscr{B}(\mathbb{R}_+) \otimes 2^E$ $(E = \{0,1\}^2)$:

$$\begin{aligned}
\nu(ds,\{(0,0)\}) &= 0,\\
\nu(ds,\{(1,0)\}) &= \left(\lambda_1 \mathbb{1}_{\{s\le T_1\}} - \lambda_{12}\mathbb{1}_{\{s\le T_1\wedge T_2\}}\right) ds,\\
\nu(ds,\{(0,1)\}) &= \left(\lambda_2 \mathbb{1}_{\{s\le T_2\}} - \lambda_{12}\mathbb{1}_{\{s\le T_1\wedge T_2\}}\right) ds,\\
\nu(ds,\{(1,1)\}) &= \left(\lambda_{12}\mathbb{1}_{\{s\le T_1\wedge T_2\}}\right) ds.
\end{aligned}$$

By arguments analogous to those used in Example 2.3.4 we infer that the nonnegative measure $\nu(dt,dx)$ is $\mathbb{F}^N$ predictable. By a direct verification, a solution to the martingale problem for $\nu$ is given by the probability measure $\mathbb{P}$ defined by the distribution function

$$\mathbb{P}(T_1 \le t_1, T_2 \le t_2) = 1 - e^{-\lambda_1 t_1} - e^{-\lambda_2 t_2} + e^{-(\lambda_1-\lambda_{12})t_1 - (\lambda_2-\lambda_{12})t_2 - \lambda_{12} t_1 \vee t_2}.$$

To see this, we prove that $M_t = \mu((0,t],A) - \nu([0,t],A)$, where $\mu(dt,dx)$ is the counting measure associated to $N$, is an $\mathbb{F}^N$ martingale for every set $A$ in $2^E$. Let, for example, $A = \{(1,1)\}$. Then $\mu((0,t],(1,1)) = \mathbb{1}_{\{T_1\wedge T_2 \le t, T_1 = T_2\}}$ and by direct computation

$$\begin{aligned}
\mathbb{P}(T_1 \le t, T_2 \le t, T_1 = T_2 | \{T_1 \wedge T_2 > s\}) &= \int_s^t \lambda_{12} e^{-(\lambda_1+\lambda_2-\lambda_{12})(u-s)} du,\\
\mathbb{P}(T_1 \wedge T_2 \ge t | \{T_1 \wedge T_2 > s\}) &= e^{-(\lambda_1+\lambda_2-\lambda_{12})(t-s)}.
\end{aligned}$$

Therefore, $M_t$ satisfies

$$\begin{aligned}
&\mathbb{E}(M_t|\mathscr{F}_s^N) = M_s + \mathbb{P}(\{s < T_1 \wedge T_2 \le t, T_1 = T_2\}|\mathscr{F}_s^N) - \lambda_{12}\int_s^t \mathbb{P}(T_1 \wedge T_2 \ge u|\mathscr{F}_s^N)\,du\\
&= M_s + \int_s^t e^{-(\lambda_1+\lambda_2-\lambda_{12})(u-s)} du\, \mathbb{1}_{\{T_1\wedge T_2 > s\}} - \int_s^t e^{-(\lambda_1+\lambda_2-\lambda_{12})(u-s)} du\, \mathbb{1}_{\{T_1\wedge T_2 > s\}} = M_s.
\end{aligned}$$

$\mathbb{P}$ is the unique probability measure on $\mathscr{F}^N$ such that the canonical process $N$ is a bivariate one-point process with compensator $\nu$. Moreover, under $\mathbb{P}$, the $\mathbb{F}^N$ compensator and the $\mathbb{F}^{N^i}$ compensator of the coordinate process $N^i$, $i=1,2$, coincide, i.e., $\widetilde{\nu}^i = \nu^i$. In fact, the $\mathbb{F}^N$ compensator of $N^1$ is given by

$$\begin{aligned}\nu^1(ds,\{1\}) &= \nu(ds,\{(1,0)\}\cup\{(1,1)\}) \\ &= \lambda_1 \mathbb{1}_{\{s\le T_1\}} - \lambda_{12}\mathbb{1}_{\{s\le T_1\wedge T_2\}} + \lambda_{12}\mathbb{1}_{\{s\le T_1\wedge T_2\}} = \lambda_1\mathbb{1}_{\{s\le T_1\}},\end{aligned}$$

and $\nu^1(ds,\{x\}) = 0$ if $x \neq 1$, i.e., $\nu^1(ds,dx) = \delta_1(dx)\lambda_1\mathbb{1}_{\{s\le T_1\}}\,ds$, and similarly for $\nu^2(ds,dx)$. Therefore, for $i=1,2$, $\nu^i(ds,\{1\})$ is $\mathbb{F}^{N^i}$ predictable and $\nu^i(ds,dx)$ is as well the compensator of $N^i$ in the filtration $\mathbb{F}^{N^i}$. It is then obvious that the $\mathbb{F}^{N^i}$ characteristic of the component process is the same (as a function of trajectories) as the $\mathbb{F}^{Y^i}$ characteristic of $Y^i$, i.e., the random measure $\nu(dt,dx)$ is a semimartingale copula.

*Remark 2.3.3.* The probability measure $\mathbb{P}$ solving the martingale problem for $\nu(dt,dx)$ given in Example 2.3.5 can be constructed via the Marshall-Olkin copula between two exponential random variables with intensity $\lambda_1$ and $\lambda_2$ (see [26, Sect. 3.1.1]). By direct computation, if $t_1,t_2>0$, then $\mathbb{P}(T_1\le t_1, T_2\le\infty) = 1-e^{-\lambda_1 t_1}$ and $\mathbb{P}(T_1\le\infty, T_2\le t_2) = 1-e^{-\lambda_2 t_2}$.

*Remark 2.3.4.* Note that the nature of dependence between the coordinate processes exhibited in Example 2.3.5 is very different from that seen in Example 2.3.4. In Example 2.3.4 the $\mathbb{F}^N$ characteristics of each component may also depend functionally on the trajectories of the other components. On the other hand, in Example 2.3.5, the dependence between components is only given by the possibility of common jumps of the processes (the jump measure of the set $\{(1,1)\}$ is positive).

## 2.3.2 Consistent Semimartingale Copulae

From the above examples we can draw some useful conclusions. Given the $\mathbb{F}^X$ characteristic triple of a multivariate process $X$, the computation of the $\mathbb{F}^{X^i}$ characteristics of the components $X^i$ can be made much simpler if we construct the multivariate process in such a way that the $\mathbb{F}^X$ characteristic triple of $X^i$ is $\mathbb{F}^{X^i}$ predictable. In this case the computation of projections can be avoided. As we have seen in the examples, projecting the $\mathbb{F}^X$ characteristics in the filtration $\mathbb{F}^{X^i}$ is rather difficult, and in fact this is possible only in very simple and special cases. These observations suggest the following definition:

**Definition 2.3.3.** We say that a two dimensional semimartingale $X=(X^1,X^2)$ defined on a stochastic basis $(\Omega,\mathscr{F}^X,\mathbb{F}^X,\mathbb{P})$ is consistent with respect to $\mathbb{F}^{X^i}$ if the $\mathbb{F}^X$ characteristic triple $(B^i,C^i,\nu^i)$ and the $\mathbb{F}^{X^i}$ characteristic triple $(\widetilde{B}^i,\widetilde{C}^i,\widetilde{\nu}^i)$ of the component process $X^i$ coincide (as functions of trajectories).

Let $Y^1, Y^2$ be two $\mathbb{R}$-valued semimartingales defined on possibly different (canonical) filtered probability spaces $(\Omega_i, \mathscr{F}^{Y^i}, \mathbb{F}^{Y^i}, \mathbb{P}^i)$, $i = 1, 2$. Moreover, let $(\widehat{B}^i, \widehat{C}^i, \widehat{\nu}^i)$ denote the characteristics of $Y^i$, $i = 1, 2$, and assume that the finite dimensional distributions of $Y^i$ are uniquely determined by its characteristic triple. Let $X = (X^1, X^2)$ denote the vector valued, canonical process on the filtered canonical stochastic basis $(\Omega, \mathscr{F}^X, \mathbb{F}^X)$ where $\Omega = \Omega_1 \times \Omega_2$.

**Definition 2.3.4.** We say that a triple $(B, C, \nu)$ defined on the basis $(\Omega, \mathscr{F}^X, \mathbb{F}^X)$ is a consistent semimartingale copula for $Y^i, i = 1, 2$, if the following conditions hold:

i) there is a unique probability measure $\mathbb{P}$ on $\mathscr{F}^X$ such that the canonical process on the stochastic basis $(\Omega, \mathscr{F}^X, \mathbb{F}^X)$ is a semimartingale with characteristic triple $(B, C, \nu)$;

ii) under $\mathbb{P}$, $X$ is consistent with respect to $\mathbb{F}^{X^i}$ and $(B^i, C^i, \nu^i)$ are equal (as functions of trajectories) to $(\widehat{B}^i, \widehat{C}^i, \widehat{\nu}^i)$ for $i = 1, 2$.

Note that the difference in the definitions of semimartingale copula and consistent semimartingale copula lies in the requirement of *consistency* imposed in the latter case; as we explained earlier, the consistency property allows one to avoid computations of projections of the characteristics on smaller filtrations. We devote the rest of this section to construct examples of consistent semimartingale copulae for some important classes of semimartingales.[4]

**Copulae between pure jump Lévy processes.** We shall now provide an elementary example of a semimartingale copula that is also a Lévy copula. There is a one-to-one correspondence between a homogeneous Poisson process with values in $\mathbb{R}^2$ and a homogeneous Poisson measure on $E = \{0, 1\}^2 \setminus \{(0, 0)\}$. We let $\nu$ denote the $\mathbb{F}$ dual predictable projection of a Poisson measure $\mu$. The measure $\nu$ is a measure on a finite set, so it is uniquely determined by its values on the atoms in $E$. Therefore a Poisson process $X$ in $\mathbb{R}^2$ is uniquely determined by

$$\nu(dt, \{1, 0\}) = \lambda_{10} dt, \quad \nu(dt, \{0, 1\}) = \lambda_{01} dt, \quad \nu(dt, \{1, 1\}) = \lambda_{11} dt \tag{2.5}$$

for some positive constants $\lambda_{10}$, $\lambda_{01}$ and $\lambda_{11}$.

*Example 2.3.6.* Let us consider two Poisson processes $X^1$ and $X^2$ with values in $\mathbb{R}^1$, with intensities $\lambda_1$ and $\lambda_2$ respectively. We will show that if real numbers $\lambda_{10}, \lambda_{01}, \lambda_{11}$ satisfy

$$\begin{aligned} \lambda_1 &= \lambda_{10} + \lambda_{11}, \\ \lambda_2 &= \lambda_{01} + \lambda_{11}, \\ \lambda_{11} &\in [0, \lambda_1 \wedge \lambda_2], \end{aligned} \tag{2.6}$$

then the measure $\nu$ in (2.5) is a semimartingale copula for $X^1, X^2$. First, (2.6) implies that $\nu$ is positive. Moreover, $\nu$ defines uniquely the probability law of a Poisson random measure on $\{0, 1\}^2$. The vector Poisson process corresponding to $\nu$ can, in fact,

[4] Since all semimartingale copulae constructed below are consistent we often omit the qualifier consistent and we only talk about semimartingale copulae.

be easily constructed from a vector of three independent unit Poisson processes, say $(N^1, N^2, N^3)$, by using time-changing. If $Y_t^1 = N^1_{\lambda_{10}t} + N^2_{\lambda_{11}t}$ and $Y_t^2 = N^3_{\lambda_{01}t} + N^2_{\lambda_{11}t}$, then it is straightforward to verify that the probability law of $Y = (Y^1, Y^2)$ is a solution of the martingale problem for $\nu(dt,dx)$. Uniqueness of the martingale problem for the triple $(0,0,\nu)$ follows from [24, Chap. 4, Theorem 5]. Finally, since

$$\nu^1(dt,\{1\}) = \nu(dt,\{(1,0)\}) + \nu(dt,\{(1,1)\}) = \lambda_{10}dt + \lambda_{11}dt = \lambda_1 dt$$

and $\nu^1(dt,\{x\}) = 0, \forall x \neq 1$ (and similarly for $\nu^2(dt,\{1\})$) we conclude that $\nu(dt,dx)$ is a semimartingale copula.[5]

**Copulae between diffusion processes.** Let us consider two $\mathbb{R}$-valued diffusion processes $X_1$ and $X_2$ defined on the spaces $(\Omega_1', \mathscr{G}^1, \mathbb{P}^1)$ and $(\Omega_2', \mathscr{G}^2, \mathbb{P}^2)$, where $(\Omega'_i, \mathscr{G}^i, \mathbb{P}^i)$ supports the Standard Brownian Motion (SBM) $W^i$, $i = 1,2$, and the filtration $\widetilde{\mathbb{G}}^i$ is generated by the SBM $W^i$. Assume that the diffusions are driven by the following SDEs:

$$dX^i(t) = \mu_i(X^i(t))dt + \sigma_i(X^i(t))dW^i(t), \quad X^i(0) = x_i, \quad i = 1,2. \tag{2.7}$$

For the moment, we suppose that $X^i$, $i = 1,2$, are strong solutions. We shall relax this assumption later on. It is well known that the $\mathbb{F}^{X^i}$ characteristics of $X^i$ are determined by $\mu_i$ and $\sigma_i$.

Now, let $(\Omega', \mathscr{G}, \mathbb{G})$ be a filtered probability space supporting a two dimensional Brownian motion, $W$, where $\mathbb{G}$ is the filtration generated by $W$ and $\Omega' = \Omega_1' \times \Omega_2'$. The problem of constructing a semimartingale copula for $X^i$ is equivalent to finding functions $m = [m_1, m_2]^T : \mathbb{R}^2 \to \mathbb{R}^2$ and $\Sigma = [\sigma_{ij}] : \mathbb{R}^2 \to L(\mathbb{R}^2, \mathbb{R}^2)$ such that the $\mathbb{F}^Y$ characteristic triple of the diffusion process $Y = (Y^1, Y^2)$, solving the SDE

$$dY(t) = m(Y(t))dt + \Sigma(Y(t))dW(t), \quad Y^i(0) = x^i, \tag{2.8}$$

satisfies Definition 2.3.1.

*Remark 2.3.5.* Note that, in the diffusion case, the filtered stochastic basis $(\Omega', \mathscr{G}, \mathbb{G})$ is not constructed according to the canonical setting. In fact, in this case, the filtration $\mathbb{G}$ may be strictly larger (or smaller) than $\mathbb{F}^Y$. However, in view of Jacod and Shiryaev [20, Sect. 2.26, Theorem III 2.26], if a unique solution-measure to (2.8) exists, it is the unique probability measure on the canonical space $(\Omega, \mathscr{F}^Y, \mathbb{F}^Y)$ such that the process $Y$ has characteristic triple $(m(Y), \sigma(Y), 0)$. We can therefore construct a semimartingale copula on the stochastic basis $(\Omega', \mathscr{G}, \mathbb{G})$.

**Proposition 2.3.5.** *Suppose that the function $\Sigma$ is chosen so that a strong solution of* (2.8) *exists and*

$$\sigma_{11}^2(x,y) + \sigma_{12}^2(x,y) = \sigma_1^2(x), \quad \sigma_{21}^2(x,y) + \sigma_{22}^2(x,y) = \sigma_2^2(y), \tag{2.9}$$

[5] The first requirement in condition (iii) of Definition 2.3.1, namely that $Y$ is consistent with respect to $\mathbb{F}^{Y^i}$, $i = 1,2$, is trivially satisfied since the characteristics are deterministic.

*with*

$$\sup_y \sigma_{12}^2(x,y) \le \sigma_1^2(x) \quad \text{and} \quad \sup_x \sigma_{21}^2(x,y) \le \sigma_2^2(y). \tag{2.10}$$

*In addition, suppose that the function $m$ satisfies*

$$m_1(x,y) = \mu_1(x), \quad m_2(x,y) = \mu_2(y). \tag{2.11}$$

*Then the processes $Y^i$ are diffusion processes, and $m = [m_1, m_2]^T$ and $\Sigma = [\sigma_{ij}]$ define a semimartingale copula for $X^1, X^2$, which we term a diffusion copula.*

The next proposition is the counterpart of Proposition 2.3.5 when the system of SDEs (2.8) does not necessarily admit a unique strong solution. We now only assume that the coefficients of Eq. (2.7) satisfy all conditions needed for existence of a weak solution.

**Proposition 2.3.6 (see [4, 32]).** *Let $Y$ be a (weak) solution of the SDE* (2.8). *Suppose that the function $\Sigma$ is measurable and satisfies the conditions (2.9) with*

$$\sup_y \sigma_{12}^2(x,y) < \sigma_1^2(x) \quad \text{and} \quad \sup_x \sigma_{21}^2(x,y) < \sigma_2^2(y). \tag{2.12}$$

*In addition, suppose that the function $m$ satisfies* (2.11). *Then the processes $Y^i$ are diffusion processes, and $m = [m_1, m_2]^T$ and $\Sigma = [\sigma_{ij}]$ define a semimartingale copula for $X^1, X^2$, which we term a diffusion copula.*

Note that in view of (2.11) we have no freedom in the choice of the function $m$. However, we do have freedom in the choice of $\Sigma$. Dependence between components of $Y$ is then fully described in terms of the functions $\sigma_{12}$ and $\sigma_{21}$. It is easy to verify that in the diffusion case

$$\text{Dvol}(Y_T) = \mathbb{E}\left(\int_0^T (\sigma_{11}(Y_s^1, Y_s^2)\sigma_{21}(Y_s^1, Y_s^2) + \sigma_{12}(Y_s^1, Y_s^2)\sigma_{22}(Y_s^1, Y_s^2))\, ds\right).$$

We immediately have the following bounds for the d-volume associated to $Y$:

$$0 < \text{Dvol}(Y_T) \le \mathbb{E}\left(\int_0^T (\sigma_1(Y_s^1)^2 + \sigma_2(Y_s^2)^2)\, ds\right) \tag{2.13}$$

as this condition is necessary for the diffusion matrix to be nonnegative definite (see [32] for a more detailed discussion).

**Copulae between finite Markov chains.** Since, in general, when dealing with Markov chains we are given their generator matrix, rather than their characteristic triple, we find it convenient to work with a counting measure which can be more directly related to the infinitesimal generator of the chain.

*Finite Markov chains and related random measures.* As before, let $(\Omega, \mathcal{F}^X, \mathbb{P})$ be the underlying probability space. We consider on this space a stochastic process $X = (X_t)_{t\ge 0}$ with values in a finite set $\mathcal{X} = \{1, 2, \ldots, N\} \subset \mathbb{N}$. As usual, by $\mathbb{F}^X$

we shall denote the natural filtration generated by $X$. As anticipated above, rather than looking at the jump characteristic of the chain, we are going to introduce a counting measure that can be more easily associated to its infinitesimal generator. To this end, for any two states $i,j \in \mathscr{X}$ such that $i \neq j$, we define the following $\mathbb{F}^X$-optional random measure on $[0,\infty)$:

$$N^{ij}((0,t]) = \sum_{0<s\leq t} \mathbb{1}_{\{X_{s-}=i, X_s=j\}}. \tag{2.14}$$

We shall simply write $N^{ij}(t)$ in place of $N^{ij}((0,t])$. Obviously, $N^{ij}(t)$ represents the number of jumps from state $i$ to state $j$ that $X$ executes over the time interval $(0,t]$. Let us denote by $\nu^{ij}$ the dual predictable projection, with respect to $\mathbb{F}^X$, of the random measure $N^{ij}$. We are now going to relate the collection of the random measures $\nu^{ij}$, $i,j \in \mathscr{X}$, $i \neq j$, to the infinitesimal generator of the chain $X$. To this end, let us define a matrix valued function $A$ on $[0,\infty)$ by

$$A(t) = [\lambda_{i,j}(t)]_{i,j\in\mathscr{X}}, \tag{2.15}$$

where $\lambda_{i,j}$ are real valued, locally integrable functions on $[0,\infty)$ such that for $t \geq 0$ and $i,j \in \mathscr{X}$, $i \neq j$, we have

$$\lambda_{i,j}(t) \geq 0 \quad \text{and} \quad \lambda_{i,i}(t) = -\sum_{j\neq i} \lambda_{i,j}(t).$$

Note that $\lambda_{i,j}(t)$ is the time-$t$ intensity of jump from state $i$ to state $j$. The following proposition establishes the connection between the random measures $\nu^{ij}$, $i,j \in \mathscr{X}$, $i \neq j$, and the infinitesimal generator of $X$.

**Proposition 2.3.7 (see [4]).** *The process $X$ is a Markov chain (with respect to $\mathbb{F}^X$) with infinitesimal generator $A(t) = [\lambda_{i,j}(t)]$ iff the dual predictable projections with respect to $\mathbb{F}^X$ of the counting measures $N^{ij}(dt)$, $i,j \in \mathscr{X}$, are of the form*

$$\nu^{ij}(dt) = \mathbb{1}_{\{X_{t-}=i\}} \lambda_{i,j}(t)dt. \tag{2.16}$$

*Copulae between Markov chains.* As usual, we shall only consider the case of bivariate Markov chains. The general multivariate case can be treated similarly. In the rest of this section we denote by $\mathscr{S}$ and $\mathscr{O}$ two finite sets. Let $X = (X^1, X^2)$ denote a two dimensional Markov chain on $\mathscr{X} = \mathscr{S} \times \mathscr{O}$, with generator function $A(t) = [\lambda^X_{ih,jk}(t)]_{i,j\in\mathscr{S},k,h\in\mathscr{O}}$. Assume that the following conditions hold:

$$\sum_{k\in\mathscr{O}} \lambda^X_{ih,jk}(t) = \sum_{k\in\mathscr{O}} \lambda^X_{ih',jk}(t), \quad \forall h,h' \in \mathscr{O},\ \forall i,j, \in \mathscr{S}, i \neq j, \tag{2.17a}$$

$$\sum_{j\in\mathscr{S}} \lambda^X_{ih,jk}(t) = \sum_{j\in\mathscr{S}} \lambda^X_{i'h,jk}(t), \quad \forall i,i' \in \mathscr{S},\ \forall k,h \in \mathscr{O}\ h \neq k. \tag{2.17b}$$

Intuitively, conditions (2.17a) and (2.17b) requires that the jump intensity of the component $X^1$ does not depend on the state of $X^2$ and vice versa. As shown in the following proposition, conditions (2.17a) and (2.17b) are sufficient to yield Markovianity of the components $X^i$, $i=1,2$, in the filtration $\mathbb{F}^X$. Moreover, we obtain an explicit characterization of the infinitesimal generator matrix of the components $X^i$ in terms of $A(t)$.

**Proposition 2.3.8 (see [4]).** *Suppose that conditions (2.17a) and (2.17b) hold and define*

$$f_{i,j}(t) := \sum_{k\in\mathscr{O}} \lambda^X_{ih,jk}(t), \quad i,j\in\mathscr{S}, i\neq j, \quad f_{i,i}(t) := -\sum_{j\in\mathscr{S},j\neq i} f_{i,j}(t), \quad \forall i\in\mathscr{S}, \tag{2.18}$$

*and*

$$g_{h,k}(t) := \sum_{j\in\mathscr{S}} \lambda^X_{ih,jk}(t), \quad k,h\in\mathscr{O}, h\neq k, \quad g_{h,h}(t) := -\sum_{k\in\mathscr{O},k\neq h} g_{h,k}(t), \quad \forall h\in\mathscr{O}. \tag{2.19}$$

*Then the components $X^1$ and $X^2$ of the Markov chain $X$ are Markov chains with respect to their natural filtrations with generator functions $A^1(t) = [f_{i,j}(t)]_{i,j\in\mathscr{S}}$ and $A^2(t) = [g_{h,k}(t)]_{k,h\in\mathscr{O}}$, respectively.*

In view of Proposition 2.3.8, it is clear how to construct the generator of a bivariate Markov chain whose components have prescribed infinitesimal generators.

**Corollary 2.3.1.** *Consider two Markov chains $Y_1$ and $Y_2$, with respect to their own filtrations, $\mathbb{F}^{Y^1}$ and $\mathbb{F}^{Y^2}$, with values in $\mathscr{S}$ and $\mathscr{O}$, respectively. Suppose that their generators are $A^1(t) = [\lambda^{Y^1}_{i,j}(t)]_{i,j\in\mathscr{S}}$ and $A^2(t) = [\lambda^{Y^2}_{h,k}(t)]_{h,k\in\mathscr{O}}$. Next, consider the system of equations in the unknowns $\lambda^X_{ih,jk}(t)$, where $i,j\in\mathscr{S}$, $h,k\in\mathscr{O}$ and $(i,h)\neq(j,k)$:*

$$\begin{cases} \sum_{k\in\mathscr{O}} \lambda^X_{ih,jk}(t) = \lambda^{Y^1}_{i,j}(t), & \forall h\in\mathscr{O}, \forall i,j\in\mathscr{S}, i\neq j \\ \sum_{j\in\mathscr{S}} \lambda^X_{ih,jk}(t) = \lambda^{Y^2}_{h,k}(t), & \forall i\in\mathscr{S}, \forall h,k\in\mathscr{O}, h\neq k. \end{cases} \tag{2.20}$$

*Suppose that the above system admits a solution such that the matrix function $A(t) = [\lambda^X_{ih,jk}(t)]_{i,j\in\mathscr{S},k,h\in\mathscr{O}}$, with*

$$\lambda^X_{ih,ih}(t) = -\sum_{(j,k)\in\mathscr{S}\times\mathscr{O},(j,k)\neq(i,h)} \lambda^X_{ih,jk}(t), \tag{2.21}$$

*properly defines an infinitesimal generator function of a Markov chain with values in $\mathscr{S}\times\mathscr{O}$. Consider a bivariate Markov chain $X := (X_1, X_2)$ on $\mathscr{S}\times\mathscr{O}$ with generator function $A(t)$. Then the components $X_1$ and $X_2$ are Markov chains with respect to their own filtrations, with generators $A^1(t)$ and $A^2(t)$.*

Note that, typically, system (2.20) contains many more unknowns than equations. In fact, given that the cardinalities of $\mathscr{S}$ and $\mathscr{O}$ are $K_{\mathscr{S}}$ and $K_{\mathscr{O}}$, respectively, the system consists of $K_{\mathscr{S}}(K_{\mathscr{S}}-1)+K_{\mathscr{O}}(K_{\mathscr{O}}-1)$ equations in $K_{\mathscr{S}}K_{\mathscr{O}}(K_{\mathscr{S}}K_{\mathscr{O}}-1)$ unknowns. Thus, in principle, one can create several bivariate Markov chains $X$ with given margins.

Now we embed the above results in the framework of semimartingale copulae. First note that the elements of the mark space of $X^1$, say $\mathscr{J}^1$, can be uniquely identified with the integers $l=j-i$ with $i,j\in\mathscr{S}$. Similarly, the elements of the mark space of $X^2$, say $\mathscr{J}^2$, can be uniquely identified with the integers $r=k-h$ with $h,k\in\mathscr{O}$. Thus, we can uniquely construct the jump characteristic[6] of $X^1$ from the collection of the random measures $\nu^{ij}$ in the following way:

$$\nu^1([0,t),l)=\sum_{i,j:j-i=l}\nu_t^{ij},$$

where, as usual, $\nu^{ij}(dt)=\mathbb{1}_{\{X_{t-}=i\}}\lambda_{i,j}(t)dt$. Now let $\nu$ be a random measure on $\mathscr{J}^1\times\mathscr{J}^2$ given by

$$\begin{aligned}\nu(dt,(0,0))&=0,\\ \nu(dt,(l,r))&=\sum_{i,j:j-i=l}\sum_{h,k:k-h=r}\nu^{ih,jk}(dt),\end{aligned}\qquad(2.22)$$

where, as usual, $\nu^{ih,jk}(dt)=\mathbb{1}_{\{X^1_{t-}=i,X^2_{t-}=h\}}\lambda^X_{ih,jk}(t)dt$.

**Proposition 2.3.9 (see [4]).** *Let $\nu$ be the random measure in (2.22). If, for $i,j\in\mathscr{S}$, $h,k\in\mathscr{O}$ and $(i,h)\neq(j,k)$, $\lambda^X_{ih,jk}$ is a positive solution of the system (2.20), then the triplet $(0,0,\nu)$ is a semimartingale copula for $Y^1$ and $Y^2$.*

It needs to be stressed that in the case of semimartingales that are also Markov processes, we can either apply the semimartingale copula methodology described above, or the Markov copula methodology, described in the following section, to model dependence between univariate components of a multivariate process with preservation of margins. This remark applies, in particular, to the Markov chain case of Proposition 2.3.9, as well as to the diffusion case of Proposition 2.3.6. In Sect. 2.5 we shall use semimartingale copulae for Markov chains in order to price ratings triggered step-up bonds.

## 2.4 Markov Copulae

In this section, which is based on [6, 31], we tackle the problem of defining and constructing "Markov copulae" using infinitesimal generators. First we introduce the class of *consistent* Markov processes for which we next define and construct Markov copulae.

---

[6] Analogous arguments hold for the jump characteristic of $X^2$ as well.

### 2.4.1 Consistent Markov Processes

Let $E = \times_{i=1}^{d} E_i$, where $E_i$ are locally compact separable spaces. We recall the notation:

1. For any index set $I \subset \{1,\dots,d\}$, we denote by $I^c$ its complementary set, and we write $E_I = \times_{i\in I} E_i$. For $x \in E$ we use the notation $x_I = (x_i,\, i \in I)$.
2. $B(E)$ is the space of bounded functions on $E$ endowed with the supremum norm. Likewise, $B(E_I)$ is the space of bounded functions on $E_I$.
3. For a linear operator $A \subset B(E) \times B(E)$, we denote its domain by $\mathscr{D}(A)$. For a suitably large set $D(E) \subseteq \mathscr{D}(A)$, we let $L(E) := \overline{D(E)}$[7].

To ensure regularity of the sample paths of the Markov processes under consideration, we shall assume that $D(E) \subseteq C_0(E)$ (and therefore $L(E) \subseteq C_0(E)$ as well). Additionally, we assume that $D(E)$ is the closure of a tensor product space, i.e.,

$$D(E) = D_I(E_I)\widehat{\otimes} D_{I^c}(E_{I^c}),$$

for suitable spaces $D_I(E_I) \subseteq C_0(E_I)$ and $D_{I^c}(E_{I^c}) \subseteq C_0(E_{I^c})$.[8] In addition we assume the space $L(E)$ is a separating subspace of $B(E)$. This condition as well will be satisfied in all the cases considered in this survey.

**Definition 2.4.1.** We define the following subspaces:

1. $B_I(E) := \{f \otimes \mathbb{1}_{E_{I^c}} : f \in B(E_I)\}$,
2. $D_I(E) := \{h \in B_I(E) : (h,g) \in$ the b.p. closure of $A$ for $g \in B(E)$ and $h = f \otimes \mathbb{1}_{E_{I^c}}$ for some $f \in D_I(E_I)\}$[9],
3. $L_I(E_I) := \overline{D_I(E_I)}$, where we assume that $L_I(E_I)$ is a separating subset of the space $B(E_I)$.

*Remark 2.4.1.* Note that if $(f_n^1, Af_n^1) \overset{b.p.}{\to} (f,h^1)$ and $(f_n^2, Af_n^2) \overset{b.p.}{\to} (f,h^2)$, then

$$h^1(x) = h^2(x) = \lim_{t\downarrow 0} \frac{T(t)f(x) - f(x)}{t}, \quad \forall x \in E.$$

This implies that if $(f,h^1)$ and $(f,h^2)$ both belong to the bounded pointwise closure of $A$, then $h^1 = h^2$.

Let now $\mathbb{P}$ be a probability measure on the filtered probability space $(\Omega, \mathscr{F}, \mathbb{F})$ and let $X = (X^1,\dots,X^d)$ be an $\mathbb{F}$-Markov process under $\mathbb{P}$, taking values in $E$. In general, the components of a vector $\mathbb{F}$-Markov process are not $\mathbb{F}$-Markovian themselves.

---

[7] For an operator $A \subset B(E) \times B(E)$ or a subspace $X$ of $B(E)$, the notations $\overline{A}$ and $\overline{X}$ signify that the closure is taken in the $\|\cdot\|_\infty$ norm.

[8] It is shown in [31] that this assumption will be satisfied in all the cases considered in this section.

[9] We implicitly assume that all functions of the form $f \otimes \mathbb{1}_{E_{I^c}}$, where $f \in D_I(E_I)$ belong to the bounded pointwise closure of $D(E)$ (for definition of the b.p. closure see [15]). This is the case in all our applications.

*Example 2.4.1.* Consider the vector process $Y_t := (W_t, M_t)$, where $W$ is a SBM,

$$M_t := N\left(\int_0^t \exp(\sigma W_s)\,ds\right)$$

and $N$ is a standard Poisson process. It is well known that $M$ is a Cox process, $Y$ is Markov in its natural filtration $\mathbb{F}^Y$, but the component $M$ is not $\mathbb{F}^Y$-Markov. In fact, since the filtration generated by $\int_0^t \exp(\sigma W_s)\,ds$ is contained in $\mathscr{F}_t^W$, $W$ is still an $\mathbb{F}^Y$ Brownian motion and

$$\mathbb{E}(M_{t+s}|\mathscr{F}_t^Y) = M_t + \mathbb{E}\left(\int_t^{t+s} \exp(\sigma W_u)\,du|\mathscr{F}_t^Y\right) = M_t + \exp(\sigma W_t)\frac{2\exp(\frac{\sigma^2}{2}s)-1}{\sigma^2}.$$

It turns out that the converse statement is also false, namely we construct a vector process which is not Markov itself with all components Markovian (in their natural filtration).

*Example 2.4.2.* Consider a pair of random times $\tau_1$ and $\tau_2$ with exponential distribution and intensities $\lambda_1$ and $\lambda_2$, and denote by $\mathbb{H} = \mathbb{H}^1 \vee \mathbb{H}^2$ the minimal filtration making $\tau_1$ and $\tau_2$ stopping times. It can be checked that the indicator processes $H_t^i := \mathbb{1}_{\{\tau_i \le t\}}$, $i = 1,2$, are Markovian in the respective natural filtrations $\mathbb{H}^i$, $i = 1,2$. In fact, for $t \ge s$,

$$\mathbb{P}\left(H_t^i = 0 \mid \mathscr{H}_s^i\right) = \mathbb{P}\left(\tau_i \ge t \mid \mathscr{H}_s^i\right) = (1 - H_s^i)\exp(-\lambda_i(t-s)).$$

Assume that the joint distribution of $(\tau_1, \tau_2)$ is given by a Gaussian copula, i.e.,

$$\mathbb{P}(\tau_1 \le t_1, \tau_2 \le t_2) := C(F_1(t_1), F_2(t_2)),$$

with $C(\cdot,\cdot) := \Phi_2(\Phi^{-1}(\cdot), \Phi^{-1}(\cdot))$, where $\Phi_2$ is the CDF of a bivariate Gaussian random variable with mean vector $(0,0)$ and covariance matrix $\Sigma$, $\Phi$ is the CDF of a standard Gaussian random variable, and $F_i(t_i) = 1 - \exp(\lambda_i t_i)$, $i = 1,2$, are the marginal CDFs.

Consider now the bivariate process $(H^1, H^2)$. It is not Markovian in its natural filtration $\mathbb{H}$. To see this, it is sufficient to notice that[10]

$$\begin{aligned}\mathbb{P}(H_t^1 = 1 \mid \mathscr{H}_s) &= \left(\frac{1 - \partial_2 C(F_1(t), F_2(\tau_2))}{1 - \partial_2 C(F_1(s), F_2(\tau_2))}\right)\mathbb{1}_{\{\tau_1 > s, \tau_2 \le s\}} \\ &+ \left(\frac{1 - F_1(t)}{1 - C(F_1(s), F_2(s))}\right)\mathbb{1}_{\{\tau_1 > s, \tau_2 > s\}},\end{aligned}$$

which is clearly nonmeasurable in $\sigma(H_s^1) \vee \sigma(H_s^2)$. In this example, starting from Markovian "marginal processes" $H^1$ and $H^2$, we constructed a non-Markovian vector process with given marginal laws.

[10] By $\partial_2 C(u_1, u_2)$ we denote the partial derivative of $C$ with respect to its second variable, evaluated at $u_2$.

*Remark 2.4.2.* The above examples show that requiring that the $\mathbb{F}$-Markov process $X$ has $\mathbb{F}$-Markov component $X^I$ is a stringent requirement. However, if the components of a multivariate Markov process $X$ are themselves Markovian, then one can apply the rich analytical apparatus of Markov processes to the analysis of both $X$ and its components. This observation motivates the following definition.

**Definition 2.4.2.** We say that a Markov process $X$ has the *Markovian consistency property* for $X^I$ (or briefly consistency property if $X^I$ is predetermined) if

$$\mathbb{E}(f(X^I_{t+s}) \mid \mathscr{F}_t) = \mathbb{E}(f(X^I_{t+s}) \mid X^I_t), \quad \forall f \in B(E_I), \tag{2.23}$$

where $X^I = (X^i, i \in I)$.

If, in addition, the law of $X^I$ agrees with the law of a given Markov process $Y$ taking values in $E_I$, and defined on some probability space $(\tilde{\Omega}, \tilde{\mathscr{F}}, \tilde{\mathbb{P}})$, i.e., for any positive integer $n$, any $t_1, t_2, \ldots, t_n \geq 0$, and any measurable subsets of $E_I$, $A_1, A_2, \ldots, A_n$,

$$\mathbb{P}(X^I_{t_i} \in A_i, i = 1, 2, \ldots, n) = \tilde{\mathbb{P}}(Y_{t_i} \in A_i, i = 1, 2, \ldots, n), \tag{2.24}$$

then we say that $X$ has the *Markovian consistency property* for $(X^I, Y)$.

*Remark 2.4.3.* Let $\mathbb{F}^I$ be the natural filtration of the process $X^I$ and let $\mathbb{G}$ be any filtration satisfying $\mathbb{F}^I \subseteq \mathbb{G} \subseteq \mathbb{F}$. It is an immediate consequence of (2.23) and the chain rule for conditional expectation that $X^I$ remains a Markov process with respect to $\mathbb{G}$. In other words, Markovian consistency also implies Markovianity of the component in its own filtration.

## 2.4.2 Markov Copulae: Generator Approach

In this section, which is based on [6, 31], we tackle the problem of defining and constructing "Markov copulae" using infinitesimal generators. In what follows, we provide conditions on the infinitesimal generator of $X$ that ensure that the Markov consistency property for $X^I$ holds. We can and will assume that the paths of $X$ are $\mathbb{P}$-a.s. in $\mathbb{D}_E[0,\infty)$. By $T(t)$ we denote the semigroup of operators on $B(E)$ defined by the transition function corresponding to $X$, and by $A$ its infinitesimal generator. We fix an index set $I \subset \{1, \ldots, d\}$. Proposition 2.4.1 below yields a necessary condition for Markovian consistency to hold.

*Remark 2.4.4.* For $f \in \mathscr{D}(A)$, $Af(x)$ determines the expected infinitesimal evolution of the process $f(X_t)$, given the initial state $X_t = x$. Intuitively, for $X^I$ to have the Markov property in the filtration $\mathbb{F}$, its infinitesimal probabilistic behavior should not depend on the state of the components $X_{I^c}$. In terms of the infinitesimal generator, this means that for a function $f$ which is a constant function of $x_{I^c}$, $Af(x)$ should only depend on the variables $x_I$. This intuition is formalized in the following proposition.

**Proposition 2.4.1 (see [6, 31]).** *Assume that $X$ is conservative and that its component $X^I$ is $\mathbb{F}$-Markov. If $f \in D_I(E)$ and $D(E) \times A(D(E)) \ni (f_n, Af_n) \stackrel{b.p.}{\to} (f, h)$, then $h$ belongs to $B_I(E)$.*

The next proposition gives sufficient conditions on the infinitesimal generator of a Markov process that ensure the Markovian consistency for components $X^I$. In addition, it provides an explicit characterization of the infinitesimal generator of $X^I$, which will be very helpful in the actual construction of Markov copulae.

**Proposition 2.4.2.** *Let $A$ be the infinitesimal generator of an $E$-valued Markov process $X$, and assume that $X$ is conservative. In addition, assume that*

$$\begin{aligned}&\forall g \in D_I(E_I),\ \text{there is a sequence } f_n \in D(E), \text{and } h^g \in C_0(E_I) \text{ such that}\\ &(f_n, Af_n) \stackrel{b.p.}{\to} (g \otimes \mathbb{1}_{E^{I^c}}, h^g \otimes \mathbb{1}_{E^{I^c}}).\end{aligned} \tag{2.25}$$

*Then:*
*(i) we can define an operator $(A^I, D_I(E_I))$ by*

$$A^I g = h^g, \text{ for all } g \in D_I(E_I). \tag{2.26}$$

*Assume, in addition, that $R(\lambda \cdot \mathrm{Id} - A^I)$ is dense in $L_I(E_I)$ for some $\lambda > 0$. Then:*
*(ii) $\overline{A^I}$ generates a strongly continuous contraction semigroup $T^I(t)$ on $L_I(E_I)$, and*
*(iii) $X^I$ is the unique Markov process corresponding to $T^I(t)$.*

Later in this section we shall be concerned with constructing operators $A$ satisfying Markovian consistency conditions, starting from the infinitesimal generators of the component processes $X^I$. The following corollary will be useful to this end:

**Corollary 2.4.1.** *Let $A$, $X$ and $\mathbb{F}$ be as in Proposition 2.4.2. Assume that condition (2.25) holds and that the operator $\overline{A^I}$, as defined in Proposition 2.4.2, generates a strongly continuous contraction semigroup on $L_I(E_I)$. Then $X^I$ is an $\mathbb{F}$-Markov process and $A^I$ coincides with the infinitesimal generator of $X^I$ on $D_I(E_I)$.*

Finally, we state sufficient conditions on the multivariate generator $A$ such that the component $X^I$ is Markovian with given finite dimensional distributions.

**Proposition 2.4.3.** *Let $X$ be a Markov process on $E$ with generator $A$, and let $Y$ be an $E_I$-valued Markov process, with infinitesimal generator $A^Y$. Suppose that the conditions of Proposition 2.4.2 are satisfied and define $A^I$ by (2.26). Moreover, suppose that $A^I = A^Y$ on $D_I(E_I)$. Then $X$ satisfies the Markovian consistency conditions for $(X^I, Y)$.*

In Proposition 2.4.2 we considered a vector-valued $\mathbb{F}$-Markov process and provided conditions on its generator ensuring that a given component is also Markov. Now we consider the problem from the opposite perspective. Given a collection of Markov processes, say $(Y^i)$, where $Y^i$ is $E_i$-valued, $i = 1, \ldots, d$, we want to construct a vector process $X = (X^1, X^2, \ldots, X^d)$ with values in $E = E_1 \times E_2 \times \cdots \times E_d$, that is Markov with respect to its natural filtration, say $\mathbb{F}$, and has the Markovian consistency property for $(X^i, Y^i)$, $i = 1, \ldots, d$.

Suppose that we are given a collection of operators, say $\mathscr{A} = \{(A^i, \mathscr{D}(A^i)) : i = 1, \ldots, d\}$, such that the closure of $A^i$ restricted to $D_i(E_i) \subset \mathscr{D}(A^i)$ generates a strongly continuous, positive contraction semigroup on $L_i(E_i) := \overline{D_i(E_i)} \subseteq C_0(E_i)$. We denote by $Y^i$, $i = 1, \ldots, d$, the corresponding $E_i$-valued Markov processes.

**Definition 2.4.3.** Assume $D(E) \subseteq \hat{\otimes}_{i=1}^d D_i(E_i)$. Let $\mathscr{C}^{\mathscr{A}} \subset \mathscr{L}(B(E), B(E))$ be a set of linear operators satisfying[11]:
i) For every element $A$ in $\mathscr{C}^{\mathscr{A}}$, the operator $\overline{A|_{D(E)}}$ generates a strongly continuous positive contraction semigroup on $L(E) := \overline{D(E)}$,
ii) For each $i = 1, 2, \ldots, d$, and for every $g \in D_i(E_i)$ there exists $f_n \in D(E)$ such that $(f_n, Af_n) \overset{b.p.}{\to} (g \otimes \mathbb{1}_{E_{i^c}}, A^i g \otimes \mathbb{1}_{E_{i^c}})$.

If $\mathscr{A}$ is not empty, then we call an element in $\mathscr{C}^{\mathscr{A}}$ a Markov copula for $\mathscr{A}$ with respect to $D(E)$.

*Remark 2.4.5.* The question whether $\mathscr{A}$ is nonempty is not easy. In some special cases it is shown by construction that $\mathscr{A}$ is not empty. In general, this question requires analysis of existence of appropriate solutions to an operator equation. In Sect. 2.4.3 we discuss symbolic Markov copulae. The question of existence of such copulae corresponds to the question of existence of solutions to certain functional equations.

Let us fix $i$. Then in view of Corollary 2.4.1, the process $X^i$ is Markov with respect to the natural filtration of $X$ and admits a generator $A^i$. Now by applying Proposition 2.4.3 we have

**Proposition 2.4.4.** *Let $A$ be an element of $\mathscr{C}^{\mathscr{A}}$. Then the canonical Markov process $X = (X^1, \ldots, X^d)$ corresponding to the semigroup generated by $A$ has the Markovian consistency property for $(X^i, Y^i)$, $i = 1, \ldots, d$.*

#### 2.4.2.1 Examples

In this subsection, we consider some important classes of Markov processes and provide a constructive answer to the problem introduced at the beginning of this section. Given a collection of Markov processes $Y^i$, how do we construct a multivariate process $X = (X^1, \ldots, X^d)$ that is Markov with respect to its natural filtration $\mathbb{F}$, and satisfies the Markovian consistency conditions for $(X^i, Y^i)$? We shall construct elements of the set $\mathscr{C}^{\mathscr{A}}$ in the following cases, which we consider important for applications:

1. The marginal processes $Y^i$, generated by $A^i$, are $\mathbb{R}$-valued diffusion processes;
2. The marginal processes $Y^i$, generated by $A^i$, are $\mathbb{R}$-valued pure jump Markov processes;

[11] We use the notation introduced in Definition 2.4.1.

3. The marginal processes $Y^i$, generated by $A^i$, are $\mathbb{R}$-valued diffusion modulated jump processes.

**Diffusion Processes.** We consider a collection of $d$ operators,

$$A^i f(x_i) = b_i(x_i)\partial_{x_i} f(x_i) + \frac{1}{2}\sigma_i(x_i)^2 \partial_{x_i}\partial_{x_i} f(x_i), \tag{2.27}$$

on $D_i(E_i) := C_c^\infty(E_i)$, where the coefficients $b_i(x_i)$ and $\sigma_i(x_i)$ are given functions in $C_b^2(E_i)$. We know that $A^i$ is a core of the infinitesimal generator of a Markov diffusion $Y^i$, taking values in $E_i = \mathbb{R}$ (see [15, Chap. 8, Theorem 2.1]). In this section we assume that $E = \mathbb{R}^d$.

In what follows we use the shorthand notation $I^{\{i\}^c}\hat{\otimes}A^i$ for $I^1\hat{\otimes}\ldots\hat{\otimes}A^i\hat{\otimes}\ldots\hat{\otimes}I^d$, where $I^m$ is the identity operator on the space $B(E_m)$, for $m = 1,\ldots,d$.

**Proposition 2.4.5.** *Let $A^i$ be as in (2.27) and define a linear operator $A$ on $D(E) := C_0^\infty(E)$ as*

$$Af(x) := \sum_{i=1}^{d} I^{\{i\}^c}\hat{\otimes}A^i f(x) + \sum_{i,j=1, i\neq j}^{d} \frac{1}{2} a_{ij}(x_i,x_j)\partial_{x_i}\partial_{x_j} f(x) \tag{2.28}$$

*where $a_{ij}(x_i,x_j)$ are such that $a_{ii}(x_i) = \sigma_i^2(x_i)$ and the (diffusion) matrix $\Sigma(x) = [a_{ij}(x_i,x_j)]$ is symmetric nonnegative definite and admits a square root $[\sigma_{ij}] := \Sigma^{\frac{1}{2}} \in C_b^2(E)$. Then the operator $A$ is a Markov copula for $\{(A^i,\mathscr{D}(A_i)) : i = 1,\ldots,d\}$.*

*Remark 2.4.6.* In view of Proposition 2.4.4, the (canonical) Markov process $X$, corresponding to the semigroup generated by $A$, has the Markovian consistency property for $(X^i,Y^i)$.

*Remark 2.4.7.* Note that dependence between the components $X^i$ is entirely characterized by the functions $a_{ij}(\cdot,\cdot)$, $i \neq j$. Therefore, every diffusion copula can be associated to a particular choice of the functions $a_{ij}(\cdot,\cdot)$.

**Markov jump processes: General case**. In this section, we assume that $E_i \subset \mathbb{R}$ are compact sets for all $i$. For $i = 1,\ldots,d$, we consider a family of operators on $D(E_i) := C(E_i)$ given by

$$A^i f(x_i) = \eta^i(x_i)\int_{E_i} (f(z_i) - f(x_i))\, \nu^i(x_i,dz_i), \tag{2.29}$$

where $\eta^i(x_i)$ are continuous functions, $\nu^i(x_i,dz_i) \in \mathscr{P}(E_i)$, and the mapping $x_i \to \nu^i(x_i,B)$ is continuous for all $i$ and $B \subset E_i$. In view of the discussion in [15, Chap. 8, Sect. 3], $A^i$ are the generators of pure jump Feller processes taking values in $E_i$, $i = 1,\ldots,n$.

**Proposition 2.4.6 (see [32]).** *Let $A^i$ be as in (2.29), and define an operator $A$ on $C(E)$ as*

$$Af(x) := \sum_{i=1}^{d} I^{\{i\}^c} \hat{\otimes} A^i f(x) + \sum_{S \in \mathscr{J}_2} \lambda^S(x) \int_E (f(z) - f(x))\, \nu^S(x, dz) \tag{2.30}$$
$$- \sum_{i=1}^{d} \sum_{S \in \mathscr{J}_2 : i \in S} \lambda^S(x) \int_E (f(z) - f(x))\, \nu^{\{i\}}(x, dz),$$

*where* $\mathscr{J}_n = \{S \in 2^{\{1,\dots,d\}} : \mathrm{card}(S) \geq n\}$ *and:*
*i)* $\nu^S(x, dz) \in \mathscr{P}(E)$ *is defined for* $S \in 2^{\{1,\dots,d\}} \setminus \emptyset$ *as*

$$\nu^S(x, dz) := \otimes_{i \in S} \nu^i(x_i, dz_i) \otimes_{j \in S^c} \delta_{x_j}(dz_j),$$

*ii) for any* $S \in \mathscr{J}_2$*, the functions* $\lambda^S$ *are nonnegative, continuous, and*

$$\sum_{S \in \mathscr{J}_2 : i \in S} \lambda^S(x) \leq \eta^i(x_i), \quad \forall x \in E, \forall i \in \{1, \dots, d\}. \tag{2.31}$$

*Let* $D(E) = C(E)$*. Then the operator* $A$ *is a Markov copula for* $\{(A^i, \mathscr{D}(A_i)) : i = 1, \dots, d\}$.

*Remark 2.4.8.* a) Notice that we can rewrite (2.30) in the form

$$Af(x) := \sum_{S \in \mathscr{J}_1} \lambda^S(x) \int_E (f(z) - f(x))\, \nu^S(x, dz), \tag{2.32}$$

with $\lambda^{\{i\}}(x) = \eta^i(x_i) - \sum_{S \in \mathscr{J}_2 : i \in S} \lambda^S(x)$ for all $i$. This, in particular, implies that $\sum_{S \in \mathscr{J}_1 : i \in S} \lambda^S(x) = \eta^i(x_i)$.
b) In view of Proposition 2.4.4, the Markov process $X$, corresponding to the semigroup generated by $A$, has the Markovian consistency property for $(X^i, Y^i)$, $i = 1, \dots, d$.

**Markov jump processes: Case of space homogeneous jump size distribution.** In this subsection, we consider special pure jump processes with generators $\{A^i, C(E_i)\}$ defined by

$$A^i f(x_i) = \eta^i(x_i) \int_{E_i} (f(z_i) - f(x_i))\, \nu^i(dz_i), \tag{2.33}$$

where $\eta^i(x_i)$ are continuous and $\nu^i(dz_i) \in \mathscr{P}(E_i)$. Let $D_i(E_i) = C(E_i)$. The jump distribution $\nu^i(dz_i)$ is space homogeneous (does not depend on $x$). It turns out that it is possible to construct multivariate Markov jump processes with an arbitrary jump distribution.

**Proposition 2.4.7 (see [31]).** *Let* $A^i$ *be as in (2.33) and let*

$$Af(x) := \sum_{i=1}^{d} I^{\{i\}^c} \hat{\otimes} A^i f(x) + \sum_{S\in\mathscr{J}_2} \lambda^S(x) \int_E (f(z)-f(x))\, \nu^S(dz) \qquad (2.34)$$
$$- \sum_{i=1}^{d} \sum_{S\in\mathscr{J}_2 : i\in S} \lambda^S(x) \int_E (f(z)-f(x))\, \nu^{\{i\}}(dz)$$

*be an operator on $C(E)$, where*
*i) $\nu^S(dz) \in \mathscr{P}(E)$ is defined as*

$$\nu^S(dz) := C^S(\nu^i(dz_i), i\in S) \otimes_{j\in S^c} \delta_{x_j}(z_j),$$

*for some copula function $C^S : [0,1]^S \to [0,1]$,*
*ii) the nonnegative continuous functions $\lambda^S$ are such that:*

$$\sum_{S\in\mathscr{J}_2 : i\in S} \lambda^S(x) \le \eta^i(x_i), \quad \forall x\in E, \forall i\in\{1,\dots,d\}.$$

*Let $D(E) := C(E)$. Then the operator $A$ is a Markov copula for $\{(A^i, \mathscr{D}(A_i)) : i = 1,\dots,d\}$.*

**Diffusion modulated Markov jump processes.** Let $Y$ be a diffusion process in $\mathbb{R}^n$, with infinitesimal generator $\mathscr{L}$ given by

$$\mathscr{L}f(y) = b(y)^T\nabla f(y) + \mathrm{trace}(a(y)\nabla\nabla^T)f(y),$$

where $b(\cdot)$ and $a(\cdot)$ are regular enough to ensure that $\overline{\mathscr{L}|_{C_c^\infty(E)}}$ generates a strongly continuous contraction semigroup on $C_0(\mathbb{R}^n)$. Using $\mathscr{L}$ we define a collection of $d$ operators on $C_c^{\infty,0}(\mathbb{R}^n\times E_i)$, $i=1,\dots,d$, by

$$A^i f(y,x_i) = \mathscr{L}\hat{\otimes} I^i f(y,x_i) + \tilde{A}^i f(y,x_i), \qquad (2.35)$$

where $I^i$ is the identity operator on $E_i$ (a compact subset of $\mathbb{R}$) and

$$\tilde{A}^i f(y,x_i) = \eta^i(y,x_i) \int_{E_i} (f(y,z_i) - f(y,x_i))\, \nu^i(y,dz_i), \qquad (2.36)$$

$\eta^i(\cdot,\cdot)$ is a continuous and bounded function of both arguments, and $\nu^i(y,dz_i)$ is a probability measure for every $y$, such that for every measurable set $B$ in $E_i$ the map $y\to\nu^i(y,B)$ is continuous and bounded for $i=1,2,\dots,d$.

Using our assumptions about $\mathscr{L}$, i.e., boundedness of the operators $\tilde{A}^i$, we see by [15, Chap. 1, Corollary 7.2] that for each $i$ the operator $A^i$ generates a strongly continuous semigroup of operators on $C_0(E_i\times\mathbb{R}^n)$. The functions $\eta^i$, $i=1,\dots,d$, are nonnegative, so $\tilde{A}^i$ satisfies the positive maximum principle. Hence the semigroup generated by $A^i$ is positive and contractive in view of the Ethier and Kurtz theorem [15, Chap. 1, Theorem 7.1]. Hence (2.35) is indeed the generator of a Markov process on $\mathbb{R}$.

**Proposition 2.4.8.** *Define an operator $A$ on $C_c^{\infty,0}(\mathbb{R}^n\times E)$ by*

$$
\begin{aligned}
Af(x,y) := {} & \mathscr{L}\hat{\otimes}If(y,x) + \sum_{i=1}^{d} I^{\{i\}^c}\hat{\otimes}\tilde{A}^i f(y,x) \\
& + \sum_{S\in\mathscr{J}_2} \lambda^S(y,x)\int_E (f(y,z)-f(y,x))\,\nu^S(y,dz) \\
& - \sum_{i=1}^{d}\sum_{S\in\mathscr{J}_2:i\in S} \lambda^S(y,x)\int_E (f(y,z)-f(y,x))\,\nu^{\{i\}}(y,dz),
\end{aligned}
\tag{2.37}
$$

*where*

*i)* $\nu^S(y,dz)\in\mathscr{P}(E)$ *is defined as*

$$\nu^S(y,dz) := C^S(\nu^i(y,dz_i),\, i\in S)\otimes_{j\in S^c}\delta_{x_j}(dz_j)$$

*for some copula function* $C^S : [0,1]^S \to [0,1]$,
*ii) the nonnegative continuous functions* $\lambda^S(y,x)$ *are such that*

$$\sum_{S\in\mathscr{J}_2:i\in S} \lambda^S(y,x) \le \eta^i(y,x_i), \quad \forall (x,y)\in E\times\mathbb{R}^n, \forall i\in\{1,\dots,d\}.$$

*Let* $D(E) := C_c^{\infty,0}(\mathbb{R}^n\times E)$. *Then* $A$ *is a Markov copula for* $\{(A^i,\mathscr{D}(A_i)) : i = 1,\dots,d\}$.

It is now easy to measure the dependence between the Markovian components $(Y,X^i)$. The d-volume is, in this case[12]:

$$\text{Dvol}((Y,X_T)) = \mathbb{E}\int_0^T \left(\frac{1}{2}\sum_{i,j=1}^{d}|a(Y_s)| + \sum_{S\in\mathscr{J}_2}\text{card(S)}\lambda(Y_s,X_s)\right) ds.$$

## 2.4.3 Markov Copulae: Symbolic Approach

In the previous section we have presented a construction of a copula between Markov processes in terms of infinitesimal generators. Here, based on [3], we present a *symbolic approach*, which makes use of pseudo-differential operators (PDO). This approach is more transparent and gives relatively simple conditions guaranteeing that a multivariate Markov process have Markovian components with respect to their own filtration. It also allows one to construct a Markov process with prescribed marginal laws. In this approach, to construct the symbol corresponding to a Markov copula, one just has to construct nonnegative definite functions satisfying appropriate conditions, whereas in the approach in [4] one has to construct an operator acting on functions. Thus, the symbolic approach allows one to avoid

[12] For an element $S$ of $\mathscr{S}$ (space of symmetric matrices), the norm $|S|$ is understood to be given by $\sqrt{trace(SS)}$.

using tensor products of infinitesimal generators and investigation of b.p. closure of operators, but it has some limitations which follow from Hoh's theorem.

In Sect. 2.4.3.1 we investigate the connection of Markovian consistency properties with the corresponding PDOs, in particular we study the question of constructing a multivariate Feller process with given marginal laws in terms of symbols of some related PDOs. Examples are provided in Sect. 2.4.3.2. In what follows, we shall only consider time-homogeneous Markov processes.

#### 2.4.3.1 Dependence and Symbols

Consider $X = (X^j, j = 1, \ldots, n)$, a time-homogeneous Markov process, defined on an underlying probability space $(\Omega, \mathscr{F}, \mathbb{P})$, taking values in $\mathbb{R}^n$.

As before, we are interested in the Markovian consistency properties (see Definition 2.4.2) of a Feller Markov process $X$. For simplicity of exposition we limit ourselves to one-dimensional margins, i.e., we consider $I = \{j\}$, but all results can be extended to the case of an arbitrary subset of components of the process $X$.

Therefore our first goal is to provide necessary and sufficient conditions which guarantee that the components of $X$ are Markov processes with respect to their natural filtrations. The second goal is to provide necessary and sufficient conditions which guarantee that the Markovian consistency condition holds for $(X^j, Y^j)$ for a given one-dimensional Markov process $Y^j$ defined on $(\widetilde{\Omega}, \widetilde{\mathscr{F}}, \widetilde{\mathbb{P}})$.

The Markov consistency properties, which are properties of transition probabilities, can be formulated in terms of the conditional characteristic functions of $X_t$ and $X_t^j$ defined as

$$\lambda_t(x,\xi) := \mathbb{E}\left(e^{-i(X_t - x,\xi)}\Big|X_0 = x\right), \quad \lambda_t^j(x_j,\xi_j) := \mathbb{E}\left(e^{-i(X_t^j - x_j)\xi_j}\Big|X_0^j = x_j\right),$$

$$\psi_t^j(y_j,\xi_j) := \widetilde{\mathbb{E}}\left(e^{-i\xi_j(Y_t^j - y_j)}\Big|Y_0^j = y_j\right).$$

Since there are no explicit formulae for the transition probabilities or conditional characteristic functions of general Markov processes, solving our problems in terms of the entire families $\lambda_t$, $\lambda_t^j$ and $\psi_t^j$ is quite inconvenient. The general form of conditional characteristic functions is well known only in the case of Lévy processes.[13]

So we have to solve our problems using a different language. The key observation is that the study of our problems in terms of the families of functions $\lambda_t$, $\lambda_t^j$ and $\psi_t^j$ turns out to be equivalent to the study of these problems in terms of the Markov semigroups corresponding to $X$, $X^j$ and $Y^j$. In Jacob [18] it is shown that for a family of functions $\lambda_t$ we can compute the semigroup $(T_t)_{t\geq 0}$ corresponding to $X$ in the following way:

$$T_t u(x) := \mathbb{E}^x u(X_t) = (2\pi)^{n/2} \int_{\mathbb{R}^n} e^{i(x,\xi)} \lambda_t(x,\xi) \hat{u}(\xi) d\xi,$$

[13] Note that in the case of Lévy processes these conditions can be significantly simplified in the sense that they can be reduced to considering $t = 1$ only.

where $\hat{u}$ denotes the Fourier transform of the function $u:\mathbb{R}^n\to\mathbb{R}$, that is,

$$\hat{u}(\xi):=(2\pi)^{-n/2}\int_{\mathbb{R}^n}e^{-i(x,\xi)}u(x)dx.$$

Analogous properties hold for the families $\lambda_t^j$ and $\psi_t^j$ and the corresponding semigroups, say $T_t^j$ and $S_t^j$.

Moreover, in view of the results of Courrège [9], the generator $A$ of $X$, acting on $u\in C_0^\infty(\mathbb{R}^n)$, the space of infinitely differentiable functions with compact support, has a representation

$$Au(x)=-q(x,D)u(x):=-(2\pi)^{-n/2}\int_{\mathbb{R}^n}e^{i(x,\xi)}q(x,\xi)\widehat{u}(\xi)d\xi, \tag{2.38}$$

where $q:\mathbb{R}^n\times\mathbb{R}^n\to\mathbb{C}$ is a measurable, continuous function in $\xi$ and for every $x$ the function $q(x,\cdot)$ is negative definite. In this context, the function $q(x,\xi)$ is called the symbol of the pseudo-differential operator $q(x,D)$ (cf. [19]), and it has the following form:

$$q(x,\xi)=i(b(x),\xi)+(\xi,a(x)\xi)+\int_{\mathbb{R}^n\setminus\{0\}}\left(1-e^{i(y,\xi)}+\frac{i(y,\xi)}{1+|y|^2}\right)\nu(x,dy) \tag{2.39}$$

where $a,b$ are Borel measurable functions, $b(x)\in\mathbb{R}^n$, $a(x)$ is a symmetric nonnegative definite matrix, and $\nu(x,dy)$ is a Lévy kernel. Moreover, if $q(x,\xi)$ is continuous (in all variables) then $q(x,D)$ maps $C_0^\infty(\mathbb{R}^n)$ into $C(\mathbb{R}^n)$ [19, Vol. 1, Theorem 4.5.7, p. 337]. Analogous results hold for $q^j$ and $\rho^j$. In particular, in the case of $Y^j$, the infinitesimal generator $B^j$, acting on $w\in C_0^\infty(\mathbb{R})$, satisfies

$$B^jw(x_j)=-(2\pi)^{-1/2}\int_{\mathbb{R}}e^{ix_j\xi_j}\widehat{w}(\xi_j)\rho_j(x_j,\xi_j)d\xi_j,$$

where

$$\rho_j(x_j,\xi_j)=ib_j(x_j)\xi_j+c_j(x_j)\xi_j^2+\int_{\mathbb{R}\setminus\{0\}}\left(1-e^{iz_j\xi_j}+\frac{iz_j\xi_j}{1+|z_j|^2}\right)\nu_j(x_j,dz_j). \tag{2.40}$$

So we pursue the study of our problems in terms of symbols of pseudo-differential operators. We shall adopt the following convention: We suppose that $f:\mathbb{R}\to\mathbb{R}$ and that $f_j:\mathbb{R}^n\to\mathbb{R}$ is defined by $f_j(x)=f(x_j)$. Note, however, that even though $f$ may be of compact support, $f_j$ will not be a function of compact support. Recall that by $e_j$ we denote the standard unit vector in $\mathbb{R}^n$ with 1 in the $j$-th position.

**Proposition 2.4.9.** *Let $X$ be a Feller process with symbol $q$ and the corresponding generator $A$ such that $C_0^\infty(\mathbb{R}^n)\subseteq D(A)$.*
*a) For every $w\in C_0^\infty(\mathbb{R})$,*

$$Aw_j(x)=-(2\pi)^{-1/2}\int_{\mathbb{R}}e^{ix_j\xi_j}\widehat{w}(\xi_j)q(x,Q_j(\xi))d\xi_j, \tag{2.41}$$

*where* $Q_j : \mathbb{R}^n \to \mathbb{R}^n$ *is the projection* $Q_j(\xi) := \mathbf{e}_j \xi_j$.
*b) Assume that* $X^j$ *is a one-dimensional Feller process with symbol* $q_j$ *and that*

$$q(x, e_j \xi_j) = q_j(x_j, \xi_j) \quad \textit{for all } x \in \mathbb{R}^n \textit{ and } \xi_j \in \mathbb{R}. \tag{2.42}$$

*Then for every* $w \in C_0^\infty(\mathbb{R})$,

$$A w_j(x) = A_j w(x_j), \ \ \forall x \in \mathbb{R}^n, \tag{2.43}$$

*where* $A$ *is the generator of* $X$, *and* $A_j$ *is the generator of* $X^j$.

The following proposition gives a necessary condition for Markovian consistency for $X^j$.

**Proposition 2.4.10.** *Let* $X$ *be a Feller process with symbol* $q$. *If* $X^j$ *is a* 1*-dimensional Feller processes with symbol* $q_j$, *then* (2.42) *holds.*

Before we consider the question of sufficiency of condition (2.42), we introduce conditions, due to Hoh [16], guaranteeing that the pseudo-differential operator $-q(x,D)$ has a unique extension which generates a Feller semigroup. Let $\psi : \mathbb{R}^n \to \mathbb{R}$ be a continuous negative definite function such that for some positive constants $r$ and $c$ we have $\psi(\xi) \geq c|\xi|^r$ for $|\xi| \geq 1$. We define

$$\lambda(\xi) := (1 + \psi(\xi))^{1/2}$$

and let $M$ be the smallest integer such that $M > (\frac{n}{r} \vee 2) + n$, and set $k = 2M + 1 - n$. For a continuous negative definite symbol $q : \mathbb{R}^n \times \mathbb{R} \to \mathbb{C}$ we make the following assumptions:
**C0:** the function $q$ is continuous in both variables;
**C1:** the map $x \mapsto q(x, \xi)$ is $k$ times continuously differentiable and

$$\left| \partial_x^\beta q(x, \xi) \right| \leq c \lambda^2(\xi), \quad \beta \in \mathbb{N}_0^n, |\beta| \leq k;$$

**C2:** for some strictly positive function $\gamma : \mathbb{R}^n \to \mathbb{R}$,

$$q(x, \xi) \geq \gamma(x) \lambda^2(\xi) \quad \text{for} \quad |\xi| \geq 1, x \in \mathbb{R}^n;$$

**C3:**

$$\sup_{x \in \mathbb{R}^n} |q(x, \xi)| \underset{\xi \to 0}{\longrightarrow} 0.$$

It is proved in Hoh [16, Theorem 5.24, p. 82] that under C0–C3 the pseudo-differential operator $-q(x,D) : C_0^\infty(\mathbb{R}^n) \to C^\infty(\mathbb{R}^n)$ has an extension which generates a Feller semigroup given by $P_t f(x) = \mathbb{E}^x f(X_t)$, where $\mathbb{E}^x$ is expectation with respect to the solution of the associated well-posed martingale problem starting at $x$.

**Theorem 2.4.1.** *Let* $X$ *be a Feller process with symbol* $q$. *Assume that*
***C4:*** *the function* $q(x, e_j \xi_j)$ *as a function of* $x$ *depends only on* $x_j$
*and denote*

$$\tilde{q}_j(x_j,\xi_j) := q(x,e_j\xi_j). \tag{2.44}$$

*If $\tilde{q}_j$ satisfies conditions C0–C3, then the component $X^j$ of $X$ is a Feller process with generator given by the symbol $\tilde{q}_j$.*

The above theorem demonstrates that (2.42) (with $q_j = \tilde{q}_j$) is a sufficient condition for Markov consistency to hold, provided that $\tilde{q}_j$ satisfies conditions C0–C3.

**Corollary 2.4.2.** *Let $X$ be a Feller process with symbol $q$ satisfying conditions C0–C4. Then the component $X^j$ of $X$ is a Feller process with generator given by the symbol $\tilde{q}_j$.*

Now we formulate an answer to the problem of Markovian consistency conditions for $(X^i,Y^i)$ which combines the results of Proposition 2.4.10 and Theorem 2.4.1.

**Theorem 2.4.2.** *Let $X = (X^1,\dots,X^n)$ be a Feller process with symbol $q$, and $Y^1,\dots,Y^n$ be an $n$-tuple of one-dimensional Feller processes with symbols $\rho_1,\dots,\rho_j$. Assume that $\rho_1,\dots,\rho_j$ satisfy assumptions C0–C3. The marginal distribution of the $j$-th coordinate of $X$ is equal to the distribution of $Y^j$ given by the symbol $\rho_j$ if and only if*

$$q(x,e_j\xi_j) = \rho_j(x_j,\xi_j) \quad \text{for all } x \in \mathbb{R}^n \text{ and } \xi_j \in \mathbb{R}. \tag{2.45}$$

*Therefore the $j$-th coordinate of $X$ is a Feller process with respect to its natural filtration.*

All these considerations allow us to provide an algorithm for construction of an $n$-dimensional Feller process with given marginal distributions, and such that its components are also Feller processes. In view of Theorem 2.4.2 we can introduce

**Definition 2.4.4 (Symbolic Markov copulae).** We say that a symbol $q$ is a Markov symbolic copula for symbols $q_1,\dots,q_n$ if for all $j = 1,\dots,n$,

$$q(x,e_j\xi_j) = q_j(x_j,\xi_j). \tag{2.46}$$

Now, our aim is to give a recipe for constructing a symbol $q$, starting from given one dimensional symbols $q_1,\dots,q_n$, such that $q$ satisfies condition (2.46). Taking into account (2.39) and (2.40) we are looking for a vector function $b$ such that

$$b_j(x) = d_j(x_j), \tag{2.47}$$

a symmetric nonnegative definite matrix function $a$ such that

$$a_{jj}(x) = c_j(x_j), \tag{2.48}$$

and a Lévy measure $\nu(x,dy)$ on $\mathbb{R}^n$ such that

$$\int_{\mathbb{R}^n\setminus\{0\}} \left(1 - e^{iy_j\xi_j} + \frac{iy_j\xi_j}{1+|y|^2}\right)\nu(x,dy) = \int_{\mathbb{R}\setminus\{0\}} \left(1 - e^{iy_j\xi_j} + \frac{iy_j\xi_j}{1+|y_j|^2}\right)\nu_j(x_j,dy_j). \tag{2.49}$$

The triple $(d_j,c_j,\nu_j)$ will be called the characteristic triple for the copula $q_j$.

### 2.4.3.2 Examples

Now we present some examples to illustrate how this idea works.

*Example 2.4.3 (Product copula).* A copula for symbols $q_1,\dots,q_n$ with characteristic triples $(d_j,c_j,\nu_j)_{j=1}^n$ is a *product copula* if its characteristic triple $(b,a,\nu)$ is defined by

$$b_j(x) := d_j(x_j), \quad a_{ij}(x) := c_j(x_j)\mathbf{1}_{\{i=j\}},$$

$$\nu(x,dy) := \sum_{j=1}^{n} \otimes_{k\neq j}\delta_{\{0\}}(dy_k)\otimes\nu_j(x_j,dy_j).$$

It is easy to see that the symbol $q$ corresponding to $(b,a,\nu)$ defined above is a copula for given symbols $q_1,\dots,q_n$. It is a copula that corresponds to independent Feller processes.

*Example 2.4.4 (Diffusion copula).* Consider $n$ one-dimensional diffusion processes with symbols given by $q_j(x_j,\xi_j) = id_j(x_j)\xi_j + c_j(x_j)\xi_j^2$, where $d_i$, $c_i$ are functions such that $q_i$ satisfies C0–C3. We define $q$ by

$$q(\xi,x) = i(b(x),\xi) + (\xi,a(x)\xi),$$

where the functions $b:\mathbb{R}^n\to\mathbb{R}^n$ and $a:\mathbb{R}^n\to\mathbb{R}^n_n$ satisfy

$$b_j(x) = d_j(x_j), \qquad a_{jj}(x) = c_j(x_j) \qquad \forall j = 1,\dots,n, \tag{2.50}$$

and moreover $b$, $a$ are chosen in such a way that C0–C3 hold. Then $q$ is a symbolic copula for $q_1,\dots,q_n$.

*Example 2.4.5 (Lévy copulae).* Consider $n$ one-dimensional Lévy processes $Z_1,\dots,Z_n$ with Lévy measures $\nu_1,\dots,\nu_n$, and a Lévy copula $F$. A Lévy measure $\nu$, which is determined by the set of tail integrals defined by formula (2.1), gives a symbol that satisfies (2.49). Therefore the construction of Lévy copulae due to Kallsen and Tankov [21] provides also a construction of symbolic copulae.

*Example 2.4.6 (Poisson copula).* Consider two one-dimensional Poisson processes. Their symbols are given by $q_i(\xi_i) = (1-e^{i\xi_i})\eta_i$, where $\eta_i$ are nonnegative constants for $i=1,2$. A symbol $q$ given by

$$q(\xi_1,\xi_2) = (1-e^{i\xi_2})\lambda_{(0,1)} + (1-e^{i\xi_1})\lambda_{(1,0)} + (1-e^{i(\xi_1+\xi_2)})\lambda_{(1,1)},$$

where $\lambda_{(0,1)},\lambda_{(1,0)},\lambda_{(1,1)}$ are nonnegative constants, defines a Markov copula iff $\lambda_{(0,1)},\lambda_{(1,0)},\lambda_{(1,1)}$ satisfy the following system of linear equations:

$$\begin{cases} \lambda_{(0,1)} + \lambda_{(1,1)} = \eta_2, \\ \lambda_{(1,0)} + \lambda_{(1,1)} = \eta_1. \end{cases}$$

The above system has infinitely many solutions which can be parameterized by $\lambda_{(1,1)}$. Since we are interested in positive solutions, we restrict $\lambda_{(1,1)}$ to the interval $[0,\lambda_1\wedge\lambda_2]$. Generalization to the $n$-dimensional case is immediate.

We can generalize Example 2.4.6 by allowing $\lambda$ to depend on $x$. In this case we have to deal with generalized $n$-dimensional Markov point processes. Moreover, we can construct copulae for Markov jump processes and Markov jump processes with the possibility of common jumps with space homogeneous jump size distribution. All these examples are presented in [3].

## 2.5 Applications in Finance

We present an application to finance of the results discussed in the previous sections. Towards this end we shall borrow some results from [5]. These results illustrate an application of Markov copulae (and semimartingale copulae) to valuation of so called *rating-triggered, step-up bonds*. We refer to [5] as well as to [1, 10] for some other financial applications of Markov copulae.

Rating-triggered step-up bonds were issued by some European telecom companies in recent 10 years. These products are of interest because they offer protection against credit events other than defaults. In particular, rating-triggered corporate step-up bonds (step-up bonds for short) are corporate coupon issues for which the coupon payment depends on the issuer's credit quality: in principle, the coupon payment increases when the credit quality of the issuer declines. In practice, credit quality is reflected in credit ratings assigned to the issuer by at least one credit rating agency (Moody's-KMV or Standard & Poor's). The provisions linking the cash flows of the step-up bonds to the credit rating of the issuer have different step amounts and different rating event triggers. In some cases, a step-up of the coupon requires a downgrade to the trigger level by both rating agencies. In other cases, there are step-up triggers for actions of each rating agency. Here, a downgrade by one agency will trigger an increase in the coupon regardless of the rating from the other agency. Provisions also vary with respect to step-down features which, as the name suggests, trigger a lowering of the coupon if the company regains its original rating after a downgrade. In general, there is no step-down below the initial coupon for ratings exceeding the initial rating.

Let $R_t$ stand for some indicator of credit quality at time $t$ (note that in this case, the process $R$ may be composed of two, or more, distinct rating processes). Assume that $t_i$, $i = 1, 2, \ldots, n$, are coupon payment dates. Here we adopt the convention that coupon paid at date $t_n$ depends only on the rating history through date $t_{n-1}$, that is: $c_n = c(R_t, t \leq t_{n-1})$ are the coupon payments. In other words, we assume that no accrual convention is in force.

Assuming that the bond's notional amount is 1, the cumulative discounted cash flow of the step-up bond is (as usual we assume that the current time is 0)

$$(1 - H_T)\beta_T + \int_{(0,T]} (1 - H_u)\beta_u \, dC_u + \beta_\tau Z_\tau H_T, \tag{2.51}$$

where $C_t = \sum_{t_i \leq t} c_i$, $\tau$ is the bond's default time, $H_t = \mathbb{1}_{\tau \leq t}$, and $Z_t$ is a (predictable) recovery process.

### 2.5.1 Pricing Rating-Triggered Step-Up Bonds via Simulation

Here, using our results on Markov copulae, we shall apply a simulation approach to pricing rating-triggered step-up bonds.

Let us consider a rating-triggered step-up bond issued by an obligor $XYZ$. Recall that, typically, cash flows associated with a step-up bond depend on ratings assigned to $XYZ$ by both Moody‘s Investors Service (Moody’s in what follows) and Standard & Poor’s (S&P in what follows). Thus, a straightforward way to model joint credit migrations would be to consider a credit migration process $R$ such that $R_t = (M_t, SP_t)$, where $M_t$ and $SP_t$ denote the time $t$ credit rating assigned to $XYZ$ by Moody’s and $SP_t$, respectively. We assume that the process $M$ is a time-homogeneous Markov chain with respect to its natural filtration, under the statistical probability $\mathbb{P}$, and that its state space is $\mathscr{K}_1 = \{1, 2, \ldots, K_1\}$. Likewise, we assume that $SP$ is a time-homogeneous Markov chain with respect to its natural filtration, under the statistical probability $\mathbb{P}$, and that its state space is $\mathscr{K}_2 = \{1, 2, \ldots, K_2\}$.

Typically, we are only provided with individual statistical characteristics of each of the processes $M$ and $SP$. Thus, in a sense, we know the marginal distributions of the joint process $R$ under the *statistical* measure $\mathbb{P}$ (where $M$ and $SP$ are considered as “univariate” margins). The crucial issue is thus appropriate modeling of dependence between $M$ and $SP$. In particular, we want to model dependence, under $\mathbb{P}$, between $M$ and $SP$ so that the joint process $R$ is a time-homogeneous Markov chain, and so that the components $M$ and $SP$ are time-homogeneous Markov chains with given $\mathbb{P}$-generators, say $A^M$ and $A^{SP}$, respectively. Thus, essentially, we need to model a $\mathbb{P}$-generator matrix, say $A^R$, so that $R$ is a time-homogeneous Markov chain with $\mathbb{P}$-generator $A^R$ and that $M$ and $SP$ are time-homogeneous Markov chains with $\mathbb{P}$-generators $A^M$ and $A^{SP}$. We can of course deal with this problem using the theory of Markov copulae.

Towards this end, we fix an underlying probability space $(\Omega, \mathscr{F}, \mathbb{P})$. On this space we consider two univariate Markov chains $M$ and $SP$, with given infinitesimal $\mathbb{P}$-generators $A^M = [a^M_{ij}]$ and $A^{SP} = [a^{SP}_{hk}]$, respectively. Next, we consider the system of equations in variables $a^R_{ih,jk}$

$$\begin{cases} \sum_{k \in \mathscr{K}_2} a^R_{ih,jk} = a^M_{ij}, \ \forall i, j \in \mathscr{K}_1, i \neq j, \ \forall h \in \mathscr{K}_2, \\ \sum_{j \in \mathscr{K}_1} a^R_{ih,jk} = a^{SP}_{hk}, \ \forall h, k \in \mathscr{K}_2, h \neq k, \ \forall i \in \mathscr{K}_1. \end{cases} \tag{2.52}$$

Now, provided that the system (2.52) has a positive solution, it follows from Corollary 2.3.1 in Sect. 2.3.2 that the resulting matrix[14] $A^R = [a^R_{ih,jk}]_{i,j \in \mathscr{K}_1, h,k \in \mathscr{K}_2}$ satisfies conditions for a $\mathbb{P}$-generator matrix of a bivariate time-homogenous Markov chain, whose components take values in finite state spaces $\mathscr{K}_1$ and $\mathscr{K}_2$ with cardinalities $K_1$ and $K_2$, respectively, and, more importantly, they are Markov chains with the

[14] System (2.52) does not involve diagonal elements of $A^R$. These elements are obtained as $a^R_{ih,ih} = -\sum_{(j,k) \in \mathscr{K}_1 \times \mathscr{K}_2 \setminus \{(i,h)\}} a^R_{ih,jk}$.

same distributions as $M$ and $SP$ under $\mathbb{P}$. Thus, indeed, the system (2.52) essentially serves as a Markov copula between the Markovian margins $M$, $SP$ and the bivariate Markov chain $R$.

Note that the system (2.52) contains many more variables than equations. Thus, one can create several bivariate Markov chains $R$ with the given margins $M$ and $SP$. In financial applications this feature leaves a lot of flexibility for various modeling options and for calibration of the model. For example, as observed by Lando and Mortensen [23], although the ratings assigned by S&P and Moody's to the same company do not necessarily coincide, split ratings are rare and are usually only observed in short time intervals. This feature can easily be modelled using the Markovian copula system (2.52) via imposing side constraints for the unknowns $a^R_{ih,jk}$. In order to model such observed behavior of the joint rating process, we thus impose additional constraints on the variables in the system (2.52). Specifically, we postulate that

$$a^R_{ih,jk} = \begin{cases} 0 & \text{if } i \neq j \text{ and } h \neq k \text{ and } j \neq k, \\ \alpha \min(a^M_{ij}, a^{SP}_{hk}) & \text{if } i \neq j \text{ and } h \neq k \text{ and } j = k, \end{cases} \tag{2.53}$$

where $\alpha \in [0,1]$ is a modeling parameter. Using constraints (2.53) we can easily solve system (2.52) (in this case the system actually becomes fully decoupled) and we can obtain the generator of the joint process. The interpretation of constraints (2.53) is the following: The components $M$ and $SP$ of the process $R$ migrate according to their marginal laws, but they tend to join, that is, they tend to both take the same value. The strength of that tendency is measured by the parameter $\alpha$. When $\alpha = 0$ then, in fact, the two components are independent processes; when $\alpha = 1$ the intensity of both components migrating simultaneously to the same rating category is maximum (given the specified functional form for the intensities of common jumps).

For pricing purposes the statistical probability measure is changed to the *EMM*. Typically, the Radon-Nikodym density is chosen in such a way that the resulting (risk-neutral) default probabilities are consistent with the term structure of CDS spreads. In addition, we require that the process $R$, which is Markovian under the statistical measure, is also Markovian under the pricing measure.

We recall that $A^R = [a^R_{ih,jk}]$ is the generator of $R$ under the statistical measure $\mathbb{P}$. In view of Corollary 4.1 in [5], given a vector $h = [h_{11}, \cdots, h_{K_1K_2}] \in \mathbb{R}^{K_1K_2}$, we can change the statistical measure $\mathbb{P}$ to an equivalent "risk-neutral" measure $\mathbb{Q}$ in such a way that $R$ is a time-homogeneous Markov chain under $\mathbb{Q}$, and its $\mathbb{Q}$-infinitesimal generator is given by

$$\widetilde{A}^R = [\widetilde{a}_{ih,jk}],$$

where

$$\widetilde{a}_{ih,jk} = a_{ih,jk}\frac{h_{jk}}{h_{ih}}, \quad \forall i,j \in \mathscr{K}_1, \quad \forall h,k \in \mathscr{K}_2, \quad (i,h) \neq (j,k)$$

and

$$\widetilde{a}_{ih,ih} = - \sum_{(j,k)\in\mathscr{K}_1\times\mathscr{K}_2\setminus\{(i,h)\}} a_{ih,jk}\frac{h_{jk}}{h_{ih}}.$$

*Remark 2.5.1.* Note that, although the change of measure preserves the Markov property of the joint process $R$, its components may not be Markov (in their natural filtration) under the new probability measure. This however is not an issue for us, as all we need to conduct computations is the Markov property of the joint process $R$ under the new measure.

An arbitrary choice of the vector $h$ may lead to a heavy parametrization of the pricing model. We suggest that the vector $h_{ij}$ be chosen as follows:

$$h_{ij} = \exp(\alpha_1 i + \alpha_2 j), \quad \forall i \in \mathscr{K}_1, \forall j \in \mathscr{K}_2,$$

where $\alpha_1$ and $\alpha_2$ are parameters to be calibrated. It turns out, as the calibration results provided below indicate, that this is a good choice.

### 2.5.2 *Model Calibration and Pricing*

The model is fully specified by three parameters, namely $\alpha$, $\alpha_1$, $\alpha_2$, which are calibrated to market data.

Let us consider a vanilla bond, which is equivalent[15] to the given step-up bond. One would presume that the price of a step-up bond is equal to the price of the equivalent vanilla bond plus the (positive) value of the step-up provision. In general, equivalent vanilla bonds are not traded on the market. However, their price can be synthesized by applying a standard bootstrapping-interpolation procedure to the market prices of traded vanilla bonds. Surprisingly, the value of the step-up provision is often negligible or even negative. This was already noted by some recent empirical literature (cf. e.g. [23]), which provides strong evidence that the market typically "underprices" step-up bonds. These findings suggest that step-up bond investors are more risk averse than vanilla bond investors. In particular, on the theoretical level, this means that the pricing kernel implied by step-up bonds prices should be different from that implied by vanilla bonds. For calibration purposes, this implies that the model parameters, or at least those relative to credit migrations, should not be calibrated to vanilla bond prices. Nevertheless, such data provides useful information. In particular, under the assumptions given below, vanilla bond prices can be used to compute a term structure of firm-specific, liquidity-adjusted, discount factors (risk-free rate + liquidity spread).

[15] By equivalent, we mean a coupon bearing bond, backed by the same company, whose all provisions, other than the step-up provision, are identical to those of the given step-up bond. That is, maturity and coupon dates are the same, and the coupons of the equivalent bond are equal to the fixed coupons of the step-up bond. In addition, credit risk is the same and liquidity risk is comparable. The term vanilla means that the step-up provision is not present.

Our first assumption is that the vanilla bond market assesses likelihood of the default event in the same way as the CDS (Credit Default Swap) market.[16] Our second assumption is that liquidity risk is priced identically by the step-up and vanilla bond markets.

Given the above, we can apply a standard bootstrapping-interpolation procedure to a pool of reference bonds[17] to obtain a term-structure of firm specific, liquidity adjusted, zero-coupons. The straightforward procedure is briefly described below. We are given a set of $J$ reference bonds with associated cash flows $CF^j_{t^j_i}$, $j = 1, \ldots, J$, and coupon dates $t^j_0 = 0, \ldots, t^j_N = T^j$ such that $T^1 < T^2 < \cdots < T^J$. The cash flows are then adjusted by the default probability implied by the CDS spreads. Let $\tau$ denote the default time of the relevant obligor. Then the default adjusted cash flows are $\widetilde{CF}^j_{t^j_i} = CF^j_{t^j_i}\mathbb{Q}(\tau > t^j_i)$. The interpolation-bootstrapping procedure is now applied to the reference bonds with default-risk adjusted cash flows, so that the resulting discount factors account only for the firm specific liquidity spread.[18] At this point, the price of an arbitrary step-up bond can be computed by simulating the evolution of the joint rating process and the relative discounted cash flows.[19] The model parameters, $\alpha$, $\alpha_1$, $\alpha_2$ are calibrated to step-up bond prices.

#### 2.5.2.1 Calibration Results

We shall present now some calibration results. The bond data, obtained from Bloomberg's Corporate Bonds section, is relative to mid market quotes on April 5, 2006.

We calibrated the model parameters to a DT (Deutsche Telecom) step-up issue described in Table 2.1:

**Table 2.1** DT step-up issue on April 5, 2006.

| | |
|---|---|
| ISIN | $XS0132407957$ |
| Maturity | 07/11/11 |
| Coupon | $6\frac{5}{8}$ Annual |
| Step provision | $\begin{cases} +50 \text{ bps, if both downgraded below single Aaa3/A-;} \\ -50 \text{ bps, if both subsequently upgraded above Baa1/BBB+} \end{cases}$ |

[16] This is not necessary since default risk can be inferred from yield spreads in the bond market, but the higher liquidity of the CDS market makes it a preferable choice.

[17] We adopt here terminology from [23] to denote vanilla bonds of several maturities which have comparable liquidity and are issued by the same company as the relevant step-up bond.

[18] Plus market risk spreads other than credit spread.

[19] Simulation seems to be the only feasible computation technique, because of certain path dependencies in the payoff structure, induced by the step-down provision present in most step-up issues. Such path dependency is well explained in [23].

Given the default probability implied by the 5-y CDS spread of DT (46 bps), the liquidity adjusted discount rates are obtained using the above mentioned bootstrapping-interpolation procedure from the following pool of reference bonds (Table 2.2):

**Table 2.2** Reference bonds pool on April 5, 2006.

| ISIN | Maturity | Coupon | Mid-price |
|---|---|---|---|
| XS0141544691 | 01/22/07 | $5\frac{1}{4}$ | 1.015698 |
| DE0002317807 | 05/20/08 | $5\frac{1}{4}$ | 1.031821 |
| XS0242840345 | 02/02/09 | 3 | 0.979798 |
| XS0217817112 | 04/22/09 | 3 | 0.978352 |
| XS0210319090 | 01/19/10 | $3\frac{1}{4}$ | 0.976716 |
| XS0210318795 | 01/19/15 | 4 | 0.960349 |

The calibration results are given in Table 2.3:

**Table 2.3** Calibration results.

| | Model price | Market price |
|---|---|---|
| Bond price | 1.11705 | 1.11705 |
| Step-up provision | 0.00574 | – |

We remark that, since our calibration problem is overdetermined (three parameters are calibrated to one piece of data), the value of the step-up provision is not uniquely defined. This can be easily overcome by calibrating the model to more step-up issues of different maturities and/or provisions.

#### 2.5.2.2 Valuation of Step-Up Bonds

Using the calibrated model, we price selected issues of DT step-up bonds; we refer to Tables 2.4 and 2.5 for the description of the bonds.

**Table 2.4** DT step-up issue $XS0113709264$ on April 5, 2006.

| ISIN | $XS0113709264$ |
|---|---|
| Maturity | 07/06/10 |
| Coupon | $6\frac{5}{8}$ Annual |
| Step provision | +50 bps, if both downgraded below single Aaa3/A-;<br>−50 bps, if both subsequently upgraded above Baa1/BBB+ |

**Table 2.5** DT step-up issue $XS0155788150$ on April 5, 2006.

| ISIN | $XS0155788150$ |
|---|---|
| Maturity | 10/07/09 |
| Coupon | $6\frac{1}{2}$ Annual |
| Step provision | $\begin{cases} +50 \text{ bps, if both downgraded below Baa1/BBB+;} \\ -50 \text{ bps, if both subsequently upgraded above Baa2/BBB} \end{cases}$ |

Table 2.6 presents the pricing results as well as the corresponding market quotes. The results are very satisfactory, indicating that the model is robust and prices consistently across maturities and step-up provisions.

**Table 2.6** Pricing results using calibrated model.

| | Mkt price/Model price | |
|---|---|---|
| ISIN | XS0113709264 | XS0155788150 |
| Bond price | 1.10105/1.103546 | 1.08435/1.08685 |
| Step-up provision | – /0.003752 | – /0.00215 |

## References


1. Assefa, S., Bielecki, T.R., Crépey, S., Jeanblanc, M.: CVA computation for counter party risk assessment and hedging in credit portfolios. Working Paper (2009)
2. Bauerle, N., Blatter, A., Muller, A.: Dependence properties and comparison results for Lévy processes. Math. Methods Operat. Res. **67**(1), 161–186 (2008)
3. Bielecki, T.R., Jakubowski, J., Niewęgłowski, M.: Study of dependence for some stochastic processes part II: Symbolic Markov copulae. Submitted (2009)
4. Bielecki, T.R., Jakubowski, J., Vidozzi, A., Vidozzi, L.: Study of dependence for some stochastic processes. Stoch. Anal. Appl. **26**, 903–924 (2008)
5. Bielecki, T.R., Vidozzi, A., Vidozzi, L.: A Markov copulae approach to pricing and hedging of credit index derivatives and ratings triggered step-up bonds. J. Credit Risk **4**(1) (2008)
6. Bielecki, T.R., Vidozzi, A., Vidozzi, L.: Multivariate Markov processes with given Markovian margins, and Markov copulae. Preprint, Illinois Institute of Technology, Chicago, IL (2008)
7. Cherubini, U., Luciano, E., Vecchiato, W.: Copula Methods in Finance. Wiley Finance Series. Wiley, Chichester (2004)
8. Cont, R., Tankov, P.: Financial Modelling with Jump Processes. Chapman & Hall/CRC Financial Mathematics Series. Chapman and Hall/CRC, Boca Raton, FL (2004)
9. Courrège, P.: Sur la forme intégro-différentielle des opérateurs de $C_K^\infty$ dans $C$ satisfaisant au principe du maximum. Sém. Théorie du Potentiel **2** (1965/1966), 3-01–3-48
10. Crépey, S., Jeanblanc, M., Zargari, B.: CDS with counterparty risk in a Markov chain copula model with joint defaults. Working Paper (2009)
11. Darsow, W.F., Nguyen, B., Olsen, E.T.: Copulas and Markov processes. Ill. J. Math. **36**(4), 600–642 (1992)
12. El Karoui, N., Jeanblanc, M., Jiao, Y.: Successive default times. Working Paper (2009)

13. Elliott, R.J.: Finite dimensional filters related to Markov chains. In: Duncan, T.E. et al. (eds.) Stochastic Theory and Adaptive Control, Proceedings of a Workshop, Lawrence, KS, 26–28 Sept 1991. Lecture Notes in Control Information Sciences, vol. 184, pp. 142–160. Springer, Berlin (1992)
14. Embrechts, P.: Did a mathematical formula really blow up Wall Street? Presentation Slides (2009)
15. Ethier, S.N., Kurtz, T.G.: Markov Processes: Characterization and Convergence. Wiley, New York, Chichester, Brisbane, Toronto, and Singapore (1986)
16. Hoh, W.: Pseudo Differential Operators Generating Markov Processes. Habilitationsschrift, Bielefeld (1998)
17. Ibragimov, R.: Copula-based characterizations for higher order Markov processes. Econ. Theory **25**(3), 819–846 (2009)
18. Jacob, N.: Characteristic Functions and Symbols in the Theory of Lévy Processes. Potential Anal. **8**, 61–68 (1998)
19. Jacob, N.: Pseudo Differential Operators, Markov Processes. Imperial College Press, Cambridge (2005)
20. Jacod, J., Shiryaev, A.N.: Limit theorems for stochastic processes. Grundlehren der Mathematischen Wissenschaften, vol. 288. Springer, Berlin (1987)
21. Kallsen, J., Tankov, P.: Characterization of dependence of multidimensional Lévy processes using Lévy copulas. J. Multivar. Anal. **97** (2006)
22. Lagerås, A.N.: Copulas for Markovian dependence. Bernoulli **16**(2), 331–342 (2010)
23. Lando, D., Mortensen, A.: On the pricing of step-up bonds in the European telecom sector. J. Credit Risk **1**(1), 71–110 (Winter 2004/05)
24. Last, G., Brandt, A.: Marked point processes on the real line. The dynamic approach. Probability and Its Applications. Springer, New York, NY (1995)
25. Li, D.X.: On default correlation: a copula function approach. J. Fixed Income **9**, 43–54 (2000)
26. Nelsen, R.B.: An introduction to copulas. Lecture Notes in Statistics, vol. 139. Springer, New York, NY (1999)
27. Protter, P.E.: Stochastic integration and differential equations. Application of Mathematics. Springer, Berlin, Heidelberg, New York (2004)
28. Salmon, F.: Recipe for disaster: the formula that killed Wall Street. Wired Mag **17**(3) (2009)
29. Tankov, P.: Simulation and option pricing in Lévy copula model. Working paper available from http://people.math.jussieu.fr/tankov/
30. Tankov, P.: Dependence structure of spectrally positive multidimensional Lévy processes (2003). Working paper available from http://people.math.jussieu.fr/tankov/
31. Vidozzi, A.: Two essays in mathematical finance. Doctoral Dissertation, Illinois Institute of Technology, Chicago, IL (2009)
32. Vidozzi, L.: Two essays on multivariate stochastic processes and applications to credit risk modeling. Doctoral Dissertation, Illinois Institute of Technology, Chicago, IL (2009)
33. Whitehouse, M.: How a formula ignited market that burned some big investors. Wall St. J. **12**(18) (2005), 1

# Chapter 3
# Copula Estimation

Barbara Choroś, Rustam Ibragimov and Elena Permiakova

**Abstract** This chapter provides a survey of estimation methods for copula models. We review parametric, semiparametric and nonparametric approaches to inference on copulas for random samples with dependent components and copula-based time series. Among other topics, the survey discusses several problems of robust statistical analysis for copula models.

## 3.1 Introduction

In this chapter, we provide a brief survey of estimation procedures for copula models. Depending on the assumptions made on copula models considered, the data generating process and an approach to inference, the procedures lead to parametric, semi-parametric and non-parametric copula inference methods for i.i.d. observations (random samples) of random vectors with dependent components and copula-based time series. We review parametric, semiparametric and nonparametric approaches to inference on copulas for random samples with dependent marginals and copula-based time series and also discuss several problems of robust estimation in these frameworks.

Barbara Choroś
Institute for Statistics and Econometrics, Humboldt-Universität zu Berlin, Berlin, Germany
e-mail: barbara.choros@wiwi.hu-berlin.de

Rustam Ibragimov
Department of Economics, Harvard University, Cambridge, MA, USA
e-mail: ribragim@fas.harvard.edu

Elena Permiakova
N. G. Chebotarev Research Institute of Mathematics and Mechanics, Kazan State University, Kazan, Russia
e-mail: epermiakova@gmail.com

P. Jaworski et al. (eds.), *Copula Theory and Its Applications*, Lecture Notes in Statistics 198,
DOI 10.1007/978-3-642-12465-5_3, 

The survey is organized as follows. Section 3.2 discusses parametric (Sect. 3.2.1), semi-parametric (Sect. 3.2.2) and non-parametric (Sect. 3.2.3) estimation methods for copulas that characterize dependence among the components of random vectors: the inference procedures are based on i.i.d. vector observations. Section 3.3 considers the case of copula-based time series. In particular, Sect. 3.3.1 discusses copula-based characterizations of (higher-dimensional) Markov processes, and Sect. 3.3.4 reviews weak dependence properties of copula-based time series. Sections 3.3.2 and 3.3.3 discuss parametric, semiparametric and nonparametric estimation methods for copulas in the time series context. Section 3.4 discusses some further copula inference approaches motivated, in part, by robustness considerations. Section 3.5 reviews empirical applications of copula estimation methods and discusses pricing Collateralized Debt Obligations (CDO) using different copula models.

## 3.2 Copula Estimation: Random Samples with Dependent Marginals

### *3.2.1 Parametric Models: Maximum Likelihood Methods and Inference from Likelihoods for Margins*

Section 10.1 in [29] provides an excellent treatment and review of maximum likelihood (ML) estimation of parameters in multivariate copula models and computationally attractive parametric inference procedures for them.

Consider the key relation in copula theory given by formula (11) in [1] under absolute continuity assumptions. In the notations of Sklar's theorem in [1], the density $f$ of the $d$-dimensional d.f. $F$ with univariate margins $F_1, F_2, ..., F_d$ and the corresponding univariate densities $f_1, f_2, ..., f_d$ can be represented as

$$f(x_1, x_2, ..., x_d) = c(F_1(x_1), F_2(x_2), ..., F_d(x_d)) \prod_{i=1}^{d} f_i(x_i), \tag{3.1}$$

where $c(u_1, u_2, ..., u_d) = \frac{\partial^d C(u_1, u_2, ..., u_d)}{\partial u_1 \partial u_2 ... \partial u_d}$ is the density of the $d$-dimensional copula $C(u_1, u_2, ..., u_d; \theta)$ in (11) in [17].

Representation (3.1) implies the following decomposition for the log-likelihood function $L = \sum_{j=1}^{n} \log f(x_1^{(j)}, x_2^{(j)}, ..., x_d^{(j)})$ of a random sample of (i.i.d.) vectors $x^{(j)} = (x_1^{(j)}, x_2^{(j)}, ..., x_d^{(j)})$, $j = 1, 2, ..., n$, with the density $f$ :

$$L = \underbrace{L_C}_{\text{dependence}} + \underbrace{\sum_{i=1}^{d} L_i}_{\text{marginals}}, \tag{3.2}$$

where $L_C = \sum_{j=1}^{n} \log c(F_1(x_1^{(j)}), F_2(x_2^{(j)}), ..., F_d(x_d^{(j)}))$ is the log-likelihood contribution in from dependence structure in data represented by the copula $C$ and $L_i = \sum_{j=1}^{n} \log f_i(x_i^{(j)})$, $i = 1, 2, ..., d$, are the log-likelihood contributions from each margin: observe that $\sum_{i=1}^{d} L_i$ in (3.2) is exactly the log-likelihood of the sample under the independence assumption.

Suppose that the copula $C$ belongs to a family of copulas indexed by a (vector) parameter $\theta$ : $C = C(u_1, u_2, ..., u_d; \theta)$ and the margins $F_i$ and the corresponding univariate densities $f_i$ are indexed by (vector) parameters $\alpha_i$ : $F_i = F_i(x_i; \alpha_i)$, $f_i = f_i(x_i; \alpha_i)$. The maximum likelihood estimator – MLE – ($\hat{\alpha}_1^{\text{MLE}}, \hat{\alpha}_2^{\text{MLE}}, ...,$ $\hat{\alpha}_d^{\text{MLE}}, \hat{\theta}_d^{\text{MLE}}$) of the model parameters $(\alpha_1, \alpha_2, ..., \alpha_d, \theta)$ corresponds to simultaneous maximization of the log-likelihood $L$ in (3.2):

$$
\begin{aligned}
(\hat{\alpha}_1^{\text{MLE}}, \hat{\alpha}_2^{\text{MLE}}, ..., \hat{\alpha}_d^{\text{MLE}}, \hat{\theta}_d^{\text{MLE}}) = \\
\arg\max_{\alpha_1,\alpha_2,...,\alpha_d,\theta} L(\alpha_1, \alpha_2, ..., \alpha_d, \theta) = \\
\arg\max_{\alpha_1,\alpha_2,...,\alpha_d,\theta} L_C(\alpha_1, \alpha_2, ..., \alpha_d, \theta) + \sum_{i=1}^{d} L_i(\alpha_i) = \\
\arg\max_{\alpha_1,\alpha_2,...,\alpha_d,\theta} \sum_{j=1}^{n} \log c(F_1(x_1^{(j)}; \alpha_1), F_2(x_2^{(j)}; \alpha_2), ..., \\
F_d(x_d^{(j)}; \alpha_d); \theta) + \sum_{i=1}^{d} \sum_{j=1}^{n} \log f_i(x_i^{(j)}; \alpha_i).
\end{aligned} \tag{3.3}
$$

Section 10.1 in [29] (see also [30, 46]) discusses a computationally attractive alternative to the maximum likelihood (ML) estimation involving simultaneous maximization over the dependence ($\theta$) and marginal $(\alpha_1, \alpha_2, ..., \alpha_d)$ parameters in (3.3). This estimation approach is motivated by decomposition (3.2) and is referred to as the method of inference functions for margins (IFM) in [29, Sect. 10.1]. In the first stage of the inference procedure, the estimators $\hat{\alpha}_i^{\text{IFM}}$ of the parameters $\alpha_i$ are estimated from the log-likelihood $L_i$ of each margin in (3.2) and (3.3): $\hat{\alpha}_i^{\text{IFM}} = \arg\max_{\alpha_i} L_i(\alpha_i)$. That is, $(\hat{\alpha}_1^{\text{IFM}}, \hat{\alpha}_2^{\text{IFM}}, ..., \hat{\alpha}_d^{\text{IFM}})$ is defined to be the MLE of the model parameters under independence. In the second stage of the procedure, the estimator $\hat{\theta}^{\text{IFM}}$ of the copula parameter $\theta^{\text{IFM}}$ is computed by maximizing the copula likelihood contribution $L_C$ in (3.2) and (3.3) with the marginal parameters $\alpha_i$ replaced by their first-stage estimators $\hat{\alpha}_i^{\text{IFM}}$ : $\hat{\theta}^{\text{IFM}} = \arg\max_{\theta} L_C(\hat{\alpha}_1^{\text{IFM}}, \hat{\alpha}_2^{\text{IFM}}, ..., \hat{\alpha}_d^{\text{IFM}}, \theta)$. While, under regularity conditions, the MLE estimator $(\hat{\alpha}_1^{\text{MLE}}, \hat{\alpha}_2^{\text{MLE}}, ..., \hat{\alpha}_d^{\text{MLE}}, \hat{\theta}^{\text{MLE}})$ solves

$$(\partial L/\partial \alpha_1, \partial L/\partial \alpha_2, ..., \partial L/\partial \alpha_d, \partial L/\partial \theta) = 0,$$

the two-stage IFM estimator $(\hat{\alpha}_1^{\text{IFM}}, \hat{\alpha}_2^{\text{IFM}}, ..., \hat{\alpha}_d^{\text{IFM}}, \hat{\theta}^{\text{IFM}})$ solves

$$(\partial L_1/\partial \alpha_1, \partial L_2/\partial \alpha_2, ..., \partial L_d/\partial \alpha_d, \partial L/\partial \theta) = 0.$$

As discussed in [29], the MLE and IFM estimation procedures are equivalent in the special case of multivariate normal d.f.'s that have multivariate Gaussian copulas discussed in Sect. 6.1 in [17] and univariate normal margins. Naturally, however, this equivalence does not in general.

Similar to the MLE, the IFM estimator $(\hat{\alpha}_1^{\text{IFM}}, \hat{\alpha}_2^{\text{IFM}}, ..., \hat{\alpha}_d^{\text{IFM}}, \hat{\theta}_d^{\text{IFM}})$ is consistent and asymptotically normal under the usual regularity conditions (see [43]) for the multivariate model and for each of its margins. However, estimation of the corresponding covariance matrices is difficult both analytically and numerically due to the need to compute many derivatives, and jackknife and related methods may be used in inference (see [29]). As discussed in [29, Sect. 10.1], efficiency comparisons based on estimation of the asymptotic covariance matrices and Monte-Carlo simulation for different dependence models suggest that the IFM approach to inference provide a highly efficient alternative to the MLE estimation of multivariate model parameters.

### *3.2.2 Semiparametric Estimation*

Similar to the inference procedures discussed in Sect. 3.2.1, semi-parametric estimation methods for copula parameters are usually motivated by density representations and decompositions for the log-likelihood of dependence models as in (3.1) and (3.2). In the first stage, the univariate margins $F_i$ are estimated non-parametrically, e.g., by the empirical d.f.'s $\hat{F}_i$ or their scaled versions. In the second stage, the copula parameters are estimated using maximization of the contribution to the log-likelihood function from the dependence structure in the data represented by a copula $C$ of interest.

Formally, consider, as in [21], the problem of estimation of the (vector) parameter $\theta$ of a family of $d$-dimensional copulas $C(u_1, u_2, ..., u_d, \theta)$ with the density $c(u_1, u_2, ..., u_d, \theta)$. Given non-parametric estimators $\hat{F}_i$ of the univariate margins $F_i$, it is natural to estimate the copula parameter $\theta$ as

$$\hat{\theta} = \arg\max_\theta L_C(\theta) = \arg\max_\theta \sum_{j=1}^{n} \log\, c(\hat{F}_1(x_1^{(j)}), \hat{F}_2(x_2^{(j)}), ..., \hat{F}_d(x_d^{(j)}); \theta).$$

As discussed in [21], the resulting semi-parametric estimator $\hat{\theta}$ of the dependence parameter $\theta$ is consistent and asymptotically normal under suitable regularity conditions. The authors of [21] propose a consistent estimator of the limiting variance-covariance matrix of $\hat{\theta}$. They further show that, under additional copula regularity assumptions that are satisfied for a large class of bivariate copulas, including bivariate Gaussian, Eyraud–Farlie–Gumbel–Morgenstern (EFGM), Clayton and Frank (see Sect. 6.1 and relations (19), (20) and (22) in [17]) families, the estimator $\hat{\theta}$ is fully efficient at independence. Numerical results presented in [21] demonstrate that efficiency of $\hat{\theta}$ compares favorably to the alternative semiparametric estimators in the Clayton family of copulas. In addition, according to the numerical results, the coverage probability of confidence intervals for $\theta$ based on asymptotic normality of $\hat{\theta}$ is close to the nominal level for random samples with Clayton copulas exhibiting small to medium range of dependence.

### 3.2.3 Nonparametric Inference and Empirical Copula Processes

Most of nonparametric estimation procedures for copulas are based on inversion formula (12) discussed in introductory survey [17]. In the procedures, an estimator $\hat{C}(u_1,u_2,...,u_d)$ of a $d$-copula $C(u_1,u_2,...,u_d)$ is typically given by an empirical analogue of inversion formula (12) in [17]. That is,

$$\hat{C}(u_1,u_2,...,u_d) = \hat{F}(\hat{F}_1^{-1}(u_1),\hat{F}_2^{-1}(u_2),...,\hat{F}_d^{-1}(u_d)), \tag{3.4}$$

where $\hat{F}$ is a nonparametric estimator of the $d$-dimensional d.f. $F$ and $\hat{F}_1^{-1},\hat{F}_2^{-1},...,$ $\hat{F}_d^{-1}$ are nonparametric estimators of the pseudo-inverses $F_i^{-1}(s) = \{t|F_i(t) \geq s\}$ of the univariate margins $F_1,F_2,...,F_d$.[1] Typically, $\hat{F}$ is taken to be the empirical $d$-dimensional d.f. $\hat{F}(x_1,x_2,...,x_d) = \frac{1}{T}\sum_{t=1}^{T} I(X_1 \leq x_1, X_2 \leq x_2,...,X_d \leq x_d)$ and $F_i^{-1}$ are estimated using the pseudo-inverses $\hat{F}_i^{-1}(s) = \{t|\hat{F}_i(t) \geq s\}$ of the empirical univariate d.f.'s $\hat{F}_i = \frac{1}{T}\sum_{t=1}^{T} I(X_i \leq x_i)$ or their rescaled versions (here and throughout the survey, $I(\cdot)$ denote the indicator functions).

Deheuvels [11, 12] established consistency and asymptotic normality of the empirical copula process for i.i.d. observations of random vectors with independent margins (the case of the product copula $C = \Pi$, see Sect. 3 in [17]). Gaenssler and Stute [20, Chap. 5] and Fermanian et al. [18] establish consistency and asymptotic normality of the empirical copula process for general copulas $C$ with continuous partial derivatives (see also [19]).

Fermanian et al. [18] further show that, under regularity conditions, asymptotic normality also holds for smoothed copula processes like $\hat{C}(u_1,u_2) = \hat{F}(\hat{F}_1^{-1}(u_1),$ $\hat{F}_2^{-1}(u_2))$ that are constructed using the nonparametric kernel estimators $\hat{F}(x_1,x_2) =$ $\frac{1}{T}\sum_{t=1}^{T} K(\frac{x-X_t}{a_T},\frac{y-Y_t}{a_T})$ of the joint d.f. $F$. Here $K(x_1,x_2) = \int_{-\infty}^{x_1}\int_{-\infty}^{x_2} k(u,v)dudv$ for some bivariate kernel function $k : \mathbf{R}^2 \to \mathbf{R}$ with $\int k(x,y)dxdy = 1$, and the sequence of bandwidths $a_T > 0$ satisfies $a_T \to 0$ as $T \to \infty$ (see also [5] for the normal asymptotics of smoothed copula processes based on local linear versions of $K$ rather than the kernel itself for dealing with the boundary bias).

[1] Here and throughout the survey, $F_i$ are assumed to be continuous as in (12) in [17], if not stated otherwise.

## 3.3 Copula-Based Time Series and Their Estimation

### 3.3.1 Copula-Based Characterizations for (Higher-Order) Markov Processes

Darsow et al. [10] obtained the following necessary and sufficient conditions for a time series process based on bivariate copulas to be first-order Markov (see also Sect. 6.4 in [38]). For copulas $A, B : [0,1]^2 \to [0,1]$, set

$$(A * B)(x,y) = \int_0^1 \frac{\partial A(x,t)}{\partial t} \cdot \frac{\partial B(t,y)}{\partial t} dt.$$

Further, for copulas $A : [0,1]^m \to [0,1]$ and $B : [0,1]^n \to [0,1]$, define their $\star$-product $A \star B : [0,1]^{m+n-1} \to [0,1]$ via

$$A \star B(x_1,\ldots,x_{m+n-1}) = \int_0^{x_m} \frac{\partial A(x_1,\ldots,x_{m-1},\xi)}{\partial \xi} \cdot \frac{\partial B(\xi,x_{m+1},\ldots,x_{m+n-1})}{\partial \xi} d\xi.$$

As shown in [10], the operators $*$ and $\star$ on copulas are distributive over convex combinations, associative and continuous in each place, but not jointly continuous.

Darsow et al. [10] show that the transition probabilities $P(s,x,t,A) = P(X_t \in A | X_s = x)$ of a real-valued stochastic process $\{X_t\}_{t \in T}$, $T \subseteq \mathbf{R}$, satisfy Chapman-Kolmogorov equations

$$P(s,x,t,A) = \int_{-\infty}^{\infty} P(u,\xi,t,A) P(s,x,u,d\xi)$$

for all Borel sets $A$, all $s < t$ in $T$, $u \in (s,t) \cap T$ and for almost all $x \in \mathbf{R}$ if and only if the copulas corresponding to bivariate d.f.'s of $X_t$ are such that

$$C_{st} = C_{su} * C_{ut}$$

for all $s,u,t \in T$ such that $s < u < t$. The paper [10] also shows that a real-valued stochastic process $\{X_t\}_{t \in T}$ is a first-order Markov process if and only if the copulas corresponding to the finite-dimensional d.f.'s of $\{X_t\}$ satisfy the conditions

$$C_{t_1,\ldots,t_n} = C_{t_1 t_2} \star C_{t_2 t_3} \star \ldots \star C_{t_{n-1} t_n}$$

for all $t_1,\ldots,t_n \in T$ such that $t_k < t_{k+1}$, $k = 1,\ldots,n-1$. In particular, a sequence of identically distributed r.v.'s $\{X_t\}_{t=1}^{\infty}$ is a stationary Markov process if and only if for all $n \geq 2$,

$$C_{1,\ldots,n}(u_1,\ldots,u_n) = \underbrace{C \star^k C \star^k \ldots \star^k C}_{n-k+1}(u_1,\ldots,u_n) = C^{n-k+1}(u_1,\ldots,u_n), \quad (3.5)$$

where $C$ is bivariate copula. It is natural to refer to the processes $\{X_t\}_{t=1}^{\infty}$ constructed via (3.5) as stationary Markov processes based on the copula $C$ or as stationary $C$-based Markov processes for short.

Among other results, Ibragimov [27] obtained extensions of the results in [10] that provide characterizations of higher-order Markov processes in terms of copulas corresponding to their finite-dimensional d.f.'s. The results in [27] show that a Markov process of order $k$ is fully determined by its $(k+1)$-dimensional copulas and one-dimensional marginal cdf's. The characterizations thus provide a justification for estimation of finite-dimensional copulas of time series with higher-order Markovian dependence structure. Using the results, [27] further obtains necessary and sufficient conditions for higher-order Markov processes to exhibit several additional dependence properties, such as $m$-dependence, $r$-independence or conditional symmetry. These conditions are closely related to $U$-statistics-based characterizations for joint distributions and copulas developed in [13]. Using the obtained results, [27] also presents a study of applicability and limitations of different copula families in constructing higher-order Markov processes with the above dependence properties.

The results in [10, 27] provide a copula-based approach to the analysis of higher-order Markov processes which is alternative to the conventional one based on transition probabilities. The advantage of the approach based on copulas is that it allows one to separate the study of dependence properties (e.g., $r$-independence, $m$-dependence or conditional symmetry) of the stochastic processes in consideration from the analysis of the effects of marginal distributions (say, unconditional heavy-tailedness or skewness). In particular, the results provide methods for construction of higher-order Markov processes with arbitrary one-dimensional margins that, possibly, satisfy additional dependence assumptions. These processes can be used, for instance, in the analysis of the robustness of statistical and econometric procedures to weak dependence. In addition, they provide examples of non-Markovian processes that nevertheless satisfy Chapman–Kolmogorov stochastic equations. Higher-order Markov processes with prescribed dependence properties can be constructed, for instance, using inversion inversion formula (12) in [17] for finite-dimensional cdf's of known examples of dependent time series (see the discussion in [10, 27]).

### 3.3.2 Parametric and Semiparametric Copula Estimation Methods for Markov Processes

As discussed in Sect. 10.4 in [29], under suitable regularity conditions, asymptotic results for ML estimation under i.i.d. observations (see Sect. 3.2.1) also hold for stationary Markov processes.

Chen and Fan [6] propose semiparametric estimation procedures for copula-based Markov processes. These procedures generalize the two-stage semiparametric inference approaches discussed in Sect. 3.2.2 to the time series framework.

Consider a stationary Markov process based on a bivariate copula $C = C(u_1, u_2; \theta)$ (see Sect. 3.3). Similar to Sect. 3.2.2, the semiparametric estimator $\hat{\theta}$ of the copula parameter in [6] is defined as

$$\hat{\theta} = \arg\max_{\theta} L_C(\theta) = \arg\max_{\theta} \sum_{j=1}^{n} \log\, c(\hat{F}(x_1^{(j)}), \hat{F}(x_2^{(j)}); \theta),$$

where $\hat{F}(x)$ is a nonparametric estimator of the univariate margin of $X_t$ : e.g., the rescaled empirical d.f. $\hat{F}(x) = \frac{1}{T+1}\sum_{t=1}^{T} I\{X_t \leq x\}$ as in [6]. Chen and Fan [6] show that the semiparametric estimator $\hat{\theta}$ is consistent and asymptotically normal under suitable regularity assumptions. These assumptions include a condition that the process $\{X_t\}$ is $\beta$-mixing with polynomial decay rate. Chen and Fan [6] provide copula-based sufficient conditions for the latter weak dependence assumption and further verify the assumptions implying consistency and asymptotic normality of $\hat{\theta}$ for Markov processes based on Gaussian, Clayton and Frank copulas. As discussed in [6], the asymptotic variance of $\hat{\theta}$ can be estimated using heteroskedasticity autocorrelation consistent (HAC) estimators (see, for instance, [1, 39] and Chap. 10 in [24]). Alternatively, the asymptotic distributions of the estimator can be approximated using bootstrap (see Sect. 4.3 in [6]).

In a recent paper, Chen, Wu and Yi [7] consider efficient sieve ML estimation methods for copula-based stationary Markov processes. The authors show that sieve MLE's of any smooth functionals of the copula parameter and marginal d.f. and are root-$n$ consistent, asymptotically normal and efficient; and that their sieve likelihood ratio statistics are asymptotically chi-square distributed. The numerical results in [7] further indicate that the sieve MLEs of copula parameters, the margins and the conditional quantiles all perform very well in finite samples even for Markov models based on tail dependent copulas, such as Clayton, Gumbel-Hougaard or (see Sect. 6.1 and relations (18) and (19) in [17]). In addition, in the case of Markov models generated via tail dependent copulas, the sieve MLEs have much smaller biases and smaller variances than the two-step semiparametric estimation procedures in [6] discussed above.

### *3.3.3 Nonparametric Copula Inference for Time Series*

Doukhan, Fermanian and Lang [16] discuss extensions of the results on copula processes in Sect. 3.2.3 to the time series framework. The results in [16] imply, in particular, that empirical copula process (3.4) is asymptotically Gaussian for weakly dependent vector-valued processes $\{X_t\}$, including the case of strongly mixing and $\beta$-mixing sequences with polynomial decay rates. More generally, asymptotic normality of the copula processes holds for random sequences that satisfy multivariate functional central limit theorem for empirical processes (see [15]). Applied to the vector-valued process $(X_t, X_{t+1})$ for a $C$-based stationary Markov sequence $\{X_t\}$ satisfying mixing conditions (see Sect. 3.3.1), these results imply asymptotic normality of the empirical copula process $\hat{C}(u_1,u_2) = \hat{G}(\hat{F}^{-1}(u_1),\hat{F}^{-1}(u_2))$, where $\hat{F}(x)$ is a nonparametric estimator of the univariate margin of $X_t$ and $\hat{G}$ is a nonparametric estimator of the bivariate d.f. of $(X_t, X_{t+1})$ : e.g., in the case of empirical d.f.'s, $\hat{F}(x) = \frac{1}{T}\sum_{t=1}^{T} I(X_t \leq x)$, and $\hat{G}(x) = \frac{1}{T}\sum_{t=1}^{T} I(X_t \leq x)I(X_{t+1} \leq x)$. Doukhan, Fermanian and Lang [16] further establish asymptotic normality of the smoothed copula process with kernel estimates of the multivariate d.f. and univariate margins in (3.4) for weakly dependent vector-valued sequences. Similar asymptotic results are also shown to hold for smoothed copula densities (see also, among others, [23] for the analysis of the asymptotics of kernel estimators of the copula density for i.i.d. vector observations with dependent components).

### 3.3.4 Dependence Properties of Copula-Based Time Series

As discussed in Sects. 3.3.2 and 3.3.3, consistency and asymptotic normality of estimators of copula functions and their parameters are obtained under assumptions of weak dependence in the time series considered. Among other results, Beare [2], Chen, Wu and Yi [7] and Lentzas and Ibragimov [35] provide a study of persistence properties of stationary copula-based Markov processes. Lentzas and Ibragimov [35] show via simulations that stationary Markov processes based on tail-dependent Clayton copulas (see (19) in [17]) can behave as long memory time series on the level of copulas exhibiting high persistence important for financial and economic applications. This long memory-like behavior is captured by an extremely slow decay of copula-based dependence measures between lagged values of the processes for commonly used lag numbers. The theoretical results in [35] further show that, in contrast, Gaussian and EFGM copulas (see Sect. 6.1 and (22) in [17]) always produce short memory stationary Markov processes. Beare [2] shows that a $C$-based stationary Markov process exhibits weak dependence properties, including $\alpha$- and $\beta$-mixing with exponential decay rates, if $C$ is a symmetric absolutely continuous copula with a square integrable density and the maximal correlation coefficient of $C$ is less than one. These results imply, in particular, that stationary Markov processes based on Gaussian and EFGM copulas are weakly dependent and mixing. Beare [2] also provides numerical results that suggest exponential decay in $\beta$-mixing coefficients and, thus, also in $\alpha$-mixing coefficients of Clayton copula-based stationary Markov processes. Chen, Wu and Yi [7] obtain theoretical results that show that tail-dependent Clayton, survival Clayton, Gumbel and $t$-copulas always generate Markov processes that are geometric ergodic and hence geometric $\beta$-mixing. The conclusions in [7] imply that, although, according to the numerical results in [35], Clayton copula-based Markov processes can behave like long memory time series on copula levels exhibiting high persistence for commonly used lag numbers, they are in fact weakly dependent and short memory in terms of mixing properties.

## 3.4 Further Copula Inference Methods

Naturally, applications of the copula inference methods described in Sects. 3.2.1, 3.2.2, 3.2.3, 3.3, 3.3.1, 3.2.2 and 3.3.3 require consistent estimation of the asymptotic covariance matrices in the Gaussian convergence results. As discussed in [29, Sect. 10] and Sect. 3.2.1 in this paper, this is a difficult task both analytically and numerically even in the case of i.i.d. observations of vectors with dependent margins. At the same time, the conclusions and numerical results in [2, 7, 35] reviewed in Sect. 3.3.4 indicate importance of development of robust methods for differentiating short and long memory in copula-based time series and robust inference approaches for copula models with dependent observations. As discussed in Chen and Fan [6] (see Sect. 3.3.2), in the case of time series, inference on the asymptotic covariance matrices may be based on HAC estimators and computationally expensive bootstrap procedures. As discussed in a number of works (see [33] and references therein), however, tests based on HAC estimators may have substantial size distortions in finite samples. Motivated, in part, by these results, Kiefer, Vogelsang and Bunzel [33] show that asymptotically justified inference in a linear time series regression may be based on inconsistent analogues of HAC estimators with a nondegenerate limiting distribution (see also [31, 32] for extensions of the approach). It would be interesting to use inconsistent analogues of HAC estimators similar to those considered in [31–33] in semiparametric tests on copula parameters in the time series framework (see Sect. 3.3.2).

Ibragimov and Mueller [28] develop a general strategy for conducting inference about a scalar parameter with potentially heterogenous and correlated data, when relatively little is known about the precise property of the correlations. Assume that the data can be classified in a finite number $q$ of groups that allow asymptotically independent normal inference about the scalar parameter of interest $\theta$ (e.g., a scalar component of the vector copula parameter as in the settings discussed in Sects. 3.2.1, 3.2.2 and 3.3.2). That is, there exist estimators $\hat{\theta}_j$ calculated for groups $j = 1, ..., q$, that are asymptotically normal: $\sqrt{T}(\hat{\theta}_j - \theta) \Rightarrow \mathcal{N}(\theta, \sigma_j)$, and $\hat{\theta}_j$ is asymptotically independent of $\hat{\theta}_i$ for $j \neq i$. The asymptotic normality of $\sqrt{T}(\hat{\theta}_j - \theta)$, $j = 1, ..., q$, typically follows from the same reasoning as the asymptotic normality of the full sample estimator $\hat{\theta}$ : as discussed in Sects. 3.2.1, 3.2.2, 3.2.3, 3.3, 3.3.1, 3.2.2 and 3.3.3, asymptotic Gaussianity is typically a standard result for many estimators of copula model parameters.

As follows from [28], one can perform an asymptotically valid test of level $\alpha$, $\alpha \leq 0.05$ of $H_0 : \theta = \theta_0$ against $H_1 : \theta \neq \theta_0$ by rejecting $H_0$ when $|t_\theta|$ exceeds the $(1 - \alpha/2)$ percentile of the Student-$t$ distribution with $q - 1$ degrees of freedom, where $t_\theta$ is the usual $t$-statistic

$$t_\theta = \sqrt{q}\frac{\overline{\hat{\theta}} - \theta_0}{s_{\hat{\theta}}} \tag{3.6}$$

[1] HAC = hierarchical Archimedean copula = nested Archimedean copula

with $\overline{\hat{\theta}} = q^{-1}\sum_{j=1}^{q}\hat{\theta}_j$, the sample mean of the group estimators $\hat{\theta}_j$, $j = 1,...,q$, and $s_{\hat{\theta}}^2 = (q-1)^{-1}\sum_{j=1}^{q}(\hat{\theta}_j - \overline{\hat{\theta}})^2$, the sample variance of $\hat{\theta}_j$, $j = 1,...,q$.

In other words, in general settings considered in [28], the usual $t$-tests can be used in the presence of asymptotic heteroskedasticity in group estimators as long the level of the tests is not greater than the typically used 5% threshold. Similar to applications considered in [28], the $t$-statistic-based approach to robust inference can be applied in statistical analysis of copula models with autocorrelated vector observations. As discussed in [28], the $t$-statistic-based approach provides a number of important advantages over the existing methods of inference in time series, panel, clustered and spatially correlated data. In particular, the approach can be employed when data are potentially heterogeneous and dependent in a largely unknown way, as is typically the case for copula models. In addition, the approach is simple to implement and does not need new tables of critical values. The assumptions of asymptotic normality for group estimators in the approach are explicit and easy to interpret, in contrast to conditions that imply validity of alternative procedures in many settings. Furthermore, as shown in [28], the $t$-statistic based approach to robust inference efficiently exploits the information contained in these regularity assumptions, both in the small sample settings (uniformly most powerful scale invariant test against a benchmark alternative with equal variances) and also in the asymptotic frameworks. The numerical results presented in [28] demonstrate that, for many dependence and heterogeneity settings considered in the literature and typically encountered in practice for time series, panel, clustered and spatially correlated data, the applications of the approach lead to robust tests with attractive finite sample performance. The analysis of the performance of the approach in inference on copula parameters and, more generally, dependence measures and other characteristics, appears to be an interesting problem that is left for further research.

## 3.5 Empirical Applications

Many works in the literature have focuses on empirical applications of copulas and related concepts in different fields, including economics, finance, risk management and insurance (see, among others, the review and discussion in [8, 13, and 37]). Hu [26] focuses on mixed copula modeling of dependence among financial variables, with the parameters estimated using semiparametric procedures described in Sect. 3.2.2 applied to the residuals in GARCH models fitted to data. Patton [40] applies copula structures for modeling asymmetric dependence in foreign exchange markets and, among other results, finds a dramatic decrease in the tail dependence parameters in Clayton family models for the Deutsche mark and the yen exchange rates following the introduction of the euro in January 1999. Giacomini, Härdle and Spokoiny [22] focus on modeling dependence of financial returns using copulas with time-varying adaptively estimated parameters. Lee and Long [34] consider extensions and empirical applications of multivariate GARCH models with copula-based specifications of dependence among the vectors components. Among other results, Caillault and Guégan [4] provide nonparametric estimates of tail dependence parameters for Asian financial indices. Smith [44, 45] discusses applications of copulas in sample selection and regime switching models.

Copulas have become standard tools in credit risk modelling, see [42]. Unfortunately, the current global financial crises has shown that industry models, that are using copulas to evaluate risk, can be very inadequate and should be treated with caution. However, copulas are powerful and flexible tools and there is still a lot of space for improvement.

In this section, we present applications of copula functions in pricing Collateralized Debt Obligation (CDO). A CDO is a structured financial product that enables securitization of a large portfolio of assets. The portfolio's risk is sliced into tranches and transfered to investors with different risk preferences. Each CDO tranche is defined by the detachment and attachment points which are the percentages of the portfolio losses. The losses are caused by defaults of the reference entities. The tranches of the iTraxx index are created by following levels: $0\%, 3\%, 6\%, 9\%, 12\%$, $22\%$, $100\%$. Investors agree to cover the losses that might appear in the range of a particular tranche in exchange for a fee. During the life of the contract investors are periodically paid a premium, usually once per quarter, which mainly depends on the type of the tranche and the value of the accumulated losses. The most risky tranche, called equity, bears first 3% losses of the portfolio nominal but pays the highest spread. If the losses exceed 3%, the next are absorbed by mezzanine tranches and after by senior tranches. The losses over 22% are allocated to the most senior tranche, often called super super senior. With the increase of seniority, the value of the offered premium decreases. For a survey of the CDO construction and pricing we refer to [3].

The values of the spreads of the CDO tranches depend on the joint behavior and joint distribution of the assets in the underlying pool and on their tendency to default simultaneously. The idea of modelling the joint distribution of defaults with copula functions was introduced by Li [36]. In this method a default is defined by a random variable called a time-until-default whose distribution is derived from market data. The joint distribution of default times is specified with a one parameter Gaussian copula. The Li model has been seen until now as the industry standard in the CDO valuation. However, during the financial meltdown it has been strongly criticized for its incapability of modelling joint extreme events due to tail independence of the underlying Gaussian copulas. Because of the drawbacks of the one-factor Gaussian copula model numerous new approaches have been proposed. Due to the high dimension nature of the problem most of the CDO models are fully parametric (see Sect. 3.2.1). Semi-parametric and non-parametric calibration procedures were proposed in [14, 41, and 42].

We present below the results of the empirical study conducted with the iTraxx Euro index series 8 with a maturity of 5 years. The series 8 was issued on 20 September 2007 and expires on 20 December 2012. For computations we consider the market values of the first $J = 5$ tranches, see Table 3.1, observed on 3rd August 2009. We also use spreads of 125 credit default swaps (CDS) that constitute the portfolio of iTraxx. The default dependency structure of the credits is specified with the one-parameter Gumbel and the one-parameter survival Clayton copula. The results are compared with the Gaussian copula model. The CDO valuations based on copulas of many parameters are studied in [9, 25]. The first step in CDO pricing consists of estimating the risk of the underlyings. The default times of portfolio credits are assumed to be exponentially distributed with parameters implied from the market quotes of CDS contracts. We assume the constant interest rate of 3% and the constant recovery rate of 40%. The fit of the models to market data is archived by minimizing the following function with respect to the copula parameter:

$$D(t_0) \stackrel{\text{def}}{=} \sum_{j=1}^{J} \frac{|s_j^c(t_0) - s_j^m(t_0)|}{s_j^m(t_0)} \to \min, \tag{3.7}$$

which sums the relative deviations of the model upfront fee $s_1^c$ and the spreads $s_j^c$, $j = 2, \ldots, J$, from the market values $s_j^m$, $j = 1, \ldots, J$, observed at time $t_0$. The detailed description of the calibration algorithm is provided in [9].

**Table 3.1** Results of CDO calibration on 3rd August 2009.

| Model | Parameter | $D$ | 0–3% | 3–6% | 6–9% | 9–12% | 12–22% |
|---|---|---|---|---|---|---|---|
| Market | | | 39.593 | 503.750 | 407.500 | 165.228 | 84.727 |
| Gauss | 0.2209 | 3.2620 | 28.957 | 1,284.324 | 622.758 | 313.985 | 86.025 |
| Surv. Clayton | 1.1215 | 2.6420 | 15.573 | 526.393 | 364.179 | 279.925 | 185.547 |
| Gumbel | 1.2000 | 1.1554 | 39.473 | 896.437 | 317.965 | 179.453 | 90.405 |

Table 3.1 presents the obtained results. The values of measure $D$ show that the market prices are better described by the Gumbel and by the survival Clayton copula than by the Gaussian copula. The improvement is achieved as the Gumbel and the survival Clayton copula exhibit the upper tail dependence which allows to quantify the risk that many obligors default at the same time. The Gumbel copula gives the most accurate fit to the market tranche quotes.

**Acknowledgements** Ibragimov gratefully acknowledges partial support provided by the National Science Foundation grant SES-0820124.

## References

1. Andrews, D. Heteroskedasticity and autocorrelation consistent covariance matrix estimation. Econometrica **59**, 817–858 (1991)
2. Beare, B. Copulas and temporal dependence. Econometrica 78, 395–410 (2010)
3. Bluhm, C., Overbeck, L.: Structured Credit Portfolio Analysis, Baskets and CDOs. CRC Press LLC, Boca Raton, FL (2006)

4. Caillault, C., Guégan, D.: Empirical estimation of tail dependence using copulas: application to Asian markets. Quant. Finance **5**, 489–501 (2005)
5. Chen, S.X., Huang, T.-M.: Nonparametric estimation of copula functions for dependence modelling. Canad. J. Stat. **35**, 265–282 (2007)
6. Chen, X., Fan, Y.: Estimation of copula-based semiparametric time series models. J. Econometr. **130**, 307–335 (2006)
7. Chen, X., Wu, W.B., Yi, Y.: Efficient estimation of copula-based semiparametric Markov models. Annal. Stat. **37**, 4214–4253 (2009)
8. Cherubini, U., Luciano, E., Vecchiato, W. Copula Methods in Finance. The Wiley Finance Series. Wiley, Chichester (2004)
9. Choroś, B., Härdle, W., Okhrin, O.: CDO and HAC. Discussion paper 2009–038, SFB 649, Humboldt Universität zu Berlin, Berlin (2009)
10. Darsow, W.F., Nguyen, B., Olsen, E.T.: Copulas and Markov processes. Ill. J. Math. **36**, 600–642 (1992)
11. Deheuvels, P.: La fonction de dépendance empirique et ses propriétés. Un test non paramétrique d'indépendance. Académie Royale de Belgique. Bulletin de la Classe des Sciences 5ième Série **65**, 274–292 (1979)
12. Deheuvels, P.: A Kolmogorov-Smirnov type test for independence and multiple samples. Revue Roumaine de Mathématiques Pures et Appliquées **26**, 213–226 (1981)
13. de la Peña, V.H., Ibragimov, R., Sharakhmetov, Sh.: Characterizations of joint distributions, copulas, information, dependence and decoupling, with applications to time series. 2nd Erich L. Lehmann Symposium – Optimality (J. Rojo, Ed.). IMS Lecture Notes – Monograph Series **49**, 183–209 (2006). Available at http://dx.doi.org/10.1214/074921706000000455
14. Dempster, M.A.H., Medova, E.A., Yang, S.W.: Empirical copulas for CDO tranche pricing using relative entropy. Int. J. Theor. Appl. Finance **10**, 679–701 (2007)
15. Doukhan, P.: Mixing: Properties and Examples. Lecture Notes in Statistics, vol. 85. Springer, New York, NY (1994)
16. Doukhan, P., Fermanian, J.-D., Lang, G.: An empirical central limit theorem with applications to copulas under weak dependence. Stat. Infer. Stoch. Process. **12**, 65–87 (2009)
17. Durante, F., Sempi, C.: Copula theory: an introduction. In: Jaworski, P., Durante, F., Härdle, W., Rychlik, T.: (eds.) Copula Theory and Its Applications, Proceedings of the Workshop, Warsaw, 25–26 Sept 2009, Springer, Dordrecht (2010)
18. Fermanian, J.-D., Radulovič, D., Wegkamp, M.: Weak convergence of empirical copula processes. Bernoulli **10**, 847–860 (2004)
19. Fermanian, J.D., Scaillet, O.: Nonparametric estimation of copulas for time series. J. Risk **5**, 25–54 (2003)
20. Gaenssler, P., Stute, W.: Seminar on Empirical Processes. DMV Seminar, 9. Birkhäuser, Basel (1989)
21. Genest, C., Ghoudi, K., Rivest, L.P.: A semiparametric estimation procedure of dependence parameters in multivariate families of distributions. Biometrika **82**, 543–552 (1995)
22. Giacomini, E., Härdle, W., Spokoiny, V.: Inhomogenous dependence modeling with timevarying copulae. J. Bus. Econ. Stat. **27**, 224–234 (2009)
23. Gijbels, I., Mielniczuk, J.: Estimating the density of a copula function. Commun. Stat. Theory Methods **19**, 445–464 (1990)
24. Hamilton, J.D.: Time Series Analysis. Princeton University Press, Princeton, NJ (1994)
25. Hofert, M., Scherer, M.: CDO pricing with nested Archimedean copulas. Working paper, Universität Ulm, Ulm (2008)
26. Hu, L.: Dependence patterns across financial markets: a mixed copula approach. Appl. Financ. Econ. **16**, 717–729 (2006)
27. Ibragimov, R.: Copula-based characterizations for higher-order Markov processes. Econom. Theory **25**, 819–846 (2009)
28. Ibragimov, R., Mueller, U.K.: *t*-statistic based correlation and heterogeneity robust inference. J. Bus. Econ. Stat. Forthcoming, http://pubs.amstat.org/doi/pdf/10.1198/jbes.2009.08046.

Also available as Harvard Institute of Economic Research Discussion Paper No. 2129, http://www.economics.harvard.edu/pub/hier/2007/HIER2129.pdf (2007)
29. Joe, H.: Multivariate Models and Dependence Concepts. Monographs on Statistics and Applied Probability, vol. 73. Chapman & Hall/CRC, New York, NY (2001)
30. Joe, H., Xu, J.J.: The estimation method of inference functions for margins for multivariate models. Technical Report no. 166, Department of Statistics, University of British Columbia, Vancouver, BC (1996)
31. Kiefer, N., Vogelsang, T.J.: Heteroskedasticity-autocorrelation robust testing using bandwidth equal to sample size. Econom. Theory **18**, 1350–1366 (2002)
32. Kiefer, N., Vogelsang, T.J.: A new asymptotic theory for heteroskedasticityautocorrelation robust tests. Econom. Theory **21**, 1130–1164 (2005)
33. Kiefer, N.M., Vogelsang, T.J., Bunzel, H.: Simple robust testing of regression hypotheses. Econometrica **68**, 695–714 (2000)
34. Lee, T.-H., Long, X.: Copula-based multivariate GARCH model with uncorrelated dependent errors. J. Econom. **150**, 207–218 (2009)
35. Lentzas, G., Ibragimov, R.: Copulas and long memory. Working paper, Harvard University, Cambridge, MA (2008)
36. Li, D.X.: On default correlation: a copula function approach. J. Fixed Income **9**, 43–54 (2000)
37. McNeil, A.J., Frey, R., Embrechts, P.: Quantitative Risk Management. Concepts, Techniques and Tools. Princeton University Press, Princeton, NJ (2005)
38. Nelsen, R.B.: An Introduction to Copulas. Springer Series in Statistics, 2nd edn. Springer, New York, NY (2006)
39. Newey, W., West, K.: Hypothesis testing with efficient method of moments estimation. Int. Econ. Rev. **28**, 777–787 (1987)
40. Patton, A.: Modelling asymmetric exchange rate dependence. Int. Econ. Rev. **47**, 527–556 (2006)
41. Schönbucher, P.J.: Portfolio losses and the term-structure of loss transition rates: a new methodology for the pricing of portfolio credit derivatives. Working paper, ETH Zürich, Zürich (2006)
42. Schönbucher, P.J.: Credit Derivatives Pricing Models: Model, Pricing and Implementation. Wiley, New York, NY (2003)
43. Serfling, R.J.: Approximation Theorems of Mathematical Statistics. Wiley, New York, NY (1980)
44. Smith, M.: Modelling sample selection using Archimedean copulas. Econom. J. **6**, 99–123 (2003)
45. Smith, M.: Using copulas to model switching regimes with an application to child labour. Econ. Record **81**, S47–S57 (2005)
46. Xu, J.J.: Statistical modeling and inference for multivariate and longitudinal discrete response data. Ph.D. Thesis, Department of Statistics, University of British Columbia, Vancouver, BC (1996)

# Chapter 4
# Pair-Copula Constructions of Multivariate Copulas

Claudia Czado

**Abstract** In this survey we introduce and discuss the pair-copula construction method to build flexible multivariate distributions. This class includes drawable (D), canonical (C) and regular vines developed in [4, 5]. Estimation and model selection methods are studied both in a classical as well as in a Bayesian setting. This flexible class of multivariate copulas can be applied to model complex dependencies. Literature to applications in modeling financial data as well as Bayesian belief networks are provided. It closes with a section on open problems.

## 4.1 Introduction

The famous Sklar's theorem (see [54]) allows to build multivariate distributions using a copula and marginal distributions. For the basic theory on copulas see the first chapter [14] or the books by Joe [32] and Nelsen [51]. Much emphasis has been put on the bivariate case and in [32, 51] many examples of bivariate copula families are given. However the class of multivariate copulas utilized so far has been limited. Especially financial applications need flexible multivariate dependence structures in the center of the distribution as well as in the tails. For value at risk (for a definition see [44]) calculations we need flexibility in particular in the tails. One such measure are the upper and lower tail dependence parameters (for a definition see [15]), which coincide for (reflection) symmetric distributions. For example the Gaussian copula allows for an arbitrary correlation matrix with zero tail dependence, while the multivariate $t$-copula has only a single degree of freedom parameter which drives the tail dependence parameter. Both the Gaussian and the $t$-copula are examples of elliptical copulas (see [18, 20]).

---

Claudia Czado
Department of Mathematics, Technische Universität München, Garching, Germany
e-mail: cczado@ma.tum.de

P. Jaworski et al. (eds.), *Copula Theory and Its Applications*, Lecture Notes in Statistics 198,
DOI 10.1007/978-3-642-12465-5_4, © Springer-Verlag Berlin Heidelberg 2010

In addition to elliptical copulas attention has focused on multivariate extensions of the Archimedean copulas. In this class we have fully and partially nested Archimedean copulas as discussed in [32, 52, 56]. Hierarchical Archimedean copulas are considered in [52], while multiplicative Archimedean copulas are proposed in [43, 50]. However these extensions require additional parameter restrictions and thus result in reduced flexibility for modeling dependence structures.

The first topic of this chapter is to present a general construction method for multivariate copulas using only bivariate copulas, which is called a pair-copula construction (PCC). This includes a simple derivation of PCC models such as D-vines and canonical vines. More general PCC's such as regular vines (see [4, 5, 39]) are introduced and some of their properties are discussed. The second topic is to provide statistical inference methods, when only parametric bivariate copulas are used as building blocks in a PCC model. Here we present three methods one based on stepwise estimation, one on maximum likelihood and one on a Bayesian approach. Applications of these methods in the literature will be given. The next topic involves model selection within a specified PCC model. Application areas will be discussed next. We close with further extensions and open problems.

## 4.2 Pair Copula Constructions of D-Vine, Canonical and Regular Vine Distributions

We assume that all joint, marginal and conditional distributions are absolutely continuous with corresponding densities. In 1996, Joe [31] gave the first pair-copula construction of a multivariate copula. He gave the construction in terms of distribution functions, while Bedford and Cooke (see [4, 5]) expressed these constructions in terms of densities. They organized these constructions in a graphically way involving a sequence of nested trees, which they called regular vines. They also identified two popular subclasses of PCC models, which they call D-vines and canonical vines. Their results are developed in more detail in the book by Kurowicka and Cooke (see [39]). First we present easy derivations of D-vines and canonical vines before we introduce general regular vines.

### *4.2.1 Pair-Copula Constructions of D-Vine and Canonical Vine Distributions*

The starting point for constructing multivariate distribution is the well known recursive decomposition of a multivariate density into products of conditional densities. For this let $(X_1,...,X_d)$ be a set of variables with joint distribution $F$ and density $f$, respectively. Consider the decomposition

$$\begin{aligned} f(x_1,...,x_d) &= f(x_d|x_1,\cdots,x_{d-1})f(x_1,\cdots,x_{d-1}) \\ &= \cdots = \prod_{t=2}^{d} f(x_t|x_1,\cdots,x_{t-1})\cdot f(x_1). \end{aligned} \tag{4.1}$$

Here $F(\cdot|\cdot)$ and later $f(\cdot|\cdot)$ denote conditional cdf's and densities, respectively.

As second ingredient we need Sklar's theorem for dimension $d = 2$ given by

$$f(x_1,x_2) = c_{12}(F_1(x_1),F_2(x_2))\cdot f_1(x_1)\cdot f_2(x_2), \tag{4.2}$$

where $c_{12}(\cdot,\cdot)$ is an arbitrary bivariate copula density. Using (4.2) we can express the conditional density of $X_1$ given $X_2$ as

$$f(x_1|x_2) = c_{12}(F_1(x_1),F_2(x_2))\cdot f_1(x_1). \tag{4.3}$$

For distinct indices $i,j,i_1,\cdots,i_k$ with $i<j$ and $i_1<\cdots<i_k$ we use the abbreviation

$$c_{i,j|i_1,\cdots,i_k} := c_{i,j|i_1,\cdots,i_k}(F(x_i|x_{i_1},\cdots,x_{i_k}),F(x_j|x_{i_1},\cdots,x_{i_k})). \tag{4.4}$$

Using (4.3) for the conditional distribution of $(X_1,X_t)$ given $X_2,\cdots X_{t-1}$ we can express $f(x_t|x_1,\cdots,x_{t-1})$ recursively as

$$\begin{aligned} f(x_t|x_1,\cdots,x_{t-1}) &= c_{1,t|2,\cdots,t-1}\cdot f(x_t|x_2,\cdots,x_{t-1}) \\ &= [\prod_{s=1}^{t-2} c_{s,t|s+1,\cdots,t-1}]\cdot c_{(t-1),t}\cdot f_t(x_t) \end{aligned} \tag{4.5}$$

Using (4.5) in (4.1) and $s=i, t=i+j$ it follows that

$$\begin{aligned} f(x_1,\ldots,x_d) &= [\prod_{t=2}^{d}\prod_{s=1}^{t-2} c_{s,t|s+1,\cdots,t-1}]\cdot[\prod_{t=2}^{d} c_{(t-1),t}][\prod_{k=1}^{d} f_k(x_k)] \\ &= [\prod_{j=1}^{d-1}\prod_{i=1}^{d-j} c_{i,(i+j)|(i+1)\cdots,(i+j-1)}]\cdot[\prod_{k=1}^{d} f_k(x_k)] \end{aligned} \tag{4.6}$$

Note that the decomposition (4.6) of the joint density consists of pair-copula densities $c_{i,j|i_1,\cdots,i_k}(\cdot,\cdot)$ evaluated at conditonal distribution functions $F(x_i|x_{i_1},\cdots,x_{i_k})$ and $F(x_j|x_{i_1},\cdots,x_{i_k})$ for specified indices $i,j,i_1,\cdots,i_k$ and marginal densities $f_k$. This is the reason why we call such a decomposition *pair-copula decomposition*. This class of decompositions was named by Bedford and Cooke a *D-vine distribution*.

A second class of decompositions is possible, when one applies (4.3) to the conditional distribution of $(X_{t-1},X_t)$ given $X_1,\cdots,X_{t-2}$ to express $f(x_t|x_1,\cdots,x_{t-1})$ recursively. This yields the following expression

$$f(x_t|x_1,\cdots,x_{t-1}) = c_{t-1,t|1,\cdots,t-2}\cdot f(x_t|x_1,\cdots,x_{t-2}). \tag{4.7}$$

Using (4.7) instead of (4.5) in (4.1) and setting $j = t - k, j + i = t$ results in the following decomposition

$$\begin{aligned}
f(x_1,\dots,x_d) &= f(x_1)\cdot\left[\prod_{t=2}^{d}\prod_{k=1}^{t-1} c_{t-k,t|1,\cdots,t-k-1}\cdot f(x_t)\right] \\
&= \left[\prod_{t=2}^{d}\prod_{k=1}^{t-1} c_{t-k,t|1,\cdots,t-k-1}\right]\cdot[\prod_{k=1}^{d} f_k(x_k)] \\
&= \left[\prod_{j=1}^{d-1}\prod_{i=1}^{d-j} c_{j,j+i|1,\cdots,j-1}\right]\cdot[\prod_{k=1}^{d} f_k(x_k)]. \qquad (4.8)
\end{aligned}$$

According to Bedford and Cooke this PCC is called a *canonical vine distribution*.

### 4.2.2 Regular Vines Distributions and Copulas

Bedford and Cooke in [4, 5] noticed that they can represent these pair-copula decompositions (4.6) and (4.8) graphically with a sequence of nested trees with undirected edges, which they call a vine tree. Edges in the trees denote the indices used for the conditional copula densities. Following [39] we recall for the convenience of the reader the definition of a regular vine. According to Definition 4.4 of [39] a *regular vine tree* on $d$ variables consists of connected trees $T_1,\cdots T_{d-1}$ with nodes $N_i$ and edges $E_i$ for $i = 1,\cdots,d-1$, which satisfy the following

1. $T_1$ has nodes $N_1 = \{1,\cdots,d\}$ and edges $E_1$.
2. For $i = 2,\cdots,d-1$ the tree $T_i$ has nodes $N_i = E_{i-1}$.
3. Two edges in tree $T_i$ are joined in tree $T_{i+1}$ if they share a common node in tree $T_i$

The edges in tree $T_i$ will be denoted by $jk|D$ where $j < k$ and $D$ is the conditioning set. Note that in contrast to [39] we order the conditioned set $\{j,k\}$ to make the order of the arguments in the bivariate copulas unique. If $D$ is the empty set, we denote the edge by $jk$. The notation of an edge $e$ in $T_i$ will depend on the two edges in $T_{i-1}$, which have a common node in $T_{i-1}$. Denote these edges by $a = j(a),k(a)|D(a)$ and $b = j(b),k(b)|D(b)$ with $V(a) := \{j(a),k(a),D(a)\}$ and $V(b) := \{j(b),k(b),D(b)\}$, respectively. The nodes $a$ and $b$ in tree $T_i$ are therefore joined by edge $e = j(e),k(e)|D(e)$, where

$$\begin{aligned}
j(e) &:= \min\{i : i \in (V(a)\cup V(b))\setminus D(e)\} \\
k(e) &:= \max\{i : i \in (V(a)\cup V(b))\setminus D(e)\} \\
D(e) &:= V(a)\cap V(b).
\end{aligned}$$

Two special vine tree specifications were identified by Bedford and Cooke, one they called *drawable vine trees* or short *D-vine trees*, while the other one is called *canonical vine trees* or short *C-vine trees*. They are defined as follows.

A regular vine tree is called

- *D-vine tree* if each node in $T-1$ has at most 2 edges.
- *C-vine tree* if each tree $T_i$ has a unique node with $d-i$ edges. The node with $d-1$ edges in tree $T_1$ is called the *root*.

In Fig. 4.1 a graphical representation of a D-vine tree in five dimensions is given, while in Fig. 4.2 we see a representation of a C-vine tree. For example the edge $e = 14|23$ in tree $T_3$ of Fig. 4.1 is derived from edges $a = 13|2$ with $V(a) = \{1,2,3\}$ and $b = 24|3$ with $V(b) = \{2,3,4\}$. Note that $D(e) = \{2,3\}, j(e) = 1$ and $k(e) = 4$.

To build up a statistical model on a regular vine tree with node set $\mathcal{N} := \{N_1, \cdots, N_{d-1}\}$ and edge set $\mathcal{E} := \{E_1, \cdots, E_{d-1}\}$ one associates each edge $e = j(e), k(e)|D(e)$ in $E_i$ with a bivariate copula density $c_{j(e),k(e)|D(e)}$. Let $\mathbf{X}_{D(e)}$ be the sub random vector of $\mathbf{X}$ indicated by the indices contained in $D(e)$. A *vine distribution* is defined as the distribution of the random vector $\mathbf{X} := (X_1, \cdots, X_d)$ with marginal densities $f_k, k = 1, \cdots, d$ and the conditional density of $(X_{j(e)}, X_{k(e)})$ given the variables $\mathbf{X}_{D(e)}$ specified as $c_{j(e),k(e)|D(e)}$ for the regular vine tree with node set $\mathcal{N}$ and edge set $\mathcal{E}$. In Theorem 4.2 of [39] it is proven that the joint density of $\mathbf{X}$ is uniquely determined and given by

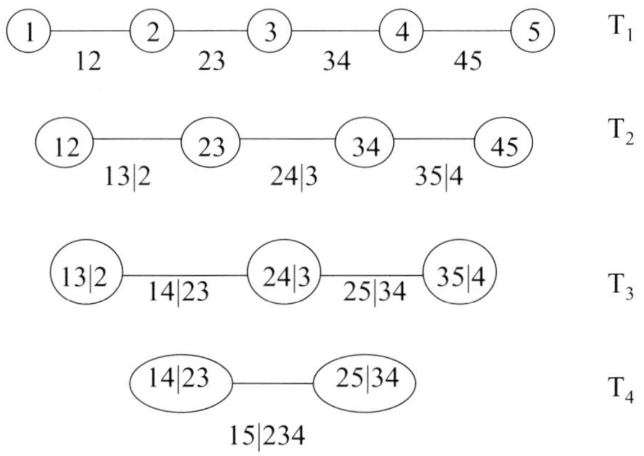


**Fig. 4.1** A D-vine tree representation for $d = 5$.

$$f(x_1,\ldots,x_d) = \prod_{r=1}^{d} f(x_r) \cdot \prod_{i=1}^{d-1} \prod_{e \in E_i} c_{j(e),k(e)|D(e)}(F(x_{j(e)}|\mathbf{x}_{D(e)}), F(x_{k(e)}|\mathbf{x}_{D(e)})), \quad (4.9)$$

where $\mathbf{x}_{D(e)}$ denotes the subvector of $\mathbf{x}$ indicated by the indices contained in $D(e)$. This is an analogue of the Hammersley-Clifford theorem for Markov random fields (see [8]) to vine distributions.

For the D-vine tree in Fig. 4.1 the corresponding vine distribution has the joint density given by

$$f(x_1,\cdots,x_5) = [\prod_{k=1}^{5} f_k(x_k)] \cdot c_{12} \cdot c_{23} \cdot c_{34} \\ \cdot\ c_{45} \cdot c_{13|2} \cdot c_{24|3} \cdot c_{35|4} \cdot c_{14|23} \cdot c_{25|34} \cdot c_{15|234}, \quad (4.10)$$

while the corresponding joint density for the C-vine distribution with tree representation displayed in Fig. 4.2 is given by

$$f(x_1,\cdots,x_5) = [\prod_{k=1}^{5} f_k(x_k)] \cdot c_{12} \cdot c_{13} \cdot c_{14} \\ \cdot\ \cdot c_{15} \cdot c_{23|1} \cdot c_{24|1} \cdot c_{25|1} \cdot c_{34|12} \cdot c_{35|12} \cdot c_{45|123}.$$

Here we used the abbreviation defined in (4.4). Comparing (4.10) to (4.6), we see that (4.10) equals (4.6) for $d = 5$. Therefore we can identify (4.6) as the joint density of a D-vine distribution. The same is true for (4.8), i.e. (4.8) is the joint density of

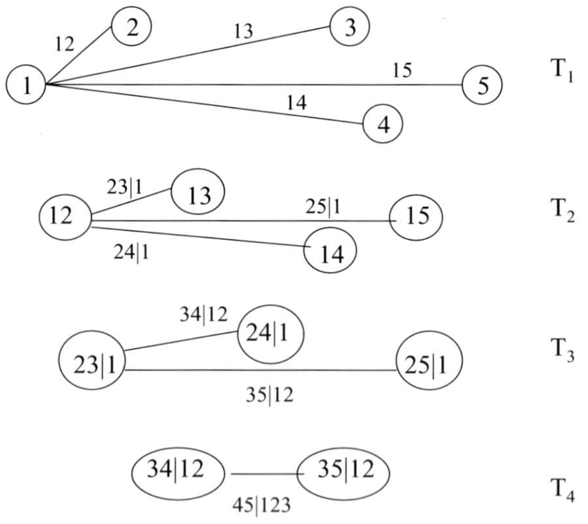


**Fig. 4.2** A C-vine tree representation for $d = 5$.

a C-vine distribution in $d$ dimensions. We can of course use vine distributions to construct copulas, by just requiring that the marginal densities in (4.9) are univariate uniform densities.

This construction of multivariate distributions and copulas is very general and flexible, since we can use any bivariate copula as building block in the PCC model. In contrast to the extended multivariate Archimedean copulas no restriction to the Archimedean pair-copulas or further parameter restrictions are necessary. In finance the most commonly used pair-copulas are the Gaussian copula, the $t$-copula, the Clayton copula and Gumbel copula (see for example [1] for definitions and properties). One limitation of the multivariate $t$-copula in financial applications is that we only have a single degree of freedom parameter which drives the tail dependence of all pairs of variables. In [2] it was first noticed that a PCC would overcome this problem and an application to financial stock data was given to demonstrate the superiority of a D-vine copula with bivariate $t$-copulas as building blocks for the PCC over a multivariate $t$-copula approach. PCC models have been compared to alternative copula based models in [7, 19] and again the PCC models performed very well among a large class of competitors.

To illustrate the model flexibility we consider a simple D-vine tree in 3 dimensions. Note that in three dimensions D-vines and C-vines coincide. The corresponding D-vine density with standard normal margins is therefore given by

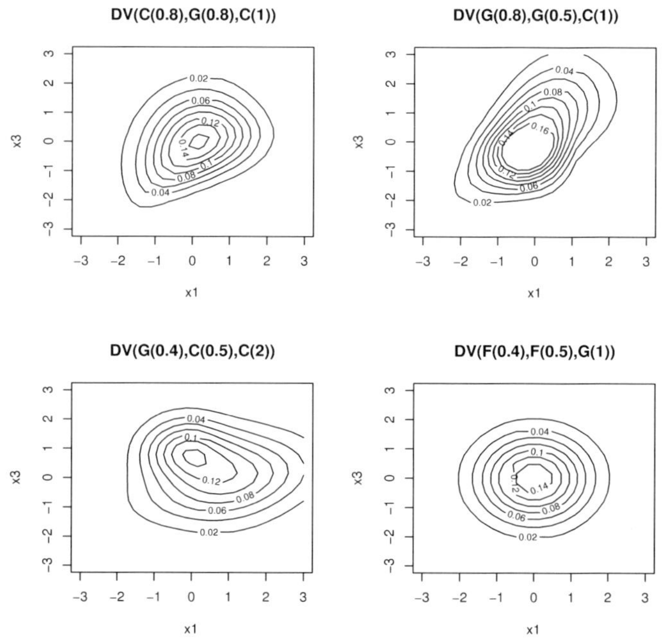


**Fig. 4.3** Density contours of $(X_1, X_3)$ for different three dimensional D-vine distributions with standard normal margins.

$$\begin{aligned} c(x_1,x_2,x_3) &= c_{12}(\Phi(x_1),\Phi(x_2))\cdot c_{23}(\Phi(x_2),\Phi(x_3)) \\ &\cdot c_{13|2}(F(x_1,|x_2),F(x_3|x_2))\cdot \phi(x_1)\cdot\phi(x_2)\cdot\phi(x_3), \end{aligned} \tag{4.11}$$

where $\Phi(\cdot)$ and $\phi(\cdot)$ denote the standard normal cdf and pdf, respectively. Here the pair-copulas $c_{12}, c_{23}$ and $c_{13|2}$ will be chosen as either a bivariate Clayton ($C(\theta)$), bivariate Gumbel ($G(\alpha)$) or bivariate Frank ($F(\eta)$) copula. The corresponding pair copula parameters are $\theta, \alpha$ and $\eta$, respectively. We will use for example the abbreviation $DV(C(0.8), G(0.8), C(1))$ to denote the D-vine copula density (4.11), where $c_{12}$ is $C(0.8)$, $c_{23}$ is $G(0.8)$ and $c_{13|2}$ is $C(1)$. The bivariate marginal density for $(X_1,X_2)$ and $(X_2,X_3)$ are directly specified, while the bivariate marginal density for $(X_1,X_3)$ needs to be computed by integrating (4.11) over the variable $x_2$. In Fig. 4.3 density contours of $(X_1,X_3)$ are plotted for four different choices. We see that a large variety of contour shapes are possible.

The tail behavior of vine copulas was investigated in [34].

In general the conditional pair-copula densities in (4.9) might depend on the conditioning values $x_{i_1},\cdots,x_{i_k}$, however in this chapter we assume the restriction that $c_{i,j|i_1,\cdots,i_k}(\cdot,\cdot)$ does not depend on $x_{i_1},\cdots,x_{i_k}$. This means that the decomposition (4.9) captures the dependency on the conditioning values solely through the arguments $F(x_i|x_{i_1},\cdots,x_{i_k})$ and $F(x_j|x_{i_1},\cdots,x_{i_k})$. In a recent paper [24], the authors investigate under which conditions the decomposition of the form (4.6) for three dimensions satisfies the above restriction as well as what are the effects if this restriction is not satisfied on the value of risk. They claim that this restriction is not so severe. Therefore we consider in the following only decompositions in which conditional pair-copula densities do not depend on the conditioning variables.

For example [13] contains a regular vine density of the form (4.9) involving foreign exchange rates. It also considers a C-vine model. Finally we want to mention that two well known multivariate copulas can be recovered using vine copulas. The first one is the multivariate Gauss copula and the second one is the multivariate $t$-copula, which was shown in detail in Sect. 2 of [13].

## 4.3 Estimation Methods for Regular Vine Copulas

For estimation of regular vine parameters Kurowicka and Cooke in [39] followed a nonstandard way involving the determinant of the correlation matrix for the random vector distributed according to a regular vine. Using bivariate normal copulas with conditional correlations in the specification of the PCC model results in a multivariate normal distribution. Here one has to use the facts that partial and conditional correlations are equal for elliptical distributions (see [3]) and that conditional distributions of normals are normal with a covariance independent of the conditioning value. Bedford and Cooke in [4] provided a one-to-one relationship between unconditional and partial correlations for Gaussian distributions. Further the determinant of the correlation matrix can be expressed in terms of partial correlations. In the case of Gaussian random vectors the distribution for the determinant of an empirical version of the correlation matrix is known (see Theorem 5.1 of [39]), however bootstrapping would be necessary for other regular vine specification to determine the distribution. Further it is unclear how useful the determinant of the induced correlation matrix is for statistical inference.

Aas et al. [2] were the first to consider more standard estimation methods such as stepwise and maximum likelihood estimation (MLE), which we will discuss now. Emphasis here is on the estimation of vine copula parameters, i.e we want to estimate the parameters of the joint density (4.9), when all marginals are uniform based on an i.i.d. sample from such a density.

Since we have an explicit expression for the joint density the likelihood is easily derived (see [2] for explicit expressions for C- and D-vine copulas). These expressions however involve conditional cdf's, for which we need expressions as well. Joe in [31] showed that for $v \in D$ and $D_{-v} := D \setminus v$

$$F(x_j|\mathbf{x}_D) = \frac{\partial C_{x_j,x_v|D_{-v}}(F(x_j|\mathbf{x}_{D_{-v}}), F(x_v|\mathbf{x}_{D_{-v}}))}{\partial F(x_v|\mathbf{x}_{D_{-v}})}. \tag{4.12}$$

For the special case where $D = \{v\}$ it follows that

$$F(x_j|x_v) = \frac{\partial C_{x_j,x_v}(F(x_j), F(x_v))}{\partial F(x_v)}.$$

In the case of uniform margins this simplifies further for a parameterized copula cdf $C_{jv}(x_j,x_v) = C_{jv}(x_j,x_v|\theta_{jv})$ to

$$h(x_j|x_v,\theta_{jv}) := \frac{\partial C_{j,v}(x_j,x_v|\theta_{jv})}{\partial x_v}. \tag{4.13}$$

We can use (4.12) to express conditional cdf's where $D$ contains more than one element. Following [13] it follows for $v \in D$

$$\begin{aligned}
F(x_j|\mathbf{x}_D) &= \int_{-\infty}^{x_j} c_{jv|D_{-v}}(F(u_j|\mathbf{x}_{D_{-v}}), F(x_v|\mathbf{x}_{D_{-v}}))\, f(u_j|\mathbf{x}_{D_{-v}})\, \mathrm{d}u_j \\
&= \int_{-\infty}^{x_j} \frac{\partial^2 C_{jv|D_{-v}}(F(u_j|\mathbf{x}_{D_{-v}}), F(x_v|\mathbf{x}_{D_{-v}}))}{\partial F(u_j|\mathbf{x}_{D_{-v}})\, \partial F(x_v|\mathbf{x}_{D_{-v}})} \frac{\partial F(u_j|\mathbf{x}_{D_{-v}})}{\partial u_j} \mathrm{d}u_j \\
&= \frac{1}{\partial F(x_v|\mathbf{x}_{D_{-v}})} \int_{-\infty}^{x_j} \underbrace{\frac{\partial^2 C_{jv|D_{-v}}(F(u_j|\mathbf{x}_{D_{-v}}), F(x_v|\mathbf{x}_{D_{-v}}))}{\partial F(u_j|\mathbf{x}_{D_{-v}})} \frac{\partial F(u_j|\mathbf{x}_{D_{-v}})}{\partial u_j}}_{\frac{\partial}{\partial u_j} C_{jv|D_{-v}}(F(u_j|\mathbf{x}_{D_{-v}}), F(x_v|\mathbf{x}_{D_{-v}}))} du_j \\
&= \frac{\partial}{\partial F(x_v|\mathbf{x}_{D_{-v}})} C_{jv|D_{-v}}(F(x_j|\mathbf{x}_{D_{-v}}), F(x_v|\mathbf{x}_{D_{-v}})) \\
&= \frac{\partial}{\partial \eta} C_{jv|D_{-v}}(F(x_j|\mathbf{x}_{D_{-v}}), \eta)|_{\eta = F(x_v|\mathbf{x}_{D_{-v}})} \\
&= h(F(x_j|\mathbf{x}_{D_{-v}})|F(x_v|\mathbf{x}_{D_{-v}}), \theta_{jv|D_{-v}}).
\end{aligned}$$

This shows that the conditional cdf's with conditioning set $D$ can be built up recursively using the $h$-function from conditional cdf's with lower dimensional conditioning set. Overall this allows an recursive determination of the likelihood. The inverses of these $h$-functions can be used to facilitate sampling from D and C vines using a conditional approach (see [2, 40]). However the number of parameters grows quadratically in the dimension $d$, since there $d \cdot (d-1)/2$ different pair-copulas to be parameterized. Therefore it is useful to consider a stepwise estimation approach, where we estimate the parameters from the first tree to the last one sequentially. In an initial step estimate the parameters corresponding to the pair-copulas in the first tree using any method you prefer. For example the correlation parameter $\rho$ of a bivariate $t$-copula pair is estimated using Kendall's $\tau$ and in second part the degree of freedom parameter $\nu$ is maximized using the estimated $\rho$. For the copula parameters identified in the second tree, one first has to transform the data with the $h$ function required for the appropriate conditional cdf using estimated parameters from the first tree to determine realizations needed in the second tree.

For example if we want to estimate the parameters of copula $c_{13|2}$. First transform the observations $\{u_{1,t}, u_{2,t}, u_{3,t}, t = 1, \cdots, n\}$ to $u_{1|2,t} := h(u_{1,t}|u_{2,t}, \hat{\theta}_{12})$ and $u_{3|2,t} := h(u_{3,t}|u_{2,t}, \hat{\theta}_{23})$, where $\hat{\theta}_{12}$ and $\hat{\theta}_{23}$ are the estimated parameters in the first tree. Now estimate $\theta_{13|2}$ based on $\{u_{1|2,t}, u_{3|2,t}; t = 1, \cdots, n\}$. Continue sequentially with this procedure until all copula parameters of all trees are estimated. Note for trees $T_i$ with $i \geq 2$ recursive applications of the $h$ functions are needed to transform to the appropriate conditional cdf.

This stepwise estimation gives parameter estimates, but so far the asymptotic distribution of the stepwise estimates has not been determined, therefore the use of these parameters as starting values is more appropriate. In contrast MLE's of the pair-copulas are efficient under regularity conditions with asymptotic variance-covariance given by the inverse of the Fisher information matrix. However it is difficult to determine the Fisher information matrix, so one uses in general the observed Hessian matrix instead. Again the Hessian matrix corresponding to a sample from (4.9) is difficult to express analytically but simple to approximate numerically. It may however happen that this numerical approximation might not yield a positive definite variance-covariance matrix. In this case further numerical manipulations are necessary. This is the reason why in the first paper on ML estimation [2] no estimated standard errors where given. In subsequent papers (see [13, 19]) these have been added. In most papers D- and C-vine copula parameters are estimated, at the moment only [13] considers a regular vine copula.

These difficulties have been noted by Min and Czado (see [47]), which instead propose to follow a Bayesian approach. Here parameters are estimated using Markov Chain Monte Carlo (MCMC) methods (see for example [9]) and they employ the Metropolis Hastings algorithm (see [28, 46]) for D-vines with pair-copulas to be chosen as a bivariate $t$-copula. Interval estimates are provided by credible intervals. This approach can also be easily extended to credible intervals for functions of the copula parameters. Examples for such functions are tail dependence coefficients, $\lambda$-function of [21] and value at risk. In particular the $\lambda$-function can be used to assess model fit. Credible intervals for the tail dependence coefficient of pairs with bivariate $t$-copula and the corresponding $\lambda$-function are provided in [47].

## 4.4 Model Selection Among Vine Specifications

The number of different D- and C-vines is very large. In [2] Aas et al show that for a C-vine decomposition on $d$ nodes there are $d!/2$ distinct C-vine trees and this is also the number of distinct D-vine trees. For regular vine trees the number is even larger (see [49]). This means that we need additional structure to select reasonable vine trees. First it may be reasonable to restrict to C- and D-vine trees. A C-vine tree may be reasonable if a there is a variable which drives all other variables. This may be the case if one considers foreign exchange rates. In all other cases a D-vine tree may be enough to consider.

For the order in the trees corresponding to a D-vine copula Aas et al [2] put the strongest bivariate dependencies in the first tree of the D-vine tree specification. Strongest bivariate dependencies within the copula distribution might be measured by Kendall's $\tau$ or the tail dependence coefficient $\lambda$, which is a function of the chosen bivariate copula.

Another approach is to choose a vine tree distribution with the smallest partial correlation in the last tree. However this requires basically a Gaussian tree distribution, since conditional correlations are only easily estimated for a Gaussian distribution, where partial correlations and conditional correlations are equal and the partial correlations have a one-to-one correspondence to unconditional correlations. This approach for Gaussian D-vine copulas has been described for example in Chap. 5 of [41]. For regular vines this problem has been considered in [36].

Once a vine tree specification is chosen one needs to select the pair-copula terms of the vine distribution. For this Aas et al. [2] suggest to follow a stepwise approach. First they consider the pairs of variables involved in the first tree and apply a goodness-of-fit (GOF) test (see [6, 22] for a review of such tests) for each such pair when the copula family varies and pick the family which gives the best fit for this pair. Now transform the data in the way described for the stepwise estimation and continue with the conditional pairs of the next tree within the vine specification. For $K$ pair-copula families this involves fitting and testing $K \cdot d \cdot (d-1)/2$ bivariate models. Mendes et al. (see [45]) are using this stepwise approach for the analysis of Brazilian financial stocks, while Shirmacher and Shirmacher in [53] use bivariate $\chi^2$ display to select the bivariate pair copula family. While this is a feasible first approach, there are obvious problems with the choice of the pair-copulas in the higher trees, since the transformed data only gives an approximation to the conditional cdf's. This uncertainty is ignored and it increases as one moves up the different trees. In addition the critical values of the GOF tests are difficult to obtain, if the test is applied not directly on an i.i.d sample of the copula, but to rank transformed standardized residuals after applying an appropriate marginal model. In this case bootstrapping might be needed.

If one wants to avoid this stepwise approach, one could attempt to apply GOF tests directly on the full $d$ dimensional sample. However if one allows for for $K$ alternative families of pair copulas, this would involve fitting $K^{d \cdot (d-1)/2}$ models, which is excessive even when $K$ and $d$ are small. Alternatively we consider now Bayesian approaches, which can traverse large model spaces without having to visit all models.

We start with the following subproblem: Once a vine tree specification is selected one is interested in the possibility of reducing the vine distribution further by identifying (conditional) independencies present in the data. For a vine distribution with a single pair-copula family this means that we want to identify pair-copula terms, which can be replaced by a bivariate independence copula. This task is easiest accomplished by using reversible jump MCMC (RJMCMC) first discussed in [23]. This approach was followed by Min and Czado [48] who investigate D-vines with bivariate $t$-copulas. The RJMCMC algorithm is developed and implemented for arbitrary dimensions. In a second approach suggested by Smith et al. (see [55]) selection indicators for each pair-copula are introduced to select between the chosen pair-copula family and the independence copula. Again a Bayesian approach is followed and an appropriate MCMC algorithm is developed. In [55] the performance of the method is tested when selection is between independence copulas and either Gauss, Clayton or Gumbel pair copulas. It is evident that both the RJMCMC or the selection indicator approach can be extended to choose between different pair-copula families and this is topic of current research.

## 4.5 Applications of Vine Distributions

One application area of vines is to positive definite matrices and correlation matrices. We have a one-to-one relationship between partial and unconditional correlations. Therefore the values of the partial correlations are unstricted in $[-1, 1]$ and can be chosen independently, while still inducing positive definite matrices. This was used in applications to linear algebra (see [37]). Random generation of correlation matrices are considered in [33] using D-vines and more general using regular vines in [42]. Random distributions of correlation matrices are useful as prior choices in a Bayesian setup. This choice was used in [41].

Another area of applications are in the area of distributions on a directed acyclic graph (DAG) or some times also called Bayesian belief network (BBN). These distributions are specified through conditional independence statements described through the graph. For Gaussian and discrete DAG's see for example [11]. Models for variables in $[0, 1]$ which are only characterized through possibly conditional rank (Spearman's) correlations on the arcs of the DAG are called nonparametric BBN's (see [38]). In [26] connections between nonparametric BBN's and a series of D-vine distributions are established. Choose for each rank correlation a copula which realizes all rank correlations in $[-1,1]$ and where a zero rank correlation induces independence. With this copula choice it is shown in [26] that the joint distribution on the BBN is uniquely induced by the rank correlations. In the case of normal BBN's the structure of the graph can be learned by removing arcs as long as the determinant of the rank correlation matrix determined by the partial correlations is close to the empirical rank correlation matrix (see [27]). Note that partial and conditional rank correlations are equal for normal BBN's. Mixed continuous and discrete BBN's are discussed in [25].

Pair copula constructions have found their applications in the analysis of financial data. Here one starts with appropriate time series models such as ARMA-GARCH or skewed-t GARCH models. The corresponding standardized residuals of each margin are now considered as an i.i.d sample over time. The dependency across margins is modeled with a a vine distribution. Here one can follow a parametric or nonparametric approach to estimate marginal and copula parameters. The simplest estimation approach is a two step approach. For the first step estimate marginal parameters separately for each margin and form standardized residuals $r_{it}$. In the parametric approach the distribution of the standardized residuals for each margin is assumed to be known. Let $F_i(\cdot|\hat{\theta}_m)$ denote this distribution, where $\hat{\theta}_i^m$ are the estimated marginal parameters for margin $i$. Now define the probability integral transforms $u_{it} := F^{-1}(r_it|\hat{\theta}_i^m)$. For the nonparametric approach one uses the empirical distribution function of $\{r_{it}, t = 1, \cdots, T\}$ instead of $F_i(\cdot|\hat{\theta}_m)$. In both approaches the data $\mathbf{u}_t = (u_{1t}, \cdots, u_{dt}), t = 1, \cdots, T$ is assumed to be an i.i.d sample from a regular vine distribution. In a second step the copula parameters are estimated based on the data $\mathbf{u}_t, t = 1, \cdots, T$ using one of the estimation methods discussed in (4.3). This two-step estimation procedure using the nonparametric approach has been followed by [2, 13], while the parametric approach was used by [30, 45] together with the stepwise procedure for selecting and estimating the copula parameters. In [10] also a parametric two step approach was followed in a regime switching setup using the EM algorithm. A truncated C vine copula was introduced in [29] and a stepwise estimation procedure to estimate the C-vine parameters is followed. The usefulness of this formulation was demonstrated in a portfolio of over 90 stocks, however standard errors are not provided.

Joint estimation approaches of both marginal and copula parameters are rare. First examples are D-vine copulas with pair t copula pairs and AR(1) margins are investigated in [12], while Gaussian D-vines copulas with regression are treated in [41]. In both papers a Bayesian approach was followed.

Finally a first application to a geostatistical continuous Markov mesh model was provided by [35] involving a 4 dimensional D-vine.

## 4.6 Summary and Open Problems

Pair copula constructions provide a powerful tool to construct flexible multivariate distributions which can be used to model complex dependencies. Especially the modeling power of financial statistical models is enormously increased. While there has been progress in developing inference methods methods, more needs to be done in the area of estimation, model selection and adaptation of the special data structures.

In the area of estimation, the lack of standard errors for the stepwise estimators of the copula parameters is evident. Here the appropriate asymptotic theory has to be developed and fast numerical implementation has to be provided. Another estimation problem to be solved is the development of fast joint estimators of marginal and copula parameters, since the now common two step estimation procedure is not efficient. Here marginal models as applied in financial statistics need to be considered. For joint estimation the Bayesian approach seems to be the most promising. For financial applications a joint Bayesian inference provides natural tools to assess the variability of value of risk estimates.

The problem of selecting the appropriate vine tree specification and the appropriate pair copula family is a challenging problem area. While in the past the flexibility was limited, the model flexibility is now so large, that one has to consider additional structures for the selection. Restrictions such as provided by truncated canonical vines in [29] are promising but need to be fully explored. Bayesian techniques such as RJMCMC or model indicators need to be studied further in the context of selecting pair copula families. Finally the problem of providing effective non-nested model selection criteria needs to be considered.

Finally adaptation to special data structures will enhance the applicability of this model construction method. Here we name the necessity to allow for time varying copula parameters. First approaches of such models are given in [29, 30]. However only stepwise estimates without standard errors are provided, while full ML or Bayesian estimates are not investigated. In insurance applications one is often faced with multivariate counts such as claim counts or multivariate zero truncated claim severities. A sampling method involving vines for multivariate counts has been developed in [16, 17]. To model dependencies among multivariate discrete or censored variables using vine distributions is another interesting research area. Finally the development of statistical models on graphs or geostatistical structures and their inference involving the pair-copula construction method is an interesting and challenging area.

**Acknowledgements** Claudia Czado is supported by the Deutsche Forschungsgemeinschaft (CZ86 1-3). Special thanks to K. Aas and A. Frigessi who introduced me to this promising research area.

## References

1. Aas, K.: Modelling the dependence structure of financial assets: a survey of four copulas. Technical Report SAMBA/22/04, Norsk Regnesentral (2004)
2. Aas, K., Czado, C., Frigessi, A., Bakken, H.: Pair-copula constructions of multiple dependence. Insur. Math. Econ. **44**, 182–198 (2009)
3. Baba, K., Sibuya, M.: Equivalence of partial and conditional correlation coefficients. J. Japan. Stat. Soc. **35**, 1–19 (2005)
4. Bedford, T., Cooke, R.: Probability density decomposition for conditionally dependent random variables modeled by vines. Ann. Math. Artif. Intell. **32**, 245–268 (2001)
5. Bedford, T., Cooke, R.: Vines – a new graphical model for dependent random variables. Ann. Stat. **30**(4), 1031–1068 (2002)
6. Berg, D.: Copula goodness-of-fit testing: an overview and power comparison. Eur. J. Finance, **15**(7/8), 675–701 (2009)
7. Berg, D., Aas, K.: Models for construction of multivariate dependence. Eur. J. Finance, **15**(7/8), 639–65, 2009
8. Besag, J.: Spatial interaction and the statistical analysis of lattice systems. J. Roy. Stat. Soc. Ser. B **36**, 192–236 (1974)
9. Chib, S.: Markov chain Monte Carlo methods: computation and inference. In: Handbook of Econometrics, pp. 3569–3649. North-Holland, Amsterdam (2001)
10. Chollete, L., Heinen, A., Valdesogo, A.: Modeling international financial returns with a multivariate regime-switching copula. J. Financ. Econom. **7**(4) 437–480, 2009
11. Cowell, R.G., Dawid, P., Lauritzen, S.L., Spiegelhalter, D.J.: Probabilistic Networks and Experts Systems. Springer, New York, NY (1999)
12. Czado, C., Gärtner, F., Min, A.: Joint Bayesian inference of D-vines with AR(1) margins. In: Kurowicka, D., Joe, H. (eds.) Dependence Modeling-Handbook on Vine Copulas, pp. 359–330. World Scientific Publishing, Singapore (2010)
13. Czado, C., Min, A, Baumann, T., Dakovic, R.: Pair-copula constructions for modeling exchange rate dependence. Technical Report, Department of Mathematics, Technische Universität München, München (2008)
14. Durante, F., Sempi, C.: Copula theory: an introduction. In: Jaworski, P., Durante, F., Härdle, W., Rychlik, T. (eds.) Copula Theory and Its Applications, Proceedings of the Workshop, Warsaw, 25–26 Sept 2009, Springer, Dordrecht (2010)
15. Embrechts, P., Lindskog, F., McNeil, A.: Modelling dependence with copulas and applications to risk management. In: Rachev, S.T. (ed.) Handbook of Heavy Tailed Distributions in Finance, pp. 329–384. Elsevier, North-Holland, Amsterdam (2003)
16. Erhardt, V., Czado, C.: Sampling count variables with specified Pearson correlation – a comparison between a naive and a C-vine sampling approach. In: Kurowicka, D., Joe, H. (eds.)

108 Claudia Czado Dependence Modeling-Handbook on Vine Copulas, pp. 359–330, World Scientific Publishing, Singapore (2010)
17. Erhardt, V., Czado, C.: A method for approximately sampling high-dimensional count variables with prespecified Pearson correlation. Technical Report, Department of Mathematics, Technische Universität München, München (2009)
18. Fang, H.B., Fang, K., Kotz, S.: The meta-elliptical distributions with given marginals. J. Multivar. Anal. **82**(1), 1–16 (2002)
19. Fischer, M., Köck, C., Schlüter, S., Weigert, F.: An empirical analysis of multivariate copula models. Quant. Finance **9**(7), 839–854 (2009)
20. Frahm, G., Junker, M., Szimayer, A.: Elliptical copulas: applicability and limitations. Stat. Probab. Lett. **63**(3), 275–286 (2003)
21. Genest, C., Rivest, L.P.: Statistical inference procedures for bivariate Archimedean copulas. J. Am. Stat. Assoc. **88**(423), 1034–1043 (1993)
22. Genest, C., Rémillard, B., Beaudoin, D.: Omnibus goodness-of-fit tests for copulas: a review and a power study. Insur. Math. Econ. **44**, 199–213 (2009)
23. Green, P.: Reversible jump Markov chain Monte Carlo computation and Bayesian model determination. Biometrika **82**, 711–732 (1995)
24. Haff, I.H., Aas, K., Frigessi, A.: On the simplified pair-copula construction – simply useful or too simplistic? Technical Report, Norwegian Computing Center, Oslo (2009)
25. Hanea, A., Kurowicka, D.: Mixed non-parametric continuous and discrete Bayesian belief networks. In: Advances in Mathematical Modeling for Reliability. IOS Press, Amsterdam (2008)
26. Hanea, A., Kurowicka, D., Cooke, R.: Hybrid method for quantifying and analyzing Bayesian belief networks. Qual. Realiab. Eng. **22**(6), 709–729 (2006)
27. Hanea, A., Kurowicka, D., Cooke, R., Ababei, D.: Mining and visualising ordinal data with non-parametric continuous BBNs. Computational Stat. Data Anal. **54**(3), 668–687 (2010)
28. Hastings, W.: Monte Carlo sampling methods using Markov chains and their applications. Biometrika **57**, 97–109 (1970)
29. Heinen, A., Valdesogo, A.: Asymmetric CAPM dependence for large dimensions: the Canonical Vine Autoregressive Model. CORE Discussion Paper 2009069, Université catholique de Louvain, Center for Operations Research and Econometrics, Louvain (2009)
30. Heinen, A., Valdesogo, A.: Dynamic D vines. In: Kurowicka, D., Joe, H. (eds) Dependence Modeling - Handbook on Vine Copulas, World Scientific Publishing, Singapore (2010)
31. Joe, H.: Families of m-Variate Distributions with Given Margins and m(m-1)/2 Bivariate Dependence Parameters. In: Rüschendorf, L., Schweitzer, B., Taylor, M.D. (eds.) Distributions with Fixed Marginals and Related Topics, pp. 120–141, Institute of Mathematical Statistics, Beachwood (1996)
32. Joe, H.: Multivariate Models and Dependence Concepts. Chapman & Hall, London (1997)
33. Joe, H.: Generating random correlation matrices based on partial correlations. J. Multivar. Anal. **97**, 2177–2189 (2006)
34. Joe, H., Li, H., Nikoloulopoulos, A.: Tail dependence and vine copulas. Technical Report, Department of Mathematics, Washington State University, Pullman, WA (2008)
35. Stien, M., Kolbjørnsen, O.: D-vine Creation of Non-Gaussian Random Fields, Technical Report SAND/07/08, Norwegian Computing Center, Oslo
36. Kurowicka, D.: Optimal truncation of vines. In: Kurowicka, D., Joe, H. (eds) Dependence Modeling - Handbook on Vine Copulas, World Scientific Publishing, Singapore (2010)
37. Kurowicka, D., Cooke, R.: A parametrization of positive definite matrices in terms of partial correlation vines. Linear Algebra Appl. **372**, 225–251 (2003)
38. Kurowicka, D., Cooke, R.: Distribution-free continuous Bayesian belief. In: Modern Statistical and Mathematical Methods in Reliabiliy. World Scientific Publishing, Singapore (2005)
39. Kurowicka, D., Cooke, R.: Uncertainty analysis with high dimensional dependence modelling. Wiley, Chichester (2006)
40. Kurowicka, D., Cooke, R.: Sampling algorithms for generating joint uniform distributions using the vine-copula method. Comput. Stat. Data Anal. **51**(6), 2889–2906 (2007)

41. Lanzendörfer, J.N.: Joint estimation of parameters in multivariate normal regression with correlated errors using pair-copula constructions and an application to finance (2009). Diploma Thesis, Center for Mathematical Sciences, Technische Universität München, München
42. Lewandowski, D., Kurowicka, D., Joe, H.: Generating random correlation matrices based on vines and extended onion method. J. Multivar. Anal. **100**, 1989–2001 (2009)
43. Liebscher, E.: Modelling and estimation of multivariate copulas. Technical Report, Working paper, University of Applied Sciences, Merseburg (2006)
44. McNeil, A.J., Frey, R., Embrechts, P.: Quantitative Risk Management: Concepts, Techniques and Tools. Princeton University Press, Princeton, NJ (2006)
45. de Melo Mendes, B.V., Semeraro, M.M., Leal, R.C.: Pair-copulas modeling in finance. Technical Report, IM/COPPEAD, Federal University at Rio de Janeiro, Brazil (2009)
46. Metropolis, N., Rosenbluth, A.W., Rosenbluth, M.N., Teller, A.H., Teller, E.: Equations of state calculations by fast computing machines. J. Chem. Phys. **21**, 1087–1092 (1953)
47. Min, A., Czado, C.: Bayesian inference for multivariate copulas using pair-copula constructions. J. Financ. Econom. (2010). To appear, preprint available under http://www-m4.ma.tum.de/Papers/index.html
48. Min, A., Czado, C.: Bayesian model selection for multivariate copulas using pair-copula constructions. Preprint (2009)
49. Morales-Napoles, O., Cooke, R., Kurowicka, D.: About the number of vines and regular vines on n nodes. Discrete Appl. Math. Submitted (2010)
50. Morillas, P.: A method to obtain new copulas from a given one. Metrika **61**, 169–184 (2005)
51. Nelsen, R.: An Introduction to Copulas, 2nd edn. Springer, New York, NY (2006)
52. Savu, C., Trede, M.: Hierarchical Archimedean copulas. In: International Conference on High Frequency Finance, Konstanz, May (2006)
53. Schirmacher, D., Schirmacher, E.: Multivariate dependence modeling using pair-copulas. Technical Report, Society of Acturaries: 2008 Enterprise Risk Management Symposium, Chicago, 14–16 Apr (2008). http://www.soa.org/library/monographs/other-monographs/2008/april/2008-erm-toc.aspx
54. Sklar, A.: Fonctions dé repartition á n dimensions et leurs marges. Publ. Inst. Stat. Univ. Paris **8**, 229–231 (1959)
55. Smith, M., Min, A., Czado, C., Almeida, C.: Modeling longitudinal data using a pair-copula decomposition of serial dependence. J. Am. Stat. Assoc. Revision (2009)
56. Whelan, N.: Sampling from Archimedian copulas. Quant. Finance **4**, 339–352 (2004)

# Chapter 5
# Risk Aggregation

Paul Embrechts and Giovanni Puccetti

**Abstract** Quantitative Risk Management (QRM) often starts with a vector of one-period profit-and-loss random variables $\mathbf{X} = (X_1, \ldots, X_d)'$ defined on some probability space $(\Omega, \mathfrak{F}, \mathbb{P})$. Risk Aggregation concerns the study of the aggregate financial position $\Psi(\mathbf{X})$, for some measurable function $\Psi : \mathbb{R}^d \to \mathbb{R}$. A risk measure $\rho$ then maps $\Psi(\mathbf{X})$ to $\rho(\Psi(\mathbf{X})) \in \mathbb{R}$, to be interpreted as the regulatory capital needed to be able to hold the aggregate position $\Psi(\mathbf{X})$ over a predetermined fixed time period. Risk Aggregation has often been studied within the framework when only the marginal distributions $F_1, \ldots, F_d$ of the individual risks $X_1, \ldots, X_d$ are available. Recently, especially in the management of operational risk, cases in which further dependence information is available have become relevant. We introduce a general mathematical framework which interpolates between marginal knowledge $(F_1, \ldots, F_d)$ and full knowledge of $F_{\mathbf{X}}$, the distribution of $\mathbf{X}$. We illustrate the basic issues through some pedagogic examples of actuarial and financial interest. In particular, we study Risk Aggregation under different mathematical set-ups, for different aggregating functionals $\Psi$ and risk measures $\rho$, focusing on Value-at-Risk. We show how the theory of Mass Transportations and tools originally developed to solve so-called Monge-Kantorovich problems turn out to be useful in this context. Finally, we introduce some new numerical integration techniques which solve some open aggregation problems and raise new interesting research issues.

Paul Embrechts, Senior SFI-Chair
Department of Mathematics, ETH Zurich, Zurich, Switzerland
e-mail: `embrechts@math.ethz.ch`

Giovanni Puccetti
Department of Mathematics for Decisions, University of Firenze, Firenze, Italy
e-mail: `giovanni.puccetti@dmd.unifi.it`

P. Jaworski et al. (eds.), *Copula Theory and Its Applications*, Lecture Notes in Statistics 198,
DOI 10.1007/978-3-642-12465-5_5, 

## 5.1 Motivations and Preliminaries

Quantitative Risk Management (QRM) standardly concerns a vector of one-period profit-and-loss random variables $\mathbf{X} = (X_1,\dots,X_d)'$ defined on some probability space $(\Omega,\mathfrak{F},\mathbb{P})$. *Risk Aggregation* concerns the study of the aggregate financial position $\Psi(\mathbf{X})$, for some measurable function $\Psi : \mathbb{R}^d \to \mathbb{R}$.

Under the terms of the New Basel Capital Accord (Basel II), internationally active banks are required to set aside capital to offset various types of risks, i.e. market, credit and operational risk; see [4]. Under the new regulations, the vector $\mathbf{X}$ represents the profit-and-loss amounts for particular lines of risk or business, and this over a given period. A risk measure $\rho$ maps the aggregate position $\Psi(\mathbf{X})$ to $\rho(\Psi(\mathbf{X})) \in \mathbb{R}$, to be interpreted as the regulatory capital needed to be able to hold the aggregate position $\Psi(\mathbf{X})$ over this predetermined fixed period. The exact calculation of $\rho(\Psi(\mathbf{X}))$ needs the joint distribution function $F_{\mathbf{X}}$ of $\mathbf{X}$; when such information is not available, special procedures are called for, typically leading to bounds on $\rho(\Psi(\mathbf{X}))$.

Risk Aggregation has often been studied within the framework when only the marginal distributions $F_1,\dots,F_d$ of the individual risks $X_1,\dots,X_d$ are available. A multitude of statistical techniques are available for estimating the univariate (marginal) distributions. It is often more difficult to capture statistically the $d$-variate structure of dependence of the vector $\mathbf{X}$. Recently, especially in the management of operational risk, cases in which further dependence information is available have become relevant.

In the following, we introduce a general mathematical framework which interpolates between marginal knowledge $(F_1,\dots,F_d)$ and full knowledge of $F_{\mathbf{X}}$. For the purpose of this paper, we disregard the statistical uncertainty related to $F_1,\dots,F_d$ and only concentrate on the probabilistic structure.

### 5.1.1 The Mathematical Framework

We follow the mathematical setup described in [38]. Let $B = \Pi_{i=1}^d B_i$ be the product of $d$ Borel spaces with $\sigma$-algebra $\mathscr{B} = \bigotimes_{i=1}^n \mathscr{B}_i$, $\mathscr{B}_i$ being the Borel $\sigma$-algebra on $B_i$. Define $I := \{1,\dots,d\}$ and let $\xi \subset 2^I$, the power set of $I$, with $\cup_{J\in\xi} J = I$. For $J \in \xi$, let $F_J \in \mathfrak{F}(B_J)$ be a *consistent* system of probability measures on $B_J = \pi_J(B) = \Pi_{j\in J} B_j$, $\pi_J$ being the natural projection from $B$ to $B_J$ and $\mathfrak{F}(B_J)$ denoting the set of all probability measures on $B_J$. Consistency of $F_J, J \in \xi$ means that $J_1, J_2 \in \xi, J_1 \cap J_2 \neq \emptyset$ implies that

$$\pi_{J_1\cap J_2} F_{J_1} = \pi_{J_1\cap J_2} F_{J_2}.$$

Finally, we denote by

$$\mathfrak{F}_\xi = \mathfrak{F}(F_J, J \in \xi)$$

the *Fréchet class* of all probability measures on $B$ having marginals $F_J, J \in \xi$.

Consistency of $F_J, J \in \xi$, is a necessary condition to guarantee that $F_\xi$ is non-empty. When $\xi$ is *regular* (see [42]), then consistency is also sufficient. When the system $\xi$ is non-regular, e.g. $\xi = \{\{1,2\},\{2,3\},\{3,1\}\}$, the Fréchet class $\mathfrak{F}_\xi$ may be empty even with consistent marginals, as illustrated in [38].

In the following, we will consider the case $B_i = \mathbb{R}$, $B = \mathbb{R}^n$. For the sake of notational simplicity, we identify probability measures on these spaces with the corresponding distribution functions. We will study only regular systems of marginals which interpolate between two particular choices of $\xi$:

- $\xi_d = \{\{1\},\dots,\{d\}\}$, also called the *simple* system of marginals, which defines the Fréchet class
$$\mathfrak{F}_{\xi_d} = \mathfrak{F}(F_1,\dots,F_d).$$
This is the most often used marginal system in Risk Aggregation and the natural setup for the theory of copulas, as discussed in [29].
- $\xi_I = \{I\}$, also called the *trivial* system of marginals, in which
$$\mathfrak{F}_{\xi_I} = \{F_{\mathbf{X}}\},$$
where $F_{\mathbf{X}}$ is the distribution function of the vector $\mathbf{X}$. This case represents the complete dependence information about $\mathbf{X}$.

There are other important cases representing intermediate dependence information between $\xi_d$ and $\xi_I$. Relevant examples are:

- $\xi_d^M = \{\{2j-1,2j\}, j = 1,\dots,d/2\}$ ($d$ even), the *multivariate* system of marginals. This system has the role of the simple marginal system when one studies aggregation of random vectors instead of aggregation of random variables.
- $\xi_d^\star = \{\{1,j\}, j = 2,\dots,d\}$, the *star-like* system of marginals and
- $\xi_d^= = \{\{j,j+1\}, j = 1,\dots,d-1\}$, the *serial* system of marginals. These latter two systems are of particular interest when dependence from bivariate datasets is available.

When $\xi$ is a partition of $I$, i.e. when all sets $J \in \xi$ are pairwise disjoint, we speak about a *non-overlapping* system of marginals, *overlapping* otherwise. According to this definition, $\xi_d$, $\xi_I$ and $\xi_d^M$ are non-overlapping marginal systems, while $\xi_d^\star$ and $\xi_d^=$ are overlapping. We study Risk Aggregation under incomplete information frameworks in Sect. 5.2. Section 5.3 will focus instead on the problems arising within the complete information system $\xi_I$.

## 5.2 Bounds for Functions of Risks: The Coupling-Dual Approach

We will focus on those risk measures $\rho(\Psi(\mathbf{X}))$ which are representable as

$$\rho(\Psi(\mathbf{X})) = \mathbb{E}[\psi(\mathbf{X})] = \int \psi \, dF_{\mathbf{X}}, \tag{5.1}$$

for some measurable function $\psi : \mathbb{R}^d \to \mathbb{R}$. This representation includes some of the most popular risk measures, such as Value-at-Risk. Most importantly, (5.1) will allow to use the theory of *Mass Transportations* within the context of Risk Aggregation. In this section, we illustrate how to obtain bounds on $\mathbb{E}[\psi(\mathbf{X})]$ under an incomplete information setting, i.e. when the distribution function $F_{\mathbf{X}}$ of the vector $\mathbf{X}$ is not completely specified. Formally, we assume that

$$F_{\mathbf{X}} \in \mathfrak{F}_{\xi}, \text{ for a fixed } \xi \subset 2^I, \xi \neq I.$$

Since $F_{\mathbf{X}}$ is not uniquely determined, there exist an entire range of values for $\mathbb{E}[\psi(\mathbf{X})]$, which are consistent with the choice of the subgroups $J \in \xi$ of marginals. The infimum and supremum of this range are defined as

$$m_{\xi}(\psi) := \inf\left\{ \int \psi \, dF_{\mathbf{X}} : F_{\mathbf{X}} \in \mathfrak{F}_{\xi} \right\}, \tag{5.2a}$$

$$M_{\xi}(\psi) := \sup\left\{ \int \psi \, dF_{\mathbf{X}} : F_{\mathbf{X}} \in \mathfrak{F}_{\xi} \right\}. \tag{5.2b}$$

Since the Fréchet class $\mathfrak{F}_{\xi}$ is convex and the problems (5.2a) and (5.2b) are linear on $F_{\mathbf{X}}$, (5.2a) and (5.2b) both admit a dual representation. This representation is to be found in the theory of Mass Transportations.

**Theorem 5.2.1.** *Let the measurable function $\psi$ be bounded or continuous, then problems* (5.2a) *and* (5.2b) *have the following dual counterparts:*

$$m_{\xi}(\psi) = \sup\left\{ \sum_{J \in \xi} \int f_J \, dF_J : f_J \in L^1(F_J), J \in \xi \text{ with} \sum_{J \in \xi} f_J \circ \pi_J \leq \psi \right\}, \tag{5.3a}$$

$$M_{\xi}(\psi) = \inf\left\{ \sum_{J \in \xi} \int g_J \, dF_J : g_J \in L^1(F_J), J \in \xi \text{ with} \sum_{J \in \xi} g_J \circ \pi_J \geq \psi \right\}. \tag{5.3b}$$

There exist several versions of Theorem 5.2.1 which are valid under weaker assumptions and more general settings, even non-topological ones. For more details on these versions, a proof of Theorem 5.2.1 and a complete coverage of the theory of Mass Transportations, we refer to the milestone book [36] and the review paper [39].

According to [24], we call a *coupling* every random vector $\mathbf{X}$ having distribution function $F_{\mathbf{X}} \in \mathfrak{F}_{\xi}$. Moreover, we call *dual choice* for (5.3a) any family of functions $\mathbf{f} = \{f_J, J \in \xi\}$ which are admissible for (5.3a). Analogously, we define a dual choice $\mathbf{g} = \{g_J, J \in \xi\}$ for (5.3b). By Theorem 5.2.1, a coupling $\mathbf{X}$ and two dual choices $\mathbf{f}$ and $\mathbf{g}$ satisfy

$$\int \psi \, dF_{\mathbf{X}} \geq m_{\xi}(\psi) \geq \sum_{J \in \xi} \int f_J dF_J, \tag{5.4a}$$

$$\int \psi \, dF_{\mathbf{X}} \leq M_{\xi}(\psi) \leq \sum_{J \in \xi} \int g_J dF_J, \tag{5.4b}$$

for all $F_{\mathbf{X}} \in \mathfrak{F}_\xi$. A coupling and a dual choice which satisfy (5.4a) (or (5.4b)) with two equalities will be called an *optimal coupling* and a *dual solution*, respectively, since they solve problem (5.2a) (or (5.2b)).

Equations (5.4a) and (5.4b) illustrate the *coupling-dual* approach in Risk Aggregation. Problems (5.2a) and (5.2b) are in general very difficult to solve, with some exceptions illustrated in Sect. 5.2.2 below. Depending on the system $\xi$ of marginals, the concept of a copula might not be useful and it may be difficult even to identify a single coupling in $F_\xi$. When the solutions of (5.2a) and (5.2b) are unknown, any dual choice satisfying (5.4a) or (5.4b) provides a bound on $m_\xi$ or $M_\xi$.

### 5.2.1 Application 1: Bounding Value-at-Risk

We now illustrate the usefulness of the dual representations (5.3a) and (5.3b), in the case of Value-at-Risk (VaR). VaR is probably the most popular risk measure in finance and insurance, this is no doubt due to its importance within the Basel II capital-adequacy framework; see [4]. The VaR of a profit-and-loss random variable $L$ at the probability (or confidence) level $\alpha \in (0,1)$ is simply the $\alpha$-quantile of its distribution, defined as

$$\mathrm{VaR}_\alpha(L) = F_L^{-1}(\alpha) = \inf\{l \in \mathbb{R} : F_L(l) \geq \alpha\}, \tag{5.5}$$

where $F_L$ is the distribution of $L$. Under the terms of Basel II, banks often measure the risk associated with a portfolio $\mathbf{X} = (X_1, \ldots, X_d)'$ in terms of $\mathrm{VaR}_\alpha(X_1 + \cdots + X_d)$, the VaR of the sum of its marginal components. This is for example the case of operational risk; see [15]. Using our notation, we have $\rho = \mathrm{VaR}$ and $\Psi = +$, the sum operator. Typical values for $\alpha$ are $\alpha = 0.95$ or $\alpha = 0.99$, or even $\alpha = 0.999$ in the case of credit and operational risk. By (5.5), bounding the VaR of a random variable $L$ from above is equivalent to bounding from below its distribution $F_L$ or, similarly, bounding from above its tail (or survival) function $\overline{F_L} = 1 - F_L$. Roughly speaking, if VaR is used to risk measure $L$, a *higher tail function* for $L$ means a *more dangerous risk*. An alternative approach to Value-at-Risk is to be found in Sect. 8.4.4 of this volume [23].

Banks often have more precise information about the marginal distributions of $\mathbf{X}$, but less about the joint distribution $F_{\mathbf{X}}$. This then immediately translates into the incomplete information setting $\xi_d$, which defines the Fréchet class $\mathfrak{F}(F_1, \ldots, F_d)$. Within $\xi_d$, banks are typically interested on a upper bound on $\mathrm{VaR}_\alpha\left(\sum_{i=1}^d X_i\right)$, since this latter amount cannot be calculated exactly. Such a bound can be obtained by solving problem (5.2b) for a particular choice of the function $\psi$, in this case $\psi = \psi(s) = 1_{\{x_1+\cdots+x_d \geq s\}}$, for some $s \in \mathbb{R}$. Thus, we define the function $M_{\xi_d}$ as

$$M_{\xi_d}(s) = \sup\left\{\int 1_{\{x_1+\cdots+x_d \geq s\}} dF_{\mathbf{X}}(x_1, \ldots, x_d), F_{\mathbf{X}} \in \mathfrak{F}(F_1, \ldots, F_d)\right\}, s \in \mathbb{R}. \tag{5.6}$$

Note that the inequality $\geq$ in the definition of the indicator function in (5.6) is essential in order to guarantee that the supremum is attained; see Remark 3.1(ii) in [13]. With respect to any random vector $(X_1,\ldots,X_d)'$ having distribution $F_{\mathbf{X}} \in \mathfrak{F}(F_1,\ldots,F_d)$, the function $M_{\xi_d}$ obviously satisfies

$$\mathbb{P}[X_1+\cdots+X_d \geq s] \leq M_{\xi_d}(s) \text{ for all } s \in \mathbb{R}, \tag{5.7}$$

while, for its inverse $M_{\xi_d}^{-1}$, we have

$$\mathrm{VaR}_\alpha(X_1+\cdots+X_d) \leq M_{\xi_d}^{-1}(1-\alpha), \text{ for all } \alpha \in (0,1). \tag{5.8}$$

According to Theorem 5.2.1, the dual counterpart of (5.6) is given by:

$$\begin{aligned} M_{\xi_d}(s) = \inf\Bigg\{ \sum_{i=1}^d \int f_i \, dF_i : f_i \in L_1(F_i), i \in I \\ \text{s.t. } \sum_{i=1}^d f_i(x_i) \geq 1_{\{x_1+\cdots+x_d \geq s\}} \text{ for all } x_i \in \mathbb{R}, i \in I \Bigg\}. \end{aligned} \tag{5.9}$$

The dual solution for (5.9) is given in [37] for the sum of two risks ($d=2$). Independently from this, [26] provided the corresponding optimal coupling. For the sum of more than two risks, (5.9) seems to be very difficult to solve. The only explicit results known in the literature are given in [37] for the case of the sum of marginals being all uniformly or binomially distributed.

When the value of $M_\xi(s)$ is unknown, Eq. (5.4b) plays a crucial role. In fact, every dual admissible choice in (5.9) gives un upper bound on $M_\xi(s)$ which, though not sharp, is conservative from a risk management point of view. This is for instance the idea used in [13] to produce bounds on $\mathrm{VaR}_\alpha\left(\sum_{i=1}^d X_i\right)$. The following theorem is a reformulation of Theorem 4.2 in the above reference and illustrates the case of a homogeneous risk portfolio, i.e. $F_i = F$ for all $i=1,\ldots,d$.

**Theorem 5.2.2.** *Let $F$ be a continuous distribution with non-negative support. If $F_i = F, i = 1,\ldots,d$, then, for every $s \geq 0$,*

$$M_{\xi_d}(s) \leq D_{\xi_d}(s) = d \inf_{r \in [0,s/d)} \frac{\int_r^{s-(d-1)r}(1-F(x))dx}{s-dr}. \tag{5.10}$$

The infimum in (5.10) can be easily calculated numerically by finding the zero-derivative points of its argument. For $d=2$, we obtain $M_{\xi_d}(s) = D_{\xi_d}(s)$, the bound given in [37]. The idea of using dual choices to produce bounds on functions of risks was discussed further in [12] (within simple systems with non-homogeneous marginals), [14] (multivariate systems) and [16] (overlapping systems). Bounds produced by a choice of admissible dual functionals are referred to as *dual bounds*. A related study of bounds on VaR can be found in [22] and in Sect. 8.4.4 of this volume [23].

In Fig. 5.1, we plot the dual bound function $D_{\xi_d}$ for a portfolio of three ($d = 3$) Gamma-distributed risks. In the same figure, we also give the tail function of the random variable $X_1 + X_2 + X_3$ in case of comonotonic ($C_{\mathbf{X}} = M$) and independent ($C_{\mathbf{X}} = \Pi$) marginals; for this notation, see [29], Chap. 5. Note that the two tail functions cross at some threshold $\hat{s}$ and the tail function obtained under comonotonicity lies above the one obtained under independence for all $s > \hat{s}$. We will return on this later in Sect. 5.3. Table 5.1 shows the upper bounds $D_{\xi_d}^{-1}(1-\alpha)$ on the VaR of the Gamma portfolio, as well as exact quantiles in case of independence and comonotonicity. Recall that for comonotonic risks VaR is additive, see also (5.15) later in the paper. Figure 5.1 and Table 5.1 exemplify the fact that, using (5.7), (5.8) and (5.10), we have

$$\mathbb{P}[X_1 + \cdots + X_d \geq s] \leq D_{\xi_d}(s), \text{ i.e. } \mathrm{VaR}_\alpha(X_1 + \cdots + X_d) \leq D_{\xi_d}^{-1}(1-\alpha),$$

for any $(X_1, \ldots, X_d)'$ having distribution $F_{\mathbf{X}} \in \mathfrak{F}_{\xi_d}$.

We finally remark that the entire curve $D_{\xi_d}(s)$ is generally obtained within seconds, independently of the number $d$ of variables under study. In general, the computational time of dual bounds strongly depends on the number of non-homogeneous marginals.

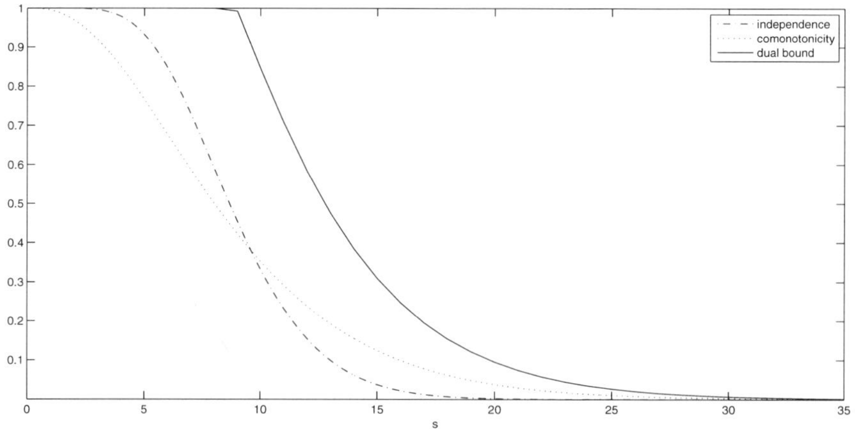


**Fig. 5.1** Plot of the tail function $\mathbb{P}[X_1 + X_2 + X_3 \geq s]$ for a $\Gamma(3,1)$-portfolio under independence and comonotonic scenarios. We also plot the upper dual bound function $D_{\xi_d}(s)$.

It is interesting to study how dual bounds vary within different marginal systems having the same univariate marginals. To this aim, we now consider $d$ risks $X_1, \ldots, X_d$ which we assume to be Pareto distributed with tail parameter $\theta$, i.e.

$$F_i(x) = \mathbb{P}[X_i \leq x] = 1 - (1+x)^{-\theta}, \; x \geq 0, \; i = 1, \ldots d. \tag{5.11}$$

**Table 5.1** $\mathrm{VaR}_\alpha(X_1+X_2+X_3)$ for a $\Gamma(3,1)$-portfolio under independence and comonotonicity, for some levels $\alpha$ of interest. We also give the corresponding upper dual bounds $D^{-1}_{\xi_d}(1-\alpha)$.

| $\alpha$ | $C_{\mathbf{X}} = \Pi$ | $C_{\mathbf{X}} = M$ | Dual bound |
|---|---|---|---|
| 0.90 | 13.00 | 15.97 | 19.80 |
| 0.95 | 14.44 | 18.89 | 22.57 |
| 0.99 | 17.41 | 25.22 | 28.67 |
| 0.999 | 21.16 | 33.69 | 36.97 |

Together with the non-overlapping marginal system $\xi_d$ studied above, we consider the overlapping star-like system $\xi_d^*$. Under $\xi_d^*$, we assume that each of the $d-1$ subvectors $(X_1,X_i), i=2,\dots,d$, is coupled by a Frank copula $C^F_\delta$ with parameter $\delta=1$. Within the system $\xi_d^*$, bounds on $\mathrm{VaR}_\alpha(\sum_{i=1}^d X_i)$ are obtained by integration of particular dual bounds in $\xi_d$. For more details on this technique, we refer to [16].

In Table 5.2, we give upper VaR limits $D^{-1}_{\xi_d^*}(\alpha)$ for Frank-Pareto portfolios of increasing dimensions. As quantile levels, we take $\alpha=0.99$ and $\alpha=0.999$. For comparison, the comonotonic quantiles are also given. Considering the absolute values reported in Table 5.2, the overlapping bounds are smaller than the corresponding bounds obtained in a non-overlapping setting. The reason is clear: switching from a non-overlapping simple system to a overlapping star-like marginal system means reducing the Fréchet class of attainable risks, i.e. having more information about the dependence structure of the portfolio $\mathbf{X}$. Formally, we have $\mathfrak{F}_{\xi_d^*} \subset \mathfrak{F}_{\xi_d}$. Under the extra information represented by $\xi_d^*$, less capital is needed to offset the underlying portfolio risk.

Detailed studies of the quality of the dual bounds have been presented in [13] for $D^{-1}_{\xi_d}$, and in [16] for $D^{-1}_{\xi_d^*}$.

**Table 5.2** Upper bounds on value-at-risk for the sum of $d$ Pareto(2)-distributed risks within the *overlapping* star-like $\xi_d^*$ and the *non-overlapping* marginal system $\xi_d$. Under the star-like system, the bivariate marginals are coupled by a Frank copula with parameter $\delta=1$.

| | $\alpha=0.99$ | | $\alpha=0.999$ | |
|---|---|---|---|---|
| $d$ | Overlapping | Non-overlapping | Overlapping | Non-overlapping |
| 3 | 29.98 | 46.70 | 95.17 | 156.98 |
| 4 | 51.82 | 70.75 | 167.24 | 248.98 |
| 5 | 78.46 | 98.44 | 253.83 | 348.55 |
| 6 | 108.99 | 129.36 | 352.62 | 458.76 |
| 7 | 143.03 | 178.20 | 463.35 | 578.66 |
| 8 | 180.12 | 218.27 | 584.19 | 707.54 |
| 9 | 220.14 | 261.00 | 712.03 | 844.81 |
| 10 | 262.83 | 306.27 | 850.30 | 990.00 |

#### 5.2.1.1 Open Problems

The search for $M_{\xi_d}(s)$, i.e. for the largest VaR over $\mathfrak{F}(F_1,\dots,F_d)$, is open when $d>2$. The proof of the optimality of the dual functionals for the case $d=2$, given in [37], is based on Strassen's theorem (see Theorem 11 in [41]). Unfortunately, Strassen's theorem does not have an obvious extension to the product of more than two marginal spaces; see [40] and references therein.

The search for $m_{\xi_d}(s)$, i.e. for the smallest VaR over $\mathfrak{F}(F_1,\dots,F_d)$, is again open when $d>2$. For general dimensions $d$, several authors have obtained an elementary lower bound for $m_{\xi_d}(s)$; see for instance [7]. In models of actuarial interest, in [15] it is shown that the last mentioned lower bound does not depend on $d$. Therefore, a better bound on $m_{\xi_d}(s)$ is needed.

Finally, VaR dual bounds of the type (5.10) are needed for more general aggregating functionals $\Psi$.

### 5.2.2 *Application 2: Supermodular Functions*

In the simple marginal setting $\xi=\xi_d$, there are some functionals $\psi$ for which the solutions of problems (5.2a) and (5.2b) are known. They form the class $\mathscr{S}_d$ of supermodular functions.

**Definition 5.2.1.** A measurable function $\psi:\mathbb{R}^d\to\mathbb{R}$ is said to be *supermodular* if

$$\psi(\mathbf{u}\wedge\mathbf{v})+\psi(\mathbf{u}\vee\mathbf{v})\geq\psi(\mathbf{u})+\psi(\mathbf{v}),\ \text{for all } \mathbf{u},\mathbf{v}\in\mathbb{R}^d,$$

where $\mathbf{u}\wedge\mathbf{v}$ is the componentwise minimum of $\mathbf{u}$ and $\mathbf{v}$, and $\mathbf{u}\vee\mathbf{v}$ is the componentwise maximum of $\mathbf{u}$ and $\mathbf{v}$.

When $d=2$, a function $c:\mathbb{R}\times\mathbb{R}\to\mathbb{R}$ is supermodular if and only if

$$\psi(x_1,y_1)+\psi(x_2,y_2)\geq\psi(x_1,y_2)+\psi(x_2,y_1),\ \text{for all } x_2\geq x_1,\ y_2\geq y_1. \tag{5.12}$$

Recall that, for any set of univariate distributions $F_1,\dots,F_d$, there exists a *comonotonic coupling* $\mathbf{X}^M$, i.e. a random vector having marginals $F_1,\dots,F_d$ and copula $M$.

**Theorem 5.2.3.** *For given univariate distributions $F_1,\dots,F_d$, denote by $\mathbf{X}^M$ a comonotonic coupling having these marginals. Let $\psi:\mathbb{R}^d\to\mathbb{R}$ be right-continuous. Then*

$$\mathbb{E}\left[\psi\left(\mathbf{X}^M\right)\right]=\sup\left\{\int\psi\,dF_{\mathbf{X}}:F_{\mathbf{X}}\in\mathfrak{F}(F_1,\dots,F_d)\right\},\ \textit{for all } F_1,\dots,F_d, \tag{5.13}$$

*if and only if $\psi\in\mathscr{S}_d$.*

*Proof.* The *if* part follows from [36, Remark 3.1.3], but many authors have derived the same result under different regularity conditions: see for instance [5, 25]. For the *only if* part, see [35]. □

The most popular supermodular function is the product $\times(\mathbf{x}) = \Pi_{i=1}^{n} x_i$. When $\psi = \times$, Theorem 5.2.3 gives the well-known result that a multivariate comonotonic distribution maximizes correlation between its marginals.

Note that Theorem 5.2.3 applies to a large class of interesting functionals, including $\psi(\mathbf{x}) = \sum_{i=1}^{d} h_i(x_i)$, where the $h_i$'s are non-decreasing (see [30]) and $\psi(\mathbf{x}) = h(\sum_{i=1}^{d} x_i)$ for $h$ non-decreasing and convex; see [27, pp. 150–155]. In insurance, $\sum_{i=1}^{d} h_i(x_i)$ and $h(\sum_{i=1}^{d} x_i)$ can be interpreted, respectively, as the risk positions for a reinsurance treaty with individual retention functions $h_i$, and a reinsurance treaty with a global retention function $h$.

We remark that the functional $\psi = 1_{\{\sum_{i=1}^{d} x_i \geq s\}}$, which defines the worst-VaR problem (5.6), is not supermodular and hence does not satisfy the assumption of Theorem 5.2.3. Hence, it may happen that a comonotonic coupling does *not* maximize the VaR of the sum of $d$ risks, as we will study in details in Sect. 5.3 below.

#### 5.2.2.1 Open Problems

For $d = 2$, the infimum in (5.13) is attained by the *countermonotonic* distribution $W(F_1, F_2)$. Since $W(F_1, \ldots, F_d)$ is not a proper distribution when $d > 2$, the search for the infimum of $\mathbb{E}[\psi(\mathbf{X})]$ among the Fréchet class $\mathfrak{F}(F_1, \ldots, F_d)$ remains open for a variety of functionals $\psi$. Especially for $\psi = \times$, Roger Nelsen (private communication) remarked that the solution of this last mentioned problem would have important consequences in the theory of dependence measures.

## 5.3 The Calculation of the Distribution of the Sum of Risks

In the trivial system of marginals $\xi = \xi_I$, we have that $\mathfrak{F}_{\xi_I} = \{F_{\mathbf{X}}\}$. This setting represents complete probabilistic information about the portfolio $\mathbf{X}$ of risks held. In fact, from a theoretical point of view, the knowledge of $F_{\mathbf{X}}$ completely determines the distribution of the random variable $\Psi(\mathbf{X})$. In practice, we will see that things are more complicated.

The system $\xi_I$ is particularly important in stress-testing, i.e. when one has different models for $F_{\mathbf{X}}$ and wants to stress-test the distribution of $\Psi(\mathbf{X})$. Especially in the context of the current (credit) crisis, financial institutions often have information on the marginal distributions of the underlying risks but want to stress-test the interdependence between these risks, for instance assuming different copula scenarios.

In the following, we will study the case of the sum of risks, i.e. $\Psi = +$. Thus, we will focus on the computation of the distribution of $\Psi(\mathbf{X}) = \sum_{i=1}^{d} X_i$, i.e.

$$\mathbb{P}[X_1 + \cdots + X_d \leq s] = \int_{\mathfrak{I}(s)} dF_{\mathbf{X}}(x_1, \ldots, x_d), s \in \mathbb{R} \tag{5.14}$$

where $\mathfrak{I}(s) = \{\mathbf{x} \in \mathbb{R}^d : \sum_{i=1}^{d} x_i \leq s\}$.

The computation of (5.14) is a rather onerous task. In the literature, there exist several methods to calculate (5.14) when the marginals $X_i$ are independent. In some rare cases, it is possible to write the integral in (5.14) in closed form. For general marginals, one can for instance rely on the Fast Fourier Transforms; see [8] and the references therein for a discussion within a risk management context.

Much less is known when the $X_i$'s are dependent. Indeed, when $\mathbf{X}$ has a general copula $C_{\mathbf{X}}$, one often has to rely on integration tools like Monte Carlo and Quasi-Monte Carlo methods. When $F_{\mathbf{X}}$ has a density function $f_{\mathbf{X}}$, these methods approximate (5.14) by the average of $f_{\mathbf{X}}$ evaluated at $M$ points $x_1,\ldots,x_M$ filling up $\mathfrak{I}(s)$ in a convenient way, i.e.

$$\int_{\mathfrak{I}(s)} dF_{\mathbf{X}}(x_1,\ldots,x_d) \simeq \frac{1}{M}\frac{s^d}{d!}\sum_{i=1}^{M} f_{\mathbf{X}}(x_i).$$

If the $x_i$'s are chosen to be (pseudo) randomly distributed, this is the *Monte Carlo* (MC) method. If the $x_i$'s are chosen as elements of a low-discrepancy sequence, this is the *Quasi-Monte Carlo* (QMC) method. A *low-discrepancy* sequence is a totally deterministic sequence of vectors that generates representative samples from a uniform distribution on given subsets. Compared to Monte Carlo methods, the advantage of using quasi-random sequences is that points cannot cluster coincidentally on some region of the set. Using Central Limit Theorem arguments, it is possible to show that traditional MC has a convergence rate of $O(M^{-1/2})$, and this independently of the number of dimensions $d$. QMC can be much faster than MC with errors approaching $O(M^{-1})$ for a smooth underlying density. For details on the theory of *rare event simulation* within MC methods, we refer the reader to the monographs [3, 18, 28]. For an introduction to QMC methods, see for instance [33]. A comprehensive overview of both methods is given in [43]. Note that all the techniques mentioned above warrant considerable expertise and, more importantly, need to be tailored to the specific problem under study. In particular, the implementation very much depends on the functional form of $f_{\mathbf{X}}$ (either direct, or through the marginals and a copula).

The re-tailoring of the rule to be iterated, from example to example, is common also to other numerical techniques for the estimation of (5.14), such as *quadrature methods*; see [6, 34] for a review. However, in the computation of multi-dimensional integrals as in (5.14), numerical quadrature rules are typically less efficient than MC and QMC.

A simple and competitive tool for the computation of the distribution function of a sum of random variables is the AEP algorithm introduced in [1]. If one knows the distribution $F_{\mathbf{X}}$ of $\mathbf{X}$, it is very easy to compute the $F_{\mathbf{X}}$-measure of hypercubes in $\mathbb{R}^d$. Thus, the authors of [1] propose a decomposition of $\mathfrak{I}(s)$ via a infinite union of (possibly overlapping) hypercubes and hence compute (5.14) in terms of the algebraic sum of the probability masses contained in them.

In the MC and QMC methods described above, the final estimates contain a source of randomness. Instead, the AEP algorithm is completely determinist because it is solely based on the geometrical properties of $\mathfrak{I}(s)$. Moreover, the accuracy of

MC and QMC methods is generally lost for problems in which the density $f_{\mathbf{X}}$ is not smooth or cannot be given in closed form, and comes at the price of an adaptation of the sampling algorithm to the specific example under study. The AEP algorithm however can handle in a uniform way any joint distribution $F_{\mathbf{X}}$ and does not require existence or smoothness of a density $f_{\mathbf{X}}$. As illustrated in [1], AEP performs better than QMC in dimensions $d = 2,3$ and slightly worse for dimensions $d = 4,5$. In these latter dimensions, however, programming a QMC sequence is much more demanding than using AEP. At the time being, AEP cannot be applied for $d > 5$ due to computational complexity (memory).

We set $d = 3$ and we use AEP to provide estimates for the tail and the quantile (VaR) function of the sum $S_3 = X_1 + X_2 + X_3$. For pedagogical reasons, we assume the marginals $F_i$ of the portfolio to be Pareto distributed with tail parameter $\theta_i > 0$. We consider the two dependence scenarios obtained by coupling the Pareto marginals either by the independent copula $C_{\mathbf{X}} = \Pi$ or via the comonotonic copula $C_{\mathbf{X}} = M$. In the following, we use the fact that VaR is additive under comonotonicity; see Prop. 3.1 in [9]. This means that, for a comonotonic vector $(X_1^M, X_2^M, X_3^M)$, we have

$$\mathrm{VaR}_\alpha(X_1^M + X_2^M + X_3^M) = \mathrm{VaR}_\alpha(X_1) + \mathrm{VaR}_\alpha(X_2) + \mathrm{VaR}_\alpha(X_3). \tag{5.15}$$

Denote by $F_\Pi$ the distribution of $S_3$ obtained under independence between the $X_i$'s and by $F_M$ the distribution of $S_3$ obtained under comonotonicity between the $X_i$'s. $\overline{F_\Pi}$ and $\overline{F_M}$, respectively, are the corresponding tail functions. We study two different cases: when the $X_i$'s have finite or infinite first moment.

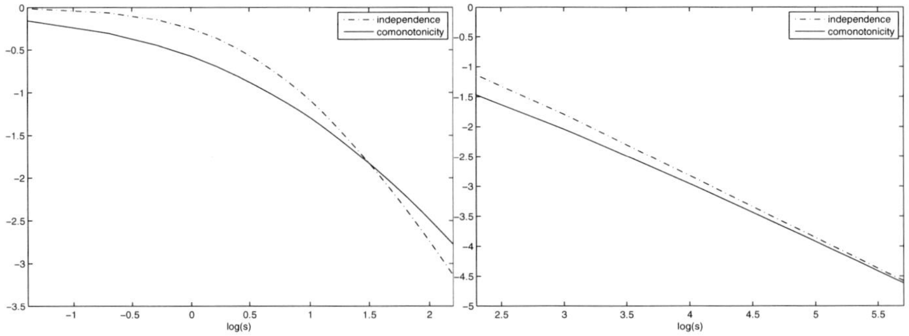


**Fig. 5.2** Log/log plots of the tail function of $X_1 + X_2 + X_3$, under independence and comonotonicity. The $X_i$'s are distributed as a Pareto(2) (*left*) and as a Pareto(1) (*right*).

*The finite-mean case.* In Fig. 5.2 (left), we plot $\overline{F_\Pi}$ and $\overline{F_M}$ when the Pareto tail parameter $\theta$ for the marginal distributions is set to 2 (the $X_i$'s have finite first moment). We note that the two curves $\overline{F_\Pi}$ and $\overline{F_M}$ cross once at some high threshold $s = \hat{s}$. For $s < \hat{s}$, we have that $\overline{F_M}(s) < \overline{F_\Pi}(s)$. Recalling (5.5), this means that

$$F_M^{-1}(\alpha) < F_{\Pi}^{-1}(\alpha), \text{ for all } \alpha < \hat{\alpha} = F_M(\hat{s}) = F_{\Pi}(\hat{s}), \tag{5.16}$$

i.e. for the lower levels $\alpha < \hat{\alpha}$, $\mathrm{VaR}_\alpha(S_3)$ is larger under independence between the $X_i$'s. For $\alpha > \hat{\alpha}$, inequality (5.16) is obviously reversed and we have that

$$F_{\Pi}^{-1}(\alpha) \leq F_M^{-1}(\alpha), \text{ for all } \alpha \geq \hat{\alpha}, \tag{5.17}$$

i.e. for the higher levels $\alpha > \hat{\alpha}$, $\mathrm{VaR}_\alpha(S_3)$ is larger under comonotonicity between the marginals. Recalling (5.15), and for the independent vector $(X_1^{\Pi}, X_2^{\Pi}, X_3^{\Pi})$, inequality (5.17) can be written as

$$\mathrm{VaR}_\alpha(X_1^{\Pi} + X_2^{\Pi} + X_3^{\Pi}) \leq \mathrm{VaR}_\alpha(X_1) + \mathrm{VaR}_\alpha(X_2) + \mathrm{VaR}_\alpha(X_3), \text{ for all } \alpha \geq \hat{\alpha}, \tag{5.18}$$

i.e. VaR is subadditive in the tail of $F_{\Pi}$. When $\theta > 1$, [19] illustrates that this tail behavior can be extended to more general dependence and marginal scenarios.

*The infinite-mean case.* Figure 5.2 (right) shows the same plot as Fig. 5.2 (left), but now the Pareto tail parameter $\theta = 1$ (the $X_i$'s have infinite first moment). We note that $\overline{F_M}(s) < \overline{F_{\Pi}}(s)$ for all $s \in \mathbb{R}$. Therefore, all the quantiles of $S_3$ under independence are larger than the corresponding quantiles under comonotonicity and inequality (5.18) is reversed:

$$\mathrm{VaR}_\alpha(X_1^{\Pi} + X_2^{\Pi} + X_3^{\Pi}) > \mathrm{VaR}_\alpha(X_1) + \mathrm{VaR}_\alpha(X_2) + \mathrm{VaR}_\alpha(X_3), \text{ for all } \alpha \in (0,1). \tag{5.19}$$

This shows that, in general, VaR *may fail to be subadditive*. Typical frameworks in which VaR shows a superadditive behavior are: marginals with infinite mean or skew distributions (as in this case) and/or marginals coupled by a non-elliptical copula; see [29]. An early interesting read on this is [11]. In [10], a mathematical summary of the issue is given within extreme value theory using the concept of multivariate regular variation.

Possible superadditivity is an important conceptual deficiency of Value-at-Risk. In fact, VaR has been heavily criticized by many authors for not being a *coherent* measure of risk; see the seminal paper [2]. Many other authors have discussed desirable properties which a general risk measure $\rho$ has to satisfy. Textbook treatments are [17, 29].

**Table 5.3** $\mathrm{VaR}_\alpha(X_1 + X_2 + X_3)$ under different dependence scenarios for three different Pareto portfolios. For a fixed level $\alpha$ and Pareto parameter $\theta$, the largest VaR value is bold-faced.

| | $X_i \backsim$ Pareto(2) | | | $X_i \backsim$ Pareto(1.3) | | | $X_i \backsim$ Pareto(1) | | |
|---|---|---|---|---|---|---|---|---|---|
| $\alpha$ | $\Pi$ | $Cl$ | $M$ | $\Pi$ | $Cl$ | $M$ | $\Pi$ | $Cl$ | $M$ |
| 0.80 | 3.92 | **4.21** | 3.71 | 8.90 | **9.36** | 7.35 | 16.69 | **17.21** | 12.00 |
| 0.90 | 5.87 | 6.45 | **6.49** | 15.36 | **16.54** | 14.63 | 33.20 | **35.05** | 27.00 |
| 0.99 | 18.37 | 19.62 | **27.00** | 84.08 | 87.34 | **100.65** | 308.21 | **315.25** | 297.00 |
| 0.999 | 55.92 | 57.37 | **91.87** | 477.44 | 481.80 | **606.28** | 3,012.97 | **3,025.00** | 2,997.00 |

Finally, in Table 5.3, we show the quantiles for $S_3$ for different levels of probabilities, under several marginal and dependence scenarios. Along with independence ($C_{\mathbf{X}} = \Pi$) and comonotonicity ($C_{\mathbf{X}} = M$), we study the case in which the copula of $\mathbf{X}$ is of Clayton type ($C_{\mathbf{X}} = Cl$). There are various points to remark about:

- The behavior of the tail function $\overline{F_{Cl}}$ of $S_3$ under the Clayton scenario is similar to the behavior of $\overline{F_\Pi}$ studied above. When $\theta > 1$, $\overline{F_{Cl}}$ and $\overline{F_M}$ cross once. This can be seen from the fact that, for $\theta = 2$ and $\theta = 1.3$, the comonotonic quantiles are smaller than the Clayton ones when the quantile level $\alpha$ is small, while they are larger when $\alpha$ is large. In this case, VaR under the Clayton scenario shows subadditivity in the tail.
  For $\theta = 1$, we have that $\overline{F_M}(s) \leq \overline{F_{Cl}}(s)$ for all $s \in \mathbb{R}$, hence VaR under the Clayton model is superadditive at all levels $\alpha$. We also note that the intersection point between the Clayton and the comonotonic curve goes to infinity as the tail parameter $\theta$ approaches 1 from above. When $\theta = 1$, the two curves do not cross.
- Since the marginal distributions of the $X_i$'s are fixed, the first moment of the sum $S_3$ does not depend on the copula $C_{\mathbf{X}}$. When $\theta > 1$, two different distributions for $S_3$ have the same *finite* mean and therefore *cannot* be *stochastically ordered*; see Sect. 1.2 in [31] for the definition of stochastic order and its properties. As a consequence, two different distributions for $S_3$ must cross. The case illustrated in Fig. 5.2, in which the intersection point is unique, is typical for two random variables which are *stop-loss ordered*; see Theorem 1.5.17 and Definition 1.5.1 in [31] (in this last reference the authors use the equivalent terminology *increasing-convex order* to indicate the stop-loss order).
  When $\theta = 1$, we have that $\mathbb{E}[S_3] = +\infty$ and it is possible that $\overline{F_M} < \overline{F_\Pi}$, i.e. the distribution of $S_3$ under independence is stochastically larger than the distribution of $S_3$ under comonotonicity, as illustrated in Fig. 5.2 (right). For general distributions, both the change of behavior with respect to stochastic dominance and superadditivity of VaR in the tail seem to be strictly related to the existence of first moments. For some further discussions on this phenomenon, see [20, 21, 32].
- For Pareto marginals of the form (5.11), the quantile function of $S_3$ can be given in closed form under the independence and comonotonic assumptions. Things are different when one assumes a Clayton-type dependence. In this latter case, the computation of the distribution and the VaRs of $S_3$ requires one of the integration techniques described above in this section. In particular, the quantiles in Table 5.3 have been obtained via AEP.

### *5.3.1 Open Problems*

In insurance and finance, there is a increasing need of software being able to compute the distribution of $\Psi(\mathbf{X})$ when the distribution of $\mathbf{X}$ is known. The authors of [1] are working on a extension of AEP to general increasing functionals $\Psi$. Moreover, efficiency of AEP for dimensions $d > 5$ needs to be improved. Finally, AEP and its competitors open the way to the computational study of large and non-homogeneous risk portfolios.

**Acknowledgements** The first author would like to thank the Swiss Finance Institute for financial support. The second author thanks both RiskLab and the FIM at the Department of Mathematics of the ETH Zurich for financial support and kind hospitality.

## References

1. Arbenz, P., Embrechts, P., Puccetti, G.: The AEP algorithm for the fast computation of the distribution of the sum of dependent random variables. Forthcoming in Bernoulli (2010)
2. Artzner, P., Delbaen, F., Eber, J.M., Heath, D.: Coherent measures of risk. Math. Finance **9**(3), 203–228 (1999)
3. Asmussen, S., Glynn, P.W.: Stochastic Simulation: Algorithms and Analysis. vol. 57, Springer, New York, NY (2007)
4. Basel Committee on Banking Supervision: International Convergence of Capital Measurement and Capital Standards. Bank for International Settlements, Basel (2006)
5. Cambanis, S., Simons, G., Stout, W.: Inequalities for $Ek(X,Y)$ when the marginals are fixed. Z. Wahrsch. Verw. Gebiete **36**(4), 285–294 (1976)
6. Davis, P.J., Rabinowitz, P.: Methods of Numerical Integration, 2nd edn. Academic Press, Orlando, FL (1984)
7. Denuit, M., Genest, C., Marceau, É.: Stochastic bounds on sums of dependent risks. Insur. Math. Econ. **25**(1), 85–104 (1999)
8. Embrechts, P., Frei, M.: Panjer recursion versus FFT for compound distributions. Math. Methods Oper. Res. **69**, 497–508 (2009)
9. Embrechts, P., Höing, A., Juri, A.: Using copulae to bound the Value-at-Risk for functions of dependent risks. Finance Stoch. **7**(2), 145–167 (2003)
10. Embrechts, P., Lambrigger, D.D., Wüthrich, M.V.: Multivariate extremes and the aggregation of dependent risks: examples and counter-examples. Extremes **12**, 107–127 (2009)
11. Embrechts, P., McNeil, A.J., Straumann, D.: Correlation and dependence in risk management: properties and pitfalls. In: Dempster, M. (ed.) Risk Management: Value at Risk and Beyond, pp. 176–223. Cambridge University Press, Cambridge (2002)
12. Embrechts, P., Puccetti, G.: Aggregating risk capital, with an application to operational risk. Geneva Risk Insur. Rev. **31**(2), 71–90 (2006)
13. Embrechts, P., Puccetti, G.: Bounds for functions of dependent risks. Finance Stoch. **10**(3), 341–352 (2006)
14. Embrechts, P., Puccetti, G.: Bounds for functions of multivariate risks. J. Mult. Anal. **97**(2), 526–547 (2006)
15. Embrechts, P., Puccetti, G.: Aggregating operational risk across matrix structured loss data. J. Oper. Risk **3**(2), 29–44 (2008)
16. Embrechts, P., Puccetti, G.: Bounds for the sum of dependent risks having overlapping marginals (2009). J. Multivar. Anal. **101**(1), 177–190 (2010)
17. Föllmer, H., Schied, A.: Stochastic Finance, 2nd edn., Walter De Gruyter, Berlin (2004)
18. Glasserman, P.: Monte Carlo Methods in Financial Engineering. Springer, New York, NY (2004)
19. Ibragimov, R.: Portfolio diversification and value at risk under thick-tailedness. Quan. Finance **9**(5), 565–580 (2009)
20. Ibragimov, R., Walden, J.: Portfolio diversification under local and moderate deviations from power laws. Insur. Math. Econ. **42**(2), 594–599 (2008)
21. Jaworski, P.: Value at risk in the presence of the power laws. Acta Phys. Polon. B **36**(8), 2575–2587 (2005)
22. Jaworski, P.: Bounds for value at risk—the approach based on copulas with homogeneous tails. Mathware Soft Comput. **15**(1), 113–124 (2008)

23. Jaworski, P.: Tail behaviour of copulas. In: Jaworski, P., Durante, F., Härdle, W., Rychlik, T. (eds.) Copula Theory and Its Applications, Proceedings of the Workshop, Warsaw, 25–26 Sept 2009. Springer, Dordrecht (2010)
24. Lindvall, T.: Lectures on the Coupling Method. Wiley, New York, NY (1992)
25. Lorentz, G.G.: An inequality for rearrangements. Am. Math. Mon. **60**, 176–179 (1953)
26. Makarov, G.D.: Estimates for the distribution function of the sum of two random variables with given marginal distributions. Theory Probab. Appl. **26**, 803–806 (1981)
27. Marshall, A.W., Olkin, I.: Inequalities: Theory of Majorization and Its Applications. Academic Press, New York, NY (1979)
28. McLeish, D.L.: Monte Carlo Simulation and Finance. Wiley, Hoboken, NJ (2005)
29. McNeil, A.J., Frey, R., Embrechts, P.: Quantitative Risk Management: Concepts, Techniques, Tools. Princeton University Press, Princeton, NJ (2005)
30. Müller, A.: Stop-loss order for portfolios of dependent risks. Insur. Math. Econ. **21**(3), 219–223 (1997)
31. Müller, A., Stoyan, D.: Comparison Methods for Stochastic Models and Risks. Wiley, Chichester (2002)
32. Nešlehová, J., Embrechts, P., Chavez-Demoulin, V.: Infinite-mean models and the LDA for operational risk. J. Oper. Risk **1**(1), 3–25 (2006)
33. Niederreiter, H.: Random Number Generation and Quasi-Monte Carlo Methods. CBMS-NSF Regional Conference Series in Applied Mathematics, vol. 63. SIAM, Philadelphia, PA (1992)
34. Press, W.H., Teukolsky, S.A., Vetterling, W.T., Flannery, B.P.: Numerical Recipes: The Art of Scientific Computing. Cambridge University Press, Cambridge (2007)
35. Puccetti, G., Scarsini, M.: Multivariate comonotonicity. J. Multivar. Anal. **101**(1), 291–304 (2010)
36. Rachev, S.T., Rüschendorf, L.: Mass Transportation Problems, vols. I–II. Springer, New York, NY (1998)
37. Rüschendorf, L.: Random variables with maximum sums. Adv. Appl. Probab. **14**(3), 623–632 (1982)
38. Rüschendorf, L.: Bounds for distributions with multivariate marginals. In: Stochastic Orders and Decision Under Risk, vol. 19, pp. 285–310. Institute of Mathematical Statistics, Hayward, CA (1991)
39. Rüschendorf, L.: Fréchet-bounds and their applications. In: Advances in Probability Distributions with Given Marginals, vol. 67, pp. 151–187. Kluwer Academic Publishers, Dordrecht (1991)
40. Shortt, R.M.: Strassen's marginal problem in two or more dimensions. Z. Wahrsch. Verw. Gebiete **64**(3), 313–325 (1983)
41. Strassen, V.: The existence of probability measures with given marginals. Ann. Math. Stat. **36**, 423–439 (1965)
42. Vorob'ev, N.N.: Consistent families of measures and their extensions. Theory Probab. Appl. **7**(2), 147–163 (1962)
43. Weinzierl, S.: Introduction to Monte Carlo Methods. Eprint arXiv:hep-ph/0006269 (2000)

# Chapter 6
# Extreme-Value Copulas

Gordon Gudendorf and Johan Segers

**Abstract** Being the limits of copulas of componentwise maxima in independent random samples, extreme-value copulas can be considered to provide appropriate models for the dependence structure between rare events. Extreme-value copulas not only arise naturally in the domain of extreme-value theory, they can also be a convenient choice to model general positive dependence structures. The aim of this survey is to present the reader with the state-of-the-art in dependence modeling via extreme-value copulas. Both probabilistic and statistical issues are reviewed, in a nonparametric as well as a parametric context.

## 6.1 Introduction

In various domains, as for example finance, insurance or environmental science, joint extreme events can have a serious impact and therefore need careful modeling. Think for instance of daily water levels at two different locations in a lake during a year. Calculation of the probability that there is a flood exceeding a certain benchmark requires knowledge of the joint distribution of maximal heights during the forecasting period. This is a typical field of application for extreme-value theory. In such situations, extreme-value copulas can be considered to provide appropriate models for the dependence structure between exceptional events.

One of the first applications of bivariate extreme-value analysis must be due to Gumbel and Goldstein [49], who analyze the maximal annual discharges of the

---

Gordon Gudendorf
Institut de statistique, Université catholique de Louvain, Louvain-la-Neuve, Belgium
e-mail: gordon.gudendorf@uclouvain.be

Johan Segers
Institut de statistique, Université catholique de Louvain, Louvain-la-Neuve, Belgium
e-mail: johan.segers@uclouvain.be

P. Jaworski et al. (eds.), *Copula Theory and Its Applications*, Lecture Notes in Statistics 198,
DOI 10.1007/978-3-642-12465-5_6, 

Ocmulgee River in Georgia at two different stations, a dataset that has been taken up again in [59]. The joint behavior of extreme returns in the foreign exchange rate market is investigated in [90], whereas the comovement of equity markets characterized by high volatility levels is studied in [68]. An application in the insurance domain can be found in [9].

Extreme-value copulas not only arise naturally in the domain of extreme events, but they can also be a convenient choice to model data with positive dependence. An advantage with respect to the much more popular class of Archimedean copulas, for instance, is that they are not symmetric. Incidentally, a hybrid class containing both the Archimedean and the extreme-value copulas as a special case are the Archimax copulas [8].

The aim of this survey is to present the reader with the state-of-the-art in dependence modeling via extreme-value copulas. Definition, origin, and basic properties of extreme-value copulas are presented in Sect. 6.2. A number of useful and popular parametric families are reviewed in Sect. 6.3. Section 6.4 provides a discussion of the most important dependence coefficients associated to extreme-value copulas. An overview of parametric and nonparametric inference methods for extreme-value copulas is given in Sect. 6.5. Finally, some further topics and pointers to the literature are gathered in Sect. 6.6.

## 6.2 Foundations

Let $X_i = (X_{i1}, \ldots, X_{id})$, $i \in \{1, \ldots, n\}$, be a sample of independent and identically distributed (iid) random vectors with common distribution function $F$, margins $F_1, \ldots, F_d$, and copula $C_F$. For convenience, assume $F$ is continuous. Consider the vector of componentwise maxima:

$$M_n = (M_{n,1}, \ldots, M_{n,d}), \qquad \text{where } M_{n,j} = \bigvee_{i=1}^{n} X_{ij}, \tag{6.1}$$

with '$\vee$' denoting maximum. Since the joint and marginal distribution functions of $M_n$ are given by $F^n$ and $F_1^n, \ldots, F_d^n$ respectively, it follows that the copula, $C_n$, of $M_n$ is given by

$$C_n(u_1, \ldots, u_d) = C_F(u_1^{1/n}, \ldots, u_d^{1/n})^n, \qquad (u_1, \ldots, u_d) \in [0,1]^d.$$

The family of extreme-value copulas arises as the limits of these copulas $C_n$ as the sample size $n$ tends to infinity.

**Definition 6.2.1.** A copula $C$ is called an *extreme-value copula* if there exists a copula $C_F$ such that

$$C_F(u_1^{1/n}, \ldots, u_d^{1/n})^n \to C(u_1, \ldots, u_d) \qquad (n \to \infty) \tag{6.2}$$

for all $(u_1,\dots,u_d) \in [0,1]^d$. The copula $C_F$ is said to be in the *domain of attraction* of $C$.

Historically, this construction dates back at least to [17, 32]. The representation of extreme-value copulas can be simplified using the concept of max-stability.

**Definition 6.2.2.** A $d$-variate copula $C$ is *max-stable* if it satisfies the relationship

$$C(u_1,\dots,u_d) = C(u_1^{1/m},\dots,u_d^{1/m})^m \tag{6.3}$$

for every integer $m \geqslant 1$ and all $(u_1,\dots,u_d) \in [0,1]^d$.

From the previous definitions, it is trivial to see that a max-stable copula is in its own domain of attraction and thus must be itself an extreme-value copula. The converse is true as well.

**Theorem 6.2.1.** *A copula is an extreme-value copula if and only if it is max-stable.*

The proof of Theorem 6.2.1 is standard: for fixed integer $m \geqslant 1$ and for $n = mk$, write

$$C_n(u_1,\dots,u_d) = C_k(u_1^{1/m},\dots,u_d^{1/m})^m.$$

Let $k$ tend to infinity on both sides of the previous display to get (6.3).

By definition, the family of extreme-value copulas coincides with the set of copulas of *extreme-value distributions*, that is, the class of limit distributions with non-degenerate margins of

$$\left(\frac{M_{n,1}-b_{n,1}}{a_{n,1}},\dots,\frac{M_{n,d}-b_{n,d}}{a_{n,d}}\right)$$

with $M_{n,j}$ as in (6.1), centering constants $b_{n,j} \in \mathbb{R}$ and scaling constants $a_{n,j} > 0$. Representations of extreme-value distributions then yield representations of extreme-value copulas. Let $\Delta_{d-1} = \{(w_1,\dots,w_d) \in [0,\infty)^d : \sum_j w_j = 1\}$ be the unit simplex in $\mathbb{R}^d$; see Fig. 6.1. The following theorem is adapted from [80], which is based in turn on [16].

**Theorem 6.2.2.** *A $d$-variate copula $C$ is an extreme-value copula if and only if there exists a finite Borel measure $H$ on $\Delta_{d-1}$, called* spectral measure, *such that*

$$C(u_1,\dots,u_d) = \exp\bigl(-\ell(-\log u_1,\dots,-\log u_d)\bigr), \qquad (u_1,\dots,u_d) \in (0,1]^d,$$

*where the* tail dependence function $\ell : [0,\infty)^d \to [0,\infty)$ *is given by*

$$\ell(x_1,\dots,x_d) = \int_{\Delta_{d-1}} \bigvee_{j=1}^{d} (w_j x_j)\, dH(w_1,\dots,w_d), \qquad (x_1,\dots,x_d) \in [0,\infty)^d. \tag{6.4}$$

*The spectral measure $H$ is arbitrary except for the $d$ moment constraints*

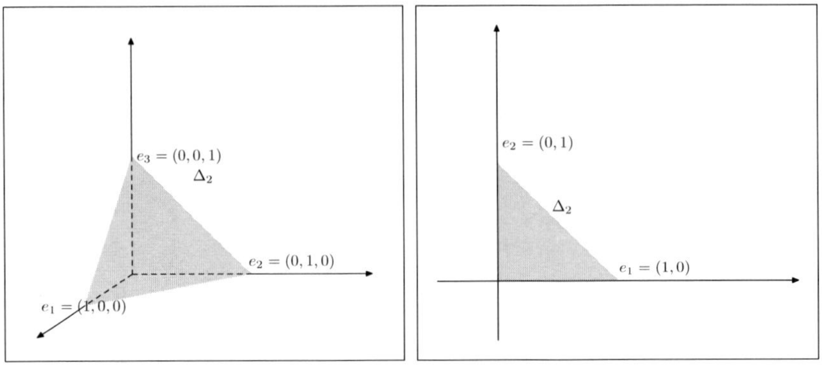


**Fig. 6.1** On the left side $\Delta_2$ is represented in $\mathbb{R}^3$, which is equivalent to the representation in $\mathbb{R}^2$ on the right side.

$$\int_{\Delta_{d-1}} w_j \, dH(w_1,\ldots,w_d) = 1, \qquad j \in \{1,\ldots,d\}. \tag{6.5}$$

The $d$ moment constraints on $H$ in (6.5) stem from the requirement that the margins of $C$ be standard uniform. They imply that $H(\Delta_{d-1}) = d$.

By a linear expansion of the logarithm and the exponential function, the domain-of-attraction Eq. (6.2) is equivalent to

$$\begin{aligned}\lim_{t \downarrow 0} t^{-1}\big(1 - C_F(1-tx_1,\ldots,1-tx_d)\big) &= -\log C(e^{-x_1},\ldots,e^{-x_d}) \\ &= \ell(x_1,\ldots,x_d)\end{aligned} \tag{6.6}$$

for all $(x_1,\ldots,x_d) \in [0,\infty)^d$; see for instance [20]. The tail dependence function $\ell$ in (6.4) is convex, homogeneous of order one, that is $\ell(cx_1,\ldots,cx_d) = c\,\ell(x_1,\ldots,x_d)$ for $c > 0$, and satisfies $\max(x_1,\ldots,x_d) \leqslant \ell(x_1,\ldots,x_d) \leqslant x_1 + \cdots + x_d$ for all $(x_1,\ldots,x_d) \in [0,\infty)^d$. By homogeneity, it is characterized by the *Pickands dependence function* $A : \Delta_{d-1} \to [1/d, 1]$, which is simply the restriction of $\ell$ to the unit simplex:

$$\ell(x_1,\ldots,x_d) = (x_1 + \cdots + x_d) A(w_1,\ldots,w_d) \qquad \text{where} \qquad w_j = \frac{x_j}{x_1 + \cdots + x_d},$$

for $(x_1,\ldots,x_d) \in [0,\infty)^d \setminus \{0\}$. The extreme-value copula $C$ can be expressed in terms of $A$ via

$$C(u_1,\ldots,u_d) = \exp\left\{ \left( \sum_{j=1}^d \log u_j \right) A\left( \frac{\log u_1}{\sum_{j=1}^d \log u_j}, \ldots, \frac{\log u_d}{\sum_{j=1}^d \log u_j} \right) \right\}.$$

The function $A$ is convex as well and satisfies $\max(w_1,\ldots,w_d) \leqslant A(w_1,\ldots,w_d) \leqslant 1$ for all $(w_1,\ldots,w_d) \in \Delta_{d-1}$. However, these properties do not characterize the class of Pickands dependence functions unless $d = 2$, see for instance the counterexample on p. 257 in [3].

In the bivariate case, we identify the unit simplex $\Delta_1 = \{(1-t,t) : t \in [0,1]\}$ in $\mathbb{R}^2$ with the interval $[0,1]$.

**Theorem 6.2.3.** *A bivariate copula $C$ is an extreme-value copula if and only if*

$$C(u,v) = (uv)^{A(\log(v)/\log(uv))}, \qquad (u,v) \in (0,1]^2 \setminus \{(1,1)\}, \tag{6.7}$$

*where $A : [0,1] \to [1/2,1]$ is convex and satisfies $t \vee (1-t) \leqslant A(t) \leqslant 1$ for all $t \in [0,1]$.*

It is worth stressing that in the bivariate case, any function $A$ satisfying the two constraints from Theorem 6.2.3 corresponds to an extreme-value copula. These functions lie in the shaded area of Fig. 6.2; in particular, $A(0) = A(1) = 1$.

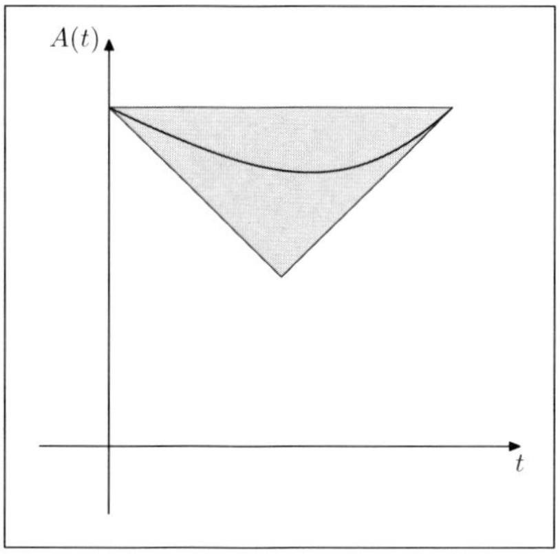


**Fig. 6.2** A typical Pickands dependence function $A$ together with the region $t \vee (1-t) \leqslant A(t) \leqslant 1$ in Theorem 6.2.3.

The upper and lower bounds for $A$ have special meanings: the upper bound $A(t) = 1$ corresponds to independence, $C(u,v) = uv$, whereas the lower bound $A(t) = t \vee (1-t)$ corresponds to perfect dependence (comonotonicity) $C(u,v) = u \wedge v$. In general, the inequality $A(t) \leqslant 1$ implies $C(u,v) \geqslant uv$, that is, extreme-value copulas are necessarily *positive quadrant dependent*.

## 6.3 Parametric Models

By Theorems 6.2.2 and 6.2.3, the class of extreme-value copulas is infinite-dimensional. Parametric submodels can be constructed in a number of ways: by calculating the limit $\ell$ in (6.6) for a given initial copula $C_F$; by specifying a spectral measure $H$; in dimension $d = 2$, by constructing a Pickands dependence function $A$. In this section, we employ the first of these methods to introduce some of the more popular families. For more extensive overviews, see e.g. [3, 65].

### 6.3.1 Logistic Model or Gumbel–Hougaard Copula

Consider the Archimedean copula

$$C_\phi(u_1,\ldots,u_d) = \phi^{\leftarrow}\big(\phi(u_1)+\cdots+\phi(u_d)\big), \qquad (u_1,\ldots,u_d)\in[0,1]^d \tag{6.8}$$

with generator $\phi : [0,1] \to [0,\infty]$ and inverse $\phi^{\leftarrow}(t) = \inf\{u \in [0,1] : \phi(u) \leqslant t\}$; the function $\phi$ should be strictly decreasing and convex and satisfy $\phi(1) = 0$, and $\phi^{\leftarrow}$ should be $d$-monotone on $(0,\infty)$, see [23, 72].

If the following limit exists,

$$\theta = -\lim_{s\downarrow 0} \frac{s\phi'(1-s)}{\phi(1-s)} \in [1,\infty] \tag{6.9}$$

then the domain-of-attraction condition (6.6) is verified for $C_F$ equal to $C_\phi$, the tail dependence function being

$$\ell(x_1,\ldots,x_d) = \begin{cases} (x_1^\theta+\cdots+x_d^\theta)^{1/\theta} & \text{if } 1\leqslant\theta<\infty, \\ x_1\vee\cdots\vee x_d & \text{if } \theta=\infty, \end{cases} \tag{6.10}$$

for $(x_1,\ldots,x_d)\in[0,\infty)^d$; see [8, 11]. The range $[1,\infty]$ for the parameter $\theta$ in (6.9) is not an assumption but rather a consequence of the properties of $\phi$. The parameter $\theta$ measures the degree of dependence, ranging from independence ($\theta = 1$) to complete dependence ($\theta = \infty$).

The extreme-value copula associated to $\ell$ in (6.10) is

$$C(u_1,\ldots,u_d) = \exp\{-\big((-\log u_1)^\theta+\cdots+(-\log u_d)^\theta\big)^{1/\theta}\},$$

known as the *Gumbel–Hougaard* or *logistic* copula. Dating back to Gumbel [46, 47], it is (one of) the oldest multivariate extreme-value models. It was discovered independently in survival analysis [15, 54]. It happens to be the only copula that is at the same time Archimedean and extreme-value [37].

The bivariate asymmetric logistic model introduced in [94] adds further flexibility to the basic logistic model. Multivariate extensions of the asymmetric logistic model were studied already in [71] and later in [14, 61]. These distributions can be generated via mixtures of certain extreme-value distributions over stable distributions, a representation that yields large possibilities for modelling that have yet begun to be explored [30, 95].

### 6.3.2 Negative Logistic Model or Galambos Copula

Let $\hat{C}_\phi$ be the survival copula of the Archimedean copula $C_\phi$ in (6.8). Specifically, if $C_\phi$ is the distribution function of the random vector $(U_1,\ldots,U_d)$, then $\hat{C}_\phi$ is the distribution function of the random vector $(1-U_1,\ldots,1-U_d)$. If the following limit exists,

$$\theta = -\lim_{s\downarrow 0} \frac{\phi(s)}{s\,\phi'(s)} \in [0,\infty] \tag{6.11}$$

then the domain-of-attraction condition (6.6) is verified for $C_F$ equal to $\hat{C}_\phi$, the tail dependence function being

$$\ell(x_1,\ldots,x_d) = \begin{cases} x_1+\cdots+x_d & \text{if } \theta = 0, \\ x_1+\cdots+x_d - \sum_{\substack{I\subset\{1,\ldots,d\}\\ |I|\geqslant 2}} (-1)^{|I|}\left(\sum_{i\in I} x_i^{-\theta}\right)^{-1/\theta} & \text{if } 0<\theta<\infty, \\ x_1\vee\cdots\vee x_d & \text{if } \theta=\infty, \end{cases}$$

for $(x_1,\ldots,x_d)\in[0,\infty)^d$; see [8, 11]. In case $0<\theta<\infty$, the sum is over all subsets $I$ of $\{1,\ldots,d\}$ of cardinality $|I|$ at least 2. The amount of dependence ranges from independence ($\theta = 0$) to complete dependence ($\theta=\infty$).

The resulting extreme-value copula is known as the *Galambos* or *negative logistic* copula, dating back to [31]. Asymmetric extensions have been proposed in [60, 61].

### 6.3.3 Hüsler–Reiss Model

For the bivariate normal distribution with correlation coefficient $\rho$ smaller than one, it is known since [87] that the marginal maxima $M_{n,1}$ and $M_{n,2}$ are asymptotically independent, that is, the domain-of-attraction condition (6.2) holds with limit copula $C(u,v)=uv$. However, for $\rho$ close to one, better approximations to the copula of $M_{n,1}$ and $M_{n,2}$ arise within a somewhat different asymptotic framework. More precisely, as in [56], consider the situation where the correlation coefficient $\rho$ associated to the bivariate Gaussian copula $C_\rho$ is allowed to change with the sample size, $\rho=\rho_n$, in such a way that $\rho_n\to 1$ as $n\to\infty$. If

$$(1-\rho_n)\log n \to \lambda^2 \in [0,+\infty] \qquad (n\to\infty),$$

then one can show that

$$C_{\rho_n}(u^{1/n},v^{1/n})^n \to C_\lambda(u,v) \qquad (n\to\infty), \qquad (u,v)\in[0,1]^2,$$

where the *Hüsler–Reiss* copula $C_\lambda$ is the bivariate extreme-value copula with Pickands dependence function

$$A(w) = (1-w)\,\Phi\left(\lambda+\frac{1}{2\lambda}\log\frac{1-w}{w}\right) + w\,\Phi\left(\lambda+\frac{1}{2\lambda}\log\frac{w}{1-w}\right)$$

for $w \in [0,1]$, with $\Phi$ representing the standard normal cumulative distribution function. The parameter $\lambda$ measures the degree of dependence, going from independence ($\lambda = \infty$) to complete dependence ($\lambda = 0$).

### 6.3.4 *The t-EV Copula*

In financial applications, the $t$-copula is sometimes preferred over the Gaussian copula because of the larger weight it assigns to the tails. The bivariate $t$-copula with $\nu > 0$ degrees of freedom and correlation parameter $\rho \in (-1,1)$ is the copula of the bivariate $t$-distribution with the same parameters and is given by

$$C_{\nu,\rho}(u,v) = \int_{-\infty}^{t_\nu^{-1}(u)} \int_{-\infty}^{t_\nu^{-1}(v)} \frac{1}{\pi\nu |P|^{1/2}} \frac{\Gamma\left(\frac{\nu}{2}+1\right)}{\Gamma\left(\frac{\nu}{2}\right)} \left(1 + \frac{x' P^{-1} x}{\nu}\right)^{-\nu/2+1} dx,$$

where $t_\nu$ represents the distribution function of the univariate $t$-distribution with $\nu$ degrees of freedom and $P$ represents the $2 \times 2$ correlation matrix with off-diagonal element $\rho$. In [19], it is shown that $C_{\nu,\rho}$ is in the domain of attraction of the bivariate extreme-value copula $C_A$ with Pickands dependence function

$$\begin{aligned} &A(w) = w t_{\nu+1}(z_w) + (1-w) t_{\nu+1}(z_{1-w}), \\ &\text{where } z_w = (1+\nu)^{1/2}[\{w/(1-w)\}^{1/\nu} - \rho](1-\rho^2)^{-1/2}, \qquad w \in [0,1]. \end{aligned} \tag{6.12}$$

This extreme-value copula was coined the *t-EV* copula. Building upon results in [41, 53], exactly the same extreme-value attractor is found in [2] for the more general class of (meta-)elliptical distributions whose generator has a regularly varying tail.

## 6.4 Dependence Coefficients

Let $(U,V)$ be a bivariate random vector with distribution function $C$, a bivariate extreme-value copula with Pickands dependence function $A$ as in (6.7). As mentioned already, the inequality $A \leqslant 1$ implies that $C(u,v) \geqslant uv$ for all $(u,v) \in [0,1]^2$, that is, $C$ is positive quadrant dependent. In fact, in [44] it was shown that extreme-value copulas are *monotone regression dependent*, that is, the conditional distribution of $U$ given $V = v$ is stochastically increasing in $v$ and *vice versa*; see also Theorem 5.2.10 in [81].

In particular, all measures of dependence of $C$ such as Kendall's $\tau$ or Spearman's $\rho_S$ must be nonnegative. The latter two can be expressed in terms of $A$ via

$$\tau = 4 \iint_{[0,1]^2} C(u,v)\, dC(u,v) - 1 = \int_0^1 \frac{t(1-t)}{A(t)}\, dA'(t),$$

$$\rho_S = 12 \iint_{[0,1]^2} uv\, dC(u,v) - 3 = 12 \int_0^1 \frac{1}{(1+A(t))^2}\, dt - 3.$$

The Stieltjes integrator $dA'(t)$ is well-defined since $A$ is a convex function on $[0,1]$; if the dependence function $A$ is twice differentiable, it can be replaced by $A''(t)\,dt$. For a proof of the identities above, see for instance [55], where it is shown that $\tau$ and $\rho_S$ satisfy $-1+\sqrt{1+3\tau} \leqslant \rho_S \leqslant \min\left(\frac{3}{2}\tau, 2\tau - \tau^2\right)$, a pair of inequalities first conjectured in [57].

The *Kendall distribution function* associated to a general bivariate copula $C$ is defined as the distribution function of the random variable $C(U,V)$, that is,

$$K(w) = \mathrm{P}[C(U,V) \leqslant w], \qquad w \in [0,1].$$

The reference to Kendall stems from the link with Kendall's $\tau$, which is given by $\tau = 4\,\mathrm{E}[C(U,V)] - 1$. For bivariate Archimedean copulas, for instance, the function $K$ not only identifies the copula [38], convergence of Archimedean copulas is actually equivalent to weak convergence of their Kendall distribution functions [10]. For bivariate extreme-value copulas, the function $K$ takes the remarkably simple form

$$K(w) = w - (1-\tau)\, w \log w, \qquad w \in [0,1], \tag{6.13}$$

as shown in [42]. In fact, in that paper the conjecture was formulated that if the Kendall distribution function of a bivariate copula is given by (6.13), then $C$ is a bivariate extreme-value copula, a conjecture which to the best of our knowledge still stands. In the same paper, Eq. (6.13) was used to formulate a test that a copula belongs to the family of extreme-value copulas; see also [40].

In the context of extremes, it is natural to study the *coefficient of upper tail dependence*. For a bivariate copula $C_F$ in the domain of attraction of an extreme-value copula with tail dependence function $\ell$ and Pickands dependence function $A$, we find

$$\begin{aligned}\lambda_U &= \lim_{u \uparrow 1} \mathrm{P}(U > u \mid V > u) = \lim_{t \downarrow 0} t^{-1}\big(2t - 1 + C(1-t, 1-t)\big) \\ &= 2 - \ell(1,1) = 2\big(1 - A(1/2)\big) \in [0,1].\end{aligned}$$

Graphically this quantity can be represented as the length between the upper boundary and the curve of the Pickands dependence function evaluated in the mid-point $1/2$, see Fig. 6.3. The coefficient $\lambda_U$ ranges from 0 ($A = 1$, independence) to 1 (complete dependence). Multivariate extensions are proposed in [67].

The related quantity $\ell(1,1) = 2A(1/2)$ is called the *extremal coefficient* in [89]. For a bivariate extreme-value copula, we find

$$\mathrm{P}(U \leqslant u, V \leqslant u) = u^{2A(1/2)}, \qquad u \in [0,1],$$

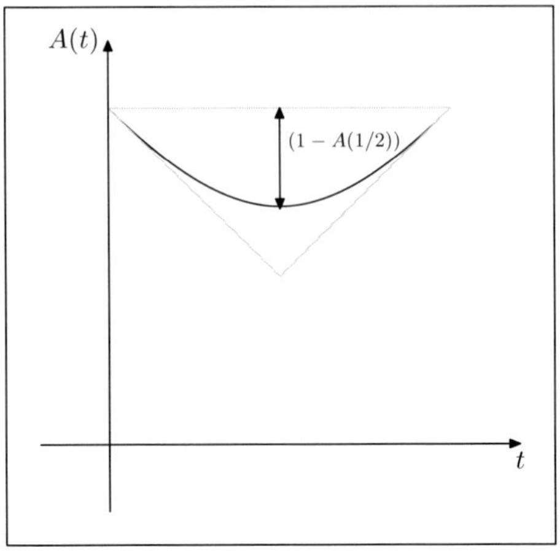


**Fig. 6.3** The coefficient of upper tail dependence $\lambda_U$ is equal to twice the length of the double arrow in the upper part of the graph.

so that $2A(1/2) \in [1,2]$ can be thought of as the (fractional) number of independent components in the copula. Multivariate extensions have been studied in [83].

For the *lower tail dependence coefficient*, the situation is trivial:

$$\lambda_L = \lim_{u \downarrow 0} \mathrm{P}(U \le u \mid V \le u) = \lim_{u \downarrow 0} u^{(2A(1/2)-1)} = \begin{cases} 0 & \text{if } A(1/2) > 1/2, \\ 1 & \text{if } A(1/2) = 1/2. \end{cases}$$

In words, except for the case of perfect dependence, $A(1/2) = 1/2$, extreme-value copulas have asymptotically independent lower tails.

## 6.5 Estimation

Let $X_i = (X_{i1}, \ldots, X_{id})$, $i \in \{1, \ldots, n\}$, be a random sample from a (continuous) distribution $F$ with margins $F_1, \ldots, F_d$ and extreme-value copula $C$:

$$F(x_1, \ldots, x_d) = C\big(F_1(x_1), \ldots, F_d(x_d)\big),$$

and $C$ as in Theorem 6.2.2. The problem considered here is statistical inference on $C$, or equivalently, on its Pickands dependence function $A$. A number of situations may arise, according to whether the extreme-value copula $C$ is completely unknown or is assumed to belong to a parametric family. In addition, the margins may be supposed to be known, parametrically modelled, or completely unknown. A survey of estimation methods in general copula models is given in Chap. 3 of this volume [12].

### 6.5.1 Parametric Estimation

Assume that the extreme-value copula $C$ belongs to a parametric family $(C_\theta : \theta \in \Theta)$ with $\Theta \subset \mathbb{R}^p$; for instance, one of the families described in Sect. 6.3. Inference on $C$ then reduces to inference on the parameter vector $\theta$. The usual way to proceed is by maximum likelihood. The likelihood is to be constructed from the copula density

$$c_\theta(u_1,\dots,u_d) = \frac{\partial^d}{\partial u_1 \cdots \partial u_d} C_\theta(u_1,\dots,u_d), \qquad (u_1,\dots,u_d) \in (0,1)^d.$$

In order for this density to exist and to be continuous, the spectral measure $H$ should be absolute continuous with continuous Radon–Nikodym derivative on all $2^d - 1$ faces of the unit simplex with respect to the Hausdorff measure of the appropriate dimension [14]. In dimension $d = 2$, the Pickands dependence function $A : [0,1] \to [1/2, 1]$ should be twice continuously differentiable on $(0,1)$, or equivalently, the spectral measure $H$ should have a continuous density on $(0,1)$ (after identification of the unit simplex in $\mathbb{R}^2$ with the unit interval).

In case the margins are unknown, they may be estimated by the (properly rescaled) empirical distribution functions

$$\hat{F}_{nj}(x) = \frac{1}{n+1} \sum_{i=1}^{n} I(X_{ij} \leqslant x), \qquad x \in \mathbb{R},\ j \in \{1,\dots,d\}. \tag{6.14}$$

(The denominator is $n+1$ rather than $n$ in order to avoid boundary effects in the pseudo-loglikelihood below.) Estimation of $\theta$ then proceeds by maximizing the pseudo-loglikelihood

$$\sum_{i=1}^{n} \log c_\theta\big(\hat{F}_{n1}(X_{i1}),\dots,\hat{F}_{nd}(X_{id})\big),$$

see [36]. The resulting estimator is consistent and asymptotically normal, and its asymptotic variance can be estimated consistently.

If the margins are modelled parametrically as well, a fully parametric model for the joint distribution $F$ arises, and the parameter vector of $F$ may be estimated by ordinary maximum likelihood. An explicit expression for the $5 \times 5$ Fisher information matrix for the bivariate distribution with Weibull margins and Gumbel copula is calculated in [74]. A multivariate extension and with arbitrary generalized extreme value margins is presented in [86].

Special attention to the boundary case of independence is given in [94]. In this case, the dependence parameter lies on the boundary of the parameter set and the Fisher information matrix is singular, implying the normal assumptions for validity of the likelihood method are no longer valid.

A robustified version of the maximum likelihood estimator is introduced in [21]. The effects of misspecification of the dependence structure are studied in [22].

### 6.5.2 Nonparametric Estimation

For simplicity, we restrict attention here to the bivariate case. For multivariate extensions, see [43, 97].

Let $(X_1,Y_1),\dots,(X_n,Y_n)$ be an independent random sample from a bivariate distribution $F$ with extreme-value copula $C$ and Pickands dependence function $A$. Assume for the moment that the marginal distribution functions $F_1$ and $F_2$ are known and put $U_i = F_1(X_i)$ and $V_i = F_2(Y_i)$ and $S_i = -\log U_i$ and $T_i = -\log V_i$. Note that $S_i$ and $T_i$ are standard exponential random variables. For $t \in [0,1]$, put

$$\xi_i(t) = \min\left(\frac{S_i}{1-t}, \frac{T_i}{t}\right),$$

with the obvious conventions for division by zero. A characterizing property of extreme-value copulas is that the distribution of $\xi_i(t)$ is exponential as well, now with mean $1/A(t)$: for $x > 0$,

$$\begin{aligned} \mathrm{P}[\xi_i(t) > x] &= \mathrm{P}[U_i < e^{-(1-t)x}, V_i < e^{-tx}] \\ &= C(e^{-(1-t)x}, e^{-tx}) = e^{-xA(t)}. \end{aligned} \tag{6.15}$$

This fact leads straightforwardly to the original Pickands estimator [80]:

$$\frac{1}{\hat{A}^P(t)} = \frac{1}{n}\sum_{i=1}^{n} \xi_i(t). \tag{6.16}$$

A major drawback of this estimator is that it does not verify any of the constraints imposed on the family of the Pickands dependence functions in Theorem 6.2.3.

Besides establishing the asymptotic properties of the original Pickands estimator, Deheuvels [18] proposed an improvement of the Pickands estimator that at least verifies the endpoint constraints $A(0) = A(1) = 1$:

$$\frac{1}{\hat{A}^D(t)} = \frac{1}{n}\sum_{i=1}^{n} \xi_i(t) - t\frac{1}{n}\sum_{i=1}^{n} \xi_i(1) - (1-t)\frac{1}{n}\sum_{i=1}^{n} \xi_i(0) + 1. \tag{6.17}$$

As shown in [85], the weights $(1-t)$ and $t$ in the Deheuvels estimator (6.17) can be understood as pragmatic choices that could be replaced by suitable weight functions $\beta_1(t)$ and $\beta_2(t)$:

$$\frac{1}{\hat{A}^D(t)} = \frac{1}{n}\sum_{i=1}^{n} \xi_i(t) - \beta_1(t)\frac{1}{n}\sum_{i=1}^{n} \xi_i(1) - \beta_2(t)\frac{1}{n}\sum_{i=1}^{n} \xi_i(0) + 1. \tag{6.18}$$

The linearity of the right-hand side of (6.17) in $\xi_i(t)$ suggests to estimate the variance-minimizing weight functions via a linear regression of $\xi_i(t)$ upon $\xi_i(0)$ and $\xi_i(1)$:

$$\xi_i(t) = \beta_0(t) + \beta_1(t)\{\xi_i(0) - 1\} + \beta_2(t)\{\xi_i(1) - 1\} + \varepsilon_i(t).$$

The estimated intercept $\hat{\beta}_0(t)$ corresponds to the minimum-variance estimator for $1/A(t)$ in the class of estimators (6.18).

In the same spirit, Hall and Tajvidi [52] proposed another approach to improve the small-sample properties of the Pickands estimator at the boundary points. For all $t \in [0,1]$ and $i \in \{1,\dots,n\}$, define

$$\bar{\xi}_i(t) = \min\left(\frac{\bar{S}_i}{1-t}, \frac{\bar{T}_i}{t}\right)$$

with

$$\bar{S}_i = \frac{S_i}{\frac{1}{n}(S_1 + \cdots + S_n)}, \qquad \bar{T}_i = \frac{T_i}{\frac{1}{n}(T_1 + \cdots + T_n)}.$$

The estimator presented in [52] is given by

$$\frac{1}{\hat{A}^{HT}(t)} = \frac{1}{n}\sum_{i=1}^{n} \bar{\xi}_i(t).$$

Not only does the estimator's construction guarantee that the endpoint conditions are verified, in addition it always verifies the constraint $\hat{A}^{HT}(t) \geqslant 1 \vee (1-t)$. Among the three nonparametric estimators mentioned so far, the Hall–Tajvidi estimator typically has the smallest asymptotic variance.

A different starting point was chosen by Capéraà, Fougères and Genest [7]: they showed that the distribution function of the random variable $Z_i = \log(U_i)/\log(U_iV_i)$ is given by

$$\mathrm{P}(Z_i \leqslant z) = z + z(1-z)\frac{A'(z)}{A(z)}, \qquad 0 \leqslant z < 1,$$

where $A'$ denotes the right-hand derivative of $A$. Solving the resulting differential equation for $A$ and replacing unknown quantities by their sample versions yields the CFG-estimator. In [85] however, it was shown that the estimator admits the simpler representation

$$\log \hat{A}^{CFG}(t) = -\frac{1}{n}\sum_{i=1}^{n}\log\xi_i(t) - (1-t)\frac{1}{n}\sum_{i=1}^{n}\log\xi_i(0) - t\frac{1}{n}\sum_{i=1}^{n}\log\xi_i(1) \tag{6.19}$$

for $t \in [0,1]$. This expression can be seen as a sample version of

$$\mathrm{E}[-\log\xi_i(t)] = \log A(t) + \gamma, \qquad t \in [0,1],$$

a relation which follows from (6.15); note that the Euler–Mascheroni constant $\gamma = 0.5772\ldots$ is equal to the mean of the standard Gumbel distribution. Again, the weights $(1-t)$ and $t$ in (6.19) can be replaced by variance-minimizing weight functions that are to be estimated from the data [43, 85]. The CFG-estimator is consistent and asymptotically normal as well, and simulations indicate that it typically performs better than the Pickands estimator and the variants by Deheuvels and Hall–Tajvidi.

Theoretical results for extreme values in the case of *unknown margins* are quite recent. To some extent, Jiménez, Villa-Deharce and Flores [59] were the first to present an in-depth treatment of this situation. However, their main theorem on uniform consistency is established under conditions that are unnecessarily restrictive. In [39], asymptotic results were established under much weaker conditions. The estimators are the same as the ones presented above, the only difference being that $U_i = F_1(X_i)$ and $V_i = F_2(Y_i)$ are replaced by

$$\hat{U}_i = \hat{F}_{n1}(X_i) = \frac{1}{n+1}\sum_{k=1}^{n} I(X_k \leqslant X_i), \qquad \hat{V}_i = \hat{F}_{n2}(Y_i) = \frac{1}{n+1}\sum_{k=1}^{n} I(Y_k \leqslant Y_i),$$

with $\hat{F}_{nj}$ as in (6.14). Observe that the resulting estimators are entirely rank-based. Contrary to the case of known margins, the endpoint-corrections are irrelevant in the sense that they do not show up in the asymptotic distribution. Again, the CFG-estimator has the smallest asymptotic variance most of the time.

The previous estimators do typically not fulfill the *shape constraints* on $A$ as given in Theorem 6.2.3. A natural way to enforce these constraints is by modifying a pilot estimate $\hat{A}$ into the convex minorant of $\big(\hat{A}(t) \vee (1-t) \vee t\big) \wedge 1$, see [18, 59, 80]. It can be shown that this transformation cannot cause the $L^\infty$ error of the estimator to increase. A different way to impose the shape constraints is by constrained spline smoothing [1, 52] or by constrained kernel estimation of the derivative of $A$ [89].

The $L^2$-viewpoint was chosen in [28]. The set $\mathscr{A}$ of Pickands dependence functions being a closed and convex subset of the space $L^2([0,1], \mathrm{d}x)$, it is possible to find for a pilot estimate $\hat{A}$ a Pickands dependence function $A \in \mathscr{A}$ that minimizes the $L^2$-distance $\int_0^1 (\hat{A} - A)^2$. By general properties of orthogonal projections, the $L^2$-error of the projected estimator cannot increase.

Finally, a nonparametric Bayesian approach has been proposed by Guillotte and Perron [45]. Driven by a nonparametric likelihood, their methodology yields an estimator with good properties: its estimation error is typically small, it automatically verifies the shape constraints, and it blends naturally with parametric likelihood methods for the margins.

## 6.6 Further Reading

About the first monograph to treat multivariate extreme-value dependence is the one by Galambos [32], with a major update in the second edition [33]. Extreme-value copulas are treated extensively in the monographs [3, 65] and briefly in the 2006 edition of Nelsen's book [73]; see also Chap. 8 in this volume [58]. The regular-variation approach to multivariate extremes is emphasized in the books by Resnick

[81, 82] and de Haan and Ferreira [50]. A highly readable introduction to extreme-value analysis is the book by Coles [13].

The first representations of bivariate extreme-value distributions are due to Finkelstein [29], Tiago de Oliveira [76], Geffroy [34, 35] and Sibuya [87]. Incidentally, the 1959 paper by Geffroy appeared in the same issue as the famous paper by Sklar [88]. The equivalence of all these representations was shown in Gumbel [48]; see also the more recent paper by Obretenov [75]. However, their representations of multivariate extreme value distributions have not enjoyed the same success as the one proposed by Pickands [80]. The domain of attraction condition seems to have been formulated for the first time by Berman [4], his standardization being to the standard exponential distribution rather than the uniform one.

A particular class of extreme-value copulas arises if the spectral measure $H$ in Theorem 6.2.2 is discrete. In that case, the stable tail dependence function $\ell$ and the Pickands dependence function $A$ are piecewise linear. In general, such distributions arise from max-linear combinations of independent random variables [83]. An early example of such a distribution is the bivariate model studied by Tiago de Oliveira in [77–79], which has a spectral measure with exactly two atoms; see also [26]. The bivariate distribution of Marshall and Olkin [70] corresponds to a spectral measure with exactly three atoms, $\{0, 1/2, 1\}$ (after identification of the unit simplex in $\mathbb{R}^2$ with the unit interval); see [69] for a multivariate extension.

Even more challenging than the estimation problem considered in Sect. 6.5 is when the random sample comes from a distribution which is merely in the domain of attraction of a multivariate extreme-value distribution. See for instance [5, 14, 26, 51, 62–64, 66] for some (semi-)parametric approaches and [1, 6, 24, 25, 27, 84] for some nonparametric ones.

For an overview of software related to extreme value analysis, see [91]. Particularly useful are the `R` packages `evd` [92], which provides algorithms for the computation, simulation [93] and estimation of certain univariate and multivariate extreme-value distributions, as well as the more general `copula` package [96].

**Acknowledgements** The authors' research was supported by IAP research network grant nr. P6/03 of the Belgian government (Belgian Science Policy) and by contract nr. 07/12/002 of the Projet d'Actions de Recherche Concertées of the Communauté française de Belgique, granted by the Académie universitaire Louvain.

## References

1. Abdous, B., Ghoudi, K.: Non-parametric estimators of multivariate extreme dependence functions. J. Nonparametr. Stat. **17**(8), 915–935 (2005)
2. Asimit, A.V., Jones, B.L.: Extreme behavior of bivariate elliptical distributions. Insur. Math. Econ. **41**, 53–61 (2007)
3. Beirlant, J., Goegebeur, Y., Segers, J., Teugels, J.: Statistics of Extremes: Theory and Applications. Wiley Series in Probability and Statistics. Wiley, Chichester (2004)
4. Berman, S.M.: Convergence to bivariate limiting extreme value distributions. Ann. Inst. Stat. Math. **13**(3), 217–223 (1961/1962)

5. Boldi, M.O., Davison, A.C.: A mixture model for multivariate extremes. J. Roy. Stat. Soc. Ser. B **69**(2), 217–229 (2007)
6. Capéraà, P., Fougères, A.L.: Estimation of a bivariate extreme value distribution. Extremes **3**, 311–329 (2000)
7. Capéraà, P., Fougères, A.L., Genest, C.: A nonparametric estimation procedure for bivariate extreme value copulas. Biometrika **84**, 567–577 (1997)
8. Capéraà, P., Fougères, A.L., Genest, C.: Bivariate distributions with given extreme value attractor. J. Multivar. Anal. **72**, 30–49 (2000)
9. Cebrián, A., Denuit, M., Lambert, P.: Analysis of bivariate tail dependence using extreme values copulas: an application to the SOA medical large claims database. Belgian Actuar. J. **3**(1), 33–41 (2003)
10. Charpentier, A., Segers, J.: Convergence of Archimedean copulas. Stat. Probab. Lett. **78**, 412–419 (2008)
11. Charpentier, A., Segers, J.: Tails of multivariate Archimedean copulas. J. Multivar. Anal. **100**, 1521–1537 (2009)
12. Choroś, B., Ibragimov, R., Permiakova, E.: Copula estimation. In: Jaworski, P., Durante, F., Härdle, W., Rychlik, T. (eds.) Copula Theory and Its Applications, Proceedings of the Workshop, Warsaw, 25–26 Sept 2009. Springer, Dordrecht (2010)
13. Coles, S.: An Introduction to Statistical Modeling of Extreme Values. Springer Series in Statistics. Springer London Ltd., London (2001)
14. Coles, S.G., Tawn, J.A.: Modelling extreme multivariate events. J. Roy. Stat. Soc. Ser. B **53**(2), 377–392 (1991)
15. Crowder, M.: A multivariate distribution with Weibull connections. J. Roy. Stat. Soc. Ser. B **51**(1), 93–107 (1989)
16. de Haan, L., Resnick, S.I.: Limit theorem for multivariate sample extremes. Zeitschrift für Wahrscheinlichkeitstheorie und verwandte Gebiete **40**, 317–337 (1977)
17. Deheuvels, P.: Probabilistic aspects of multivariate extremes. In: Tiago de Oliveira, J. (ed.) Statistical Extremes and Applications, pp. 117–130. Reidel, Dordrecht (1984)
18. Deheuvels, P.: On the limiting behavior of the Pickands estimator for bivariate extreme-value distributions. Stat. Probab. Lett. **12**(5), 429–439 (1991)
19. Demarta, S., McNeil, A.: The t-copula and related copulas. Int. Stat. Rev. **73**, 111–129 (2005)
20. Drees, H., Huang, X.: Best attainable rates of convergence for estimates of the stable tail dependence function. J. Multivar. Anal. **64**, 25–47 (1998)
21. Dupuis, D.J., Morgenthaler, S.: Robust weighted likelihood estimators with an application to bivariate extreme value problems. Can. J. Stat. **30**(1), 17–36 (2002)
22. Dupuis, D.J., Tawn, J.A.: Effects of mis-specification in bivariate extreme value problems. Extremes **4**, 315–330 (2001)
23. Durante, F., Sempi, S.: Copula theory: an introduction. In: Jaworski, P., Durante, F., Härdle, W., Rychlik, T. (eds.) Copula Theory and Its Applications, Proceedings of the Workshop, Warsaw, 25–26 Sept 2009, Springer, Dordrecht (2010)
24. Einmahl, J.H.J., de Haan, L., Li, D.: Weighted approximations to tail copula processes with application to testing the bivariate extreme value condition. Ann. Stat. **34**(4), 1987–2014 (2006)
25. Einmahl, J.H.J., de Haan, L., Piterbarg, V.I.: Nonparametric estimation of the spectral measure of an extreme value distribution. Ann. Stat. **29**(5), 1401–1423 (2001)
26. Einmahl, J.H.J., Krajina, A., Segers, J.: A method of moments estimator of tail dependence. Bernoulli **14**(4), 1003–1026 (2008)
27. Einmahl, J.H.J., Segers, J.: Maximum empirical likelihood estimation of the spectral measure of an extreme-value distribution. Ann. Stat. **37**(5B), 2953–2989 (2009)
28. Fils-Villetard, A., Guillou, A., Segers, J.: Projection estimators of Pickands dependence functions. Can. J. Stat. **36**(3), 369–382 (2008)
29. Finkelstein, B.V.: On the limiting distributions of the extreme terms of a variational series of a two-dimensional random quantity. Dokladi Akademia SSSR **91**(2), 209–211 (1953). In Russian

30. Fougères, A.L., Nolan, J.P., Rootzén, H.: Models for dependent extremes using stable mixtures. Scand. J. Stat. **36**, 42–59 (2009)
31. Galambos, J.: Order statistics of samples from multivariate distributions. J. Am. Stat. Assoc. **70**(351, Part 1), 674–680 (1975)
32. Galambos, J.: The Asymptotic Theory of Extreme Order Statistics. Wiley Series in Probability and Mathematical Statistics. Wiley, New York, NY, Chichester, Brisbane (1978)
33. Galambos, J.: The Asymptotic Theory of Extreme Order Statistics, 2nd edn. Robert E. Krieger Publishing Co. Inc., Melbourne, FL (1987)
34. Geffroy, J.: Contributions a la théorie des valeurs extrêmes. Publ. Inst. Stat. Univ. Paris **7**, 37–121 (1958)
35. Geffroy, J.: Contributions a la théorie des valeurs extrêmes. Publ. Inst. Stat. Univ. Paris **8**, 123–184 (1959)
36. Genest, C., Ghoudi, K., Rivest, L.P.: A semiparametric estimation procedure of dependence parameters in multivariate families of distributions. Biometrika **82**(3), 543–552 (1995)
37. Genest, C., Rivest, L.P.: A characterization of Gumbel's family of extreme value distributions. Stat. Probab. Lett. **8**(3), 207–211 (1989)
38. Genest, C., Rivest, L.P.: Statistical inference procedures for bivariate Archimedean copulas. J. Am. Stat. Assoc. **88**, 1034–1043 (1993)
39. Genest, C., Segers, J.: Rank-based inference for bivariate extreme-value copulas. Ann. Stat. **37**(5B), 2990–3022 (2009)
40. Ghorbal, N.B., Genest, C., Nešlehová, J.: On the Ghoudi, Khoudraji, and Rivest test for extreme-value dependence. Can. J. Stat. **37**(4), 534–552 (2009)
41. Ghoudi, B., Fougères, A.L., Ghoudi, K.: Extreme behavior for bivariate elliptical distributions. Can. J. Stat. **33**(3), 317–334 (2005)
42. Ghoudi, K., Khoudraji, A., Rivest, L.P.: Propriétés statistiques des copules de valeurs extrêmes bidimensionnelles. Can. J. Stat. **26**(1), 187–197 (1998)
43. Gudendorf, G., Segers, J.: Nonparametric estimation of an extreme-value copula in arbitrary dimensions. Technical Report DP0923, Institut de statistique, Université catholique de Louvain, Louvain-la-Neuve (2009). http://www.uclouvain.be/stat.arXiv:0910.0845v1 [math.ST]
44. Guillem, A.G.: Structure de dépendance des lois de valeurs extrêmes bivariées. C. R. Acad. Sci. Paris Série I **330** 593–596 (2000)
45. Guillotte, S., Perron, F.: A Bayesian estimator for the dependence function of a bivariate extreme-value distribution. Can. J. Stat. **36**(3), 383–396 (2008)
46. Gumbel, E.J.: Bivariate exponential distributions. J. Am. Stat. Assoc. **55**, 698–707 (1960)
47. Gumbel, E.J.: Bivariate logistic distributions. J. Am. Stat. Assoc. **56**, 335–349 (1961)
48. Gumbel, E.J.: Multivariate extremal distributions. Bull. Inst. Internat. Stat. **39**(livraison 2), 471–475 (1962)
49. Gumbel, E.J., Goldstein, N.: Analysis of empirical bivariate extremal distributions. J. Am. Stat. Assoc. **59**(307), 794–816 (1964)
50. de Haan, L., Ferreira, A.: Extreme Value Theory. Springer Series in Operations Research and Financial Engineering. Springer, New York, NY (2006)
51. de Haan, L., Neves, C., Peng, L.: Parametric tail copula estimation and model testing. J. Multivar. Anal. **99**(6), 1260–1275 (2008)
52. Hall, P., Tajvidi, N.: Distribution and dependence-function estimation for bivariate extremevalue distributions. Bernoulli **6**(5), 835–844 (2000)
53. Hashorva, E.: Extremes of asymptotically spherical and elliptical random vectors. Insur. Math. Econ. **36**, 285–302 (2005)
54. Hougaard, P.: A class of multivariate failure time distributions. Biometrika **73**(3), 671–678 (1986)
55. Hürlimann, W.: Hutchinson–Lai's conjecture for bivariate extreme value copulas. Stat. Probab. Lett. **61**, 191–198 (2003)
56. Hüsler, J., Reiss, R.: Maxima of normal random vectors: Between independence and complete dependence. Stat. Probab. Lett. **7**, 283–286 (1989)

57. Hutchinson, T., Lai, C.: Continuous Bivariate Distributions, Emphasizing Applications. Rumbsy Scientific, Adelaide (1990)
58. Jaworski, P.: Tail behaviour of copulas. In: Jaworski, P., Durante, F., Härdle, W., Rychlik, T. (eds.) Copula Theory and Its Applications, Proceedings of the Workshop, Warsaw, 25–26 Sept 2009, Springer, Dordrecht (2010)
59. Jiménez, J.R., Villa-Diharce, E., Flores, M.: Nonparametric estimation of the dependence function in bivariate extreme value distributions. J. Multivar. Anal. **76**(2), 159–191 (2001)
60. Joe, H.: Families of min-stable multivariate exponential and multivariate extreme value distributions. Stat. Probab. Lett. **9**(1), 75–81 (1990)
61. Joe, H.: Multivariate extreme-value distributions with applications to environmental data. Can. J. Stat. **22**(1) (1994)
62. Joe, H., Smith, R.L., Weissman, I.: Threshold methods for extremes. J. Roy. Stat. Soc. Ser B **54**, 171–183 (1992)
63. Klüppelberg, C., Kuhn, G., Peng, L.: Estimating the tail dependence function of an elliptical distribution. Bernoulli **13**(1), 229–251 (2007)
64. Klüppelberg, C., Kuhn, G., Peng, L.: Semi-parametric models for the multivariate tail dependence function—the asymptotically dependent case. Scand. J. Stat. **35**(4), 701–718 (2008)
65. Kotz, S., Nadarajah, S.: Extreme Value Distributions. Imperial College Press, London (2000). Theory and applications
66. Ledford, A.W., Tawn, J.A.: Statistics for near independence in multivariate extreme values. Biometrika **83**(1), 169–187 (1996)
67. Li, H.: Orthant tail dependence of multivariate extreme value distributions. J. Multivar. Anal. **100**, 243–256 (2009)
68. Longin, F., Solnik, B.: Extreme correlation of international equity markets. J. Finance **56**(2), 649–676 (2001)
69. Mai, J.F., Scherer, M.: Lévy-frailty copulas. J. Multivar. Anal. **100**(7), 1567–1585 (2009)
70. Marshall, A.W., Olkin, I.: A multivariate exponential distribution. J. Am. Stat. Assoc. **62**, 30–44 (1967)
71. McFadden, D.: Modelling the choice of residential location. In: Karlquist, A. (ed.) Spatial interaction Theory and Planning Models, pp. 75–96. North-Holland, Amsterdam (1978)
72. McNeil, A., Nešlehová, J.: Multivariate archimedean copulas, $d$-monotone functions and $\lambda_1$-norm symmetric distributions. Ann. Stat. **37**(5B), 3059–3097 (2009)
73. Nelsen, R.B.: An Introduction to Copulas, 2nd edn. Springer Series in Statistics. Springer, New York, NY (2006)
74. Oakes, D., Manatunga, A.K.: Fisher information for a bivariate extreme value distribution. Biometrika **79**(4), 827–832 (1992)
75. Obretenov, A.: On the dependence function of sibuya in multivariate extreme value theory. J. Multivar. Anal. **36**(1), 35–43 (1991)
76. Tiago de Oliveira, J.: Extremal distributions. Revista Faculdade de Ciencias de Lisboa **7**, 219–227 (1958)
77. Tiago de Oliveira, J.: Regression in the nondifferentiable bivariate extreme models. J. Am. Stat. Assoc. **69**, 816–818 (1974)
78. Tiago de Oliveira, J.: Bivariate extremes: foundations and statistics. In: Multivariate Analysis, V (Proc. Fifth Internat. Sympos., Univ. Pittsburgh, Pittsburgh, PA, 1978), pp. 349–366. North-Holland, Amsterdam (1980)
79. Tiago de Oliveira, J.: Statistical decision for bivariate extremes. In: Extreme Value Theory (Oberwolfach, 1987). Lecture Notes in Statistics, vol. 51, pp. 246–261. Springer, New York, NY (1989)
80. Pickands, J.: Multivariate extreme value distributions, Proceedings of the 43rd session of the International Statistical Institute, vol. 2 (Buenos Aires, 1981), vol. 49, pp. 859–878, 894–902 (1981). With a discussion
81. Resnick, S.I.: Extreme Values, Regular Variation and Point Processes. Springer Series in Operations Research and Financial Engineering, vol. 4. Springer, New York, NY (1987)

82. Resnick, S.I.: Heavy-Tail Phenomena. Springer Series in Operations Research and Financial Engineering. Springer, New York, NY (2007). Probabilistic and statistical modeling
83. Schlather, M., Tawn, J.A.: Inequalities for the extremal coefficients of multivariate extreme value distributions. Extremes **5**, 87–102 (2002)
84. Schmidt, R., Stadtmüller, U.: Non-parametric estimation of tail dependence. Scand. J. Stat. **33**(2), 307–335 (2006)
85. Segers, J.: Non-parametric inference for bivariate extreme-value copulas. In: Ahsanulah, M., Kirmani, S. (eds.) Extreme Value Distributions, Chap. 9, pp. 181–203. Nova Science Publishers, Inc., New York, NY (2007). Older version available as CentER DP 2004–91, Tilburg University
86. Shi, D.: Fisher information for a multivariate extreme value distribution. Biometrika **82**(3), 644–649 (1995)
87. Sibuya, M.: Bivariate extreme statistics, I. Ann. Inst. Stat. Math. **11**, 195–210 (1960)
88. Sklar, A.: Fonctions de répartition à $n$ dimensions et leurs marges. Publ. Inst. Statist. Univ. Paris **8**, 229–231 (1959)
89. Smith, R.L., Tawn, J.A., Yuen, H.K.: Statistics of multivariate extremes. Int. Stat. Rev. / Revue Internationale de Statistique **58**(1), 47–58 (1990)
90. Stărică, C.: Multivariate extremes for models with constant conditional correlations. J. Empir. Finance **6**(5), 515–553 (1999)
91. Stephenson, A., Gilleland, E.: Software for the analysis of extreme events: the current state and future directions. Extremes **8**(3), 87–109 (2005)
92. Stephenson, A.G.: evd: Extreme value distributions. R News **2**(2), June (2002). http://CRAN.R-project.org/doc/Rnews/
93. Stephenson, A.G.: Simulating multivariate extreme value analysis of logistic type. Extremes **6**, 49–59 (2003)
94. Tawn, J.A.: Extreme value theory: models and estimation. Biometrika **75**, 397–415 (1988)
95. Toulemonde, G., Guillou, A., Naveau, P., Vrac, M., Chevalier, F.: Autoregressive models for maxima and their applications to $CH_4$ and $N_2O$. Environmetrics (2009). DOI 10.1002/env.992
96. Yan, J.: Enjoy the joy of copulas: with a package copula. J. Stat. Softw. **21**(4), 1–21 (2007)
97. Zhang, D., Wells, M.T., Peng, L.: Nonparametric estimation of the dependence function for a multivariate extreme value distribution. J. Multivar. Anal. **99**(4), 577–588 (2008)

# Chapter 7
# Construction and Sampling of Nested Archimedean Copulas

Marius Hofert

**Abstract** Nested Archimedean copulas are explicit copulas which generalize Archimedean copulas to allow for asymmetries. Starting with completely monotone Archimedean generators, it is usually not clear when the corresponding Archimedean copulas can be nested to build indeed a proper copula. This article presents some results about the construction of nested Archimedean copulas. The presented construction principles are directly linked to sampling algorithms, which are also discussed in this work.

## 7.1 Introduction

Archimedean copulas are considered to be one of the most famous classes of copulas. This has several reasons. In contrast to elliptical copulas, Archimedean copulas are not constructed via Sklar's Theorem from known multivariate distributions. One rather starts with a nice functional form in terms of a one-place real function, the Archimedean generator, and asks for properties required in order to obtain a proper copula. As a result of such a construction, Archimedean copulas can be written in explicit form, one of the main advantages of this class of copulas. Another advantage of Archimedean copulas is their great flexibility in modeling different kinds of dependencies. Again in contrast to elliptical copulas, Archimedean copulas are not restricted to radial symmetry. They are therefore able to capture different lower and upper tail dependencies, often a desired feature for modeling purposes, e.g. when common losses are observed to happen with a higher probability than common profits. Further, many known copula families are Archimedean, which further emphasizes the importance of this class of copulas.

Marius Hofert
Institute of Number Theory and Probability Theory, Ulm University, 89081 Ulm, Germany
e-mail: marius.hofert@uni-ulm.de

P. Jaworski et al. (eds.), *Copula Theory and Its Applications*, Lecture Notes in Statistics 198,
DOI 10.1007/978-3-642-12465-5_7, © Springer-Verlag Berlin Heidelberg 2010

Nested Archimedean copulas only recently appeared in the literature. Although [14] already investigate dependence structures which are even more general than nested Archimedean copulas, the hierarchical functional form and the term "nesting" first appears in [13, pp. 87]. Since then, nested Archimedean copulas appear at several points in the literature, including [10, 11, 19, 26, 31] in the context of sampling algorithms, [7, 24, 25, 27] mainly in the context of statistical inference, and [20, pp. 226] in the context of risk management. Other references are [4, 5, 9, 12] specifically in the context of credit risk applications. Due to their hierarchical structure, note that some of these authors refer to nested Archimedean copulas as hierarchical Archimedean copulas.

Nested Archimedean copulas are indeed generalizations of Archimedean copulas. As they allow for asymmetries and provide more flexibility, at the same time sharing the nice properties of Archimedean copulas such as an explicit functional form, nested Archimedean copulas are an interesting class of copulas to study. They can easily be incorporated in many known models, possibly replacing symmetric copulas there. There are several situations in which proper nested Archimedean copulas should be considered more appropriate than their symmetric counterparts. First and foremost, nested Archimedean copulas are often more adequate in large-dimensional problems, e.g. modeling a portfolio in more than, say, 50 dimensions and assuming perfectly equicorrelated components usually contradicts patterns found in real-life data. It is frequently the case that some portfolio components are stronger correlated than others, being similarly affected by macroeconomic effects, political decisions, geographical regions, industrial cooperations, or consumer trends. A hierarchical structure for the portfolio model therefore often naturally emerges. Moreover, neglecting this structure is a potential source of errors, see, e.g., [12]. Note that there are other developments in copula theory that also take this into account, e.g. the pair-copula approach, see, e.g., [17] or [1].

The goal of this chapter is to give an introduction to the notion of nested Archimedean copulas and to sampling algorithms of such. As can be seen from the aforementioned references, efficient sampling algorithms for nested Archimedean copulas are of major interest, either in their own right or in applications such as credit-risk models. However, they are also linked to the construction of nested Archimedean copulas. The construction of bivariate Archimedean copulas, as well as their symmetric multivariate extensions, is well covered by the literature. To be more precise, [29] already obtained precise conditions on an Archimedean generator to lead to a proper bivariate copula. An extension of this result to multivariate Archimedean copulas was recently presented by [21]. Concerning nested Archimedean copulas, there is only a sufficient condition on the underlying Archimedean generators known such that the resulting nested structure is indeed a proper copula. This motivates the construction of nested Archimedean copulas. As such a construction is typically linked to a mixture representation, sampling routines naturally appear in the construction of nested Archimedean copulas.

This chapter is organized as follows. Section 7.2 introduces the notion of nested Archimedean copulas. Section 7.3 presents the sufficient nesting condition of [13, p. 88], also investigated by [19]. Further, we take a closer look at this condition and even slightly relax it. Based on this result, two conditions under which Archimedean generators can be mixed to build proper nested Archimedean copulas are presented in Sect. 7.4. Section 7.5 presents sampling algorithms for nested Archimedean copulas, both from a theoretical and practical point of view. Section 7.6 concludes.

## 7.2 Nested Archimedean Copulas

The study of nested Archimedean copulas is, by construction, intimately connected with the study of symmetric Archimedean copulas. The latter are therefore also addressed in what follows.

**Definition 7.2.1.** An *(Archimedean) generator* is a continuous, decreasing function $\psi : [0,\infty] \to [0,1]$ which satisfies $\psi(0) = 1$, $\psi(\infty) := \lim_{t\to\infty} \psi(t) = 0$, and which is strictly decreasing on $[0, \inf\{t : \psi(t) = 0\}]$; the set of all such functions is denoted by $\Psi$. An Archimedean generator $\psi \in \Psi$ is called *strict* if $\inf\{t : \psi(t) = 0\} = \infty$. A $d$-dimensional copula $C$ is called *Archimedean* if it permits the representation

$$C(u) = C(u;\psi) := \psi(\psi^{-1}(u_1) + \cdots + \psi^{-1}(u_d)),\ u \in I^d, \tag{7.1}$$

for some Archimedean generator $\psi \in \Psi$ with inverse $\psi^{-1} : [0,1] \to [0,\infty]$, where $\psi^{-1}(0) := \inf\{t : \psi(t) = 0\}$.

Note that the notation in terms of $\varphi = \psi^{-1}$ is also circulating in which case $\varphi$ is referred to as generator, see, e.g., [8, 28], and [22, p. 112]. For studying multivariate Archimedean copulas in more than two dimensions, it turns out to be more convenient to work with the representation (7.1) in terms of the generator $\psi$; this notation may be found e.g. in [13, p. 86] and [21].

If $C$ is defined recursively via (7.1), for $d = 2$, and via

$$C(u_1,\ldots,u_d;\psi_0,\ldots,\psi_{d-2}) := \psi_0(\psi_0^{-1}(u_1) + \psi_0^{-1}(C(u_2,\ldots,u_d;\psi_1,\ldots,\psi_{d-2}))), \tag{7.2}$$

for $d \geq 3$, $C$ is called *fully-nested Archimedean copula* with $d-1$ *nesting levels* or *hierarchies*. Combinations of fully-nested Archimedean copulas with symmetric Archimedean copulas are called *partially-nested Archimedean copulas*. Fully- and partially-nested Archimedean copulas are summarized as *nested* (or *hierarchical*) *Archimedean copulas*.

The most simple proper fully-nested Archimedean copula $C$ is obtained for $d = 3$ with two nesting levels, given by

$$\begin{aligned} C(u) &= C(u_1, C(u_2,u_3;\psi_1);\psi_0) \\ &= \psi_0\big(\psi_0^{-1}(u_1) + \psi_0^{-1}\big(\psi_1(\psi_1^{-1}(u_2) + \psi_1^{-1}(u_3))\big)\big),\ u \in I^3. \end{aligned} \tag{7.3}$$

Its structure can be depicted by a tree as given in Fig. 7.1.

In many applications, partially-nested Archimedean copulas of the form

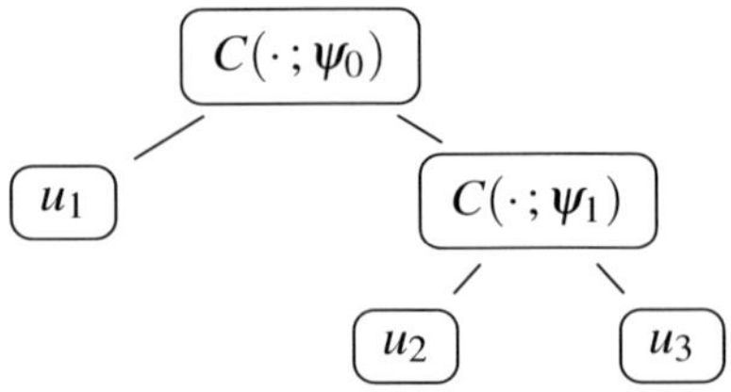


**Fig. 7.1** Tree structure of a fully-nested Archimedean copula.

$$\begin{aligned}C(u) &= C(C(u_{11},\dots,u_{1d_1};\psi_1),\dots,C(u_{S1},\dots,u_{Sd_S};\psi_S);\psi_0)\\ &= \psi_0(\psi_0^{-1}(\psi_1(\psi_1^{-1}(u_{11})+\dots+\psi_1^{-1}(u_{1d_1})))+\dots\\ &\quad+\psi_0^{-1}(\psi_S(\psi_S^{-1}(u_{S1})+\dots+\psi_S^{-1}(u_{Sd_S}))))\\ &= \psi_0\left(\sum_{s=1}^{S}\psi_0^{-1}\left(\psi_s\left(\sum_{l=1}^{d_s}\psi_s^{-1}(u_{sl})\right)\right)\right),\ u\in I^d \qquad (7.4)\end{aligned}$$

involving $S$ different *sectors* of dimensions $d_s$, $s\in\{1,\dots,S\}$, with $\sum_{s=1}^{S} d_s = d$ often arise naturally. The corresponding tree representation is given in Fig. 7.2. One can

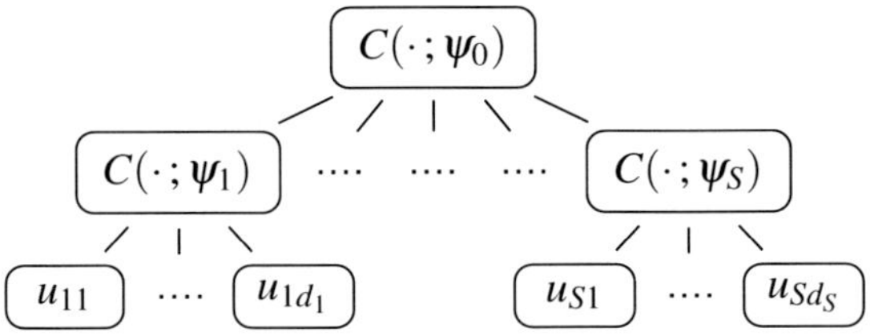


**Fig. 7.2** Tree structure of a partially-nested Archimedean copula.

think of the partially-nested Archimedean copula (7.4) as capturing the dependence among components in the same sector by some symmetric Archimedean copula and combining these sector copulas by an overall dependence structure, again of the symmetric Archimedean type. To put it in other words, all random variables belonging to the same sector $s$, $s\in\{1,\dots,S\}$, have the sector copula generated by $\psi_s$ as dependence structure, whereas random variables belonging to different sectors have the copula generated by $\psi_0$ as common dependence structure.

Note that nested Archimedean copulas are quite flexible, there are already $d!/k!$ possibilities of combining $d$ random variables according to the fully-nested Archimedean copula (7.2) with a $k$-dimensional symmetric Archimedean copula on the lowest level, $k\in\{2,\dots,d\}$. The corresponding copula has $d-k+1$ parameters i.e. possibly different dependencies among the random variables.

## 7.3 A Sufficient Nesting Condition

Concerning the construction of nested Archimedean copulas, the first natural question to ask is, under which conditions Archimedean copulas can be nested such that proper copulas result. To the best of our knowledge, there is only a sufficient condition known. For explaining it, we need several notions of monotonicity, defined as follows.

**Definition 7.3.1.** Let $a,b \in \bar{\mathbb{R}} := \mathbb{R} \cup \{-\infty,\infty\}$ with $a < b$ and $f : [a,b] \to \mathbb{R}$ with $f(-\infty) := \lim_{x\to-\infty} f(x)$ if $a = -\infty$ and $f(\infty) := \lim_{x\to\infty} f(x)$ if $b = \infty$ in which case the limits are assumed to exist in the improper sense. Then $f$ is called *absolutely monotone* on $[a,b]$ if it is continuous and admits derivatives of all orders satisfying $f^{(k)}(x) \geq 0$ for all $x \in (a,b)$ and $k \in \mathbb{N}_0$; $f$ is called *d-monotone*, $d \geq 2$, on $[a,b]$ if it is continuous and admits derivatives up to the order $d-2$ satisfying $(-1)^k f^{(k)}(x) \geq 0$ for all $k \in \{0,\ldots,d-2\}$, $x \in (a,b)$, and $(-1)^{d-2} f^{(d-2)}(x)$ is decreasing and convex on $(a,b)$; $f$ is called *completely monotone* on $[a,b]$ if it is continuous and admits derivatives of all orders satisfying $(-1)^k f^{(k)}(x) \geq 0$ for all $x \in (a,b)$ and $k \in \mathbb{N}_0$. If the respective monotonicity holds on the whole domain of $f$, the interval is dropped for simplicity.

Important properties of absolutely monotone and completely monotone functions are summarized in the following proposition. Part 1 may be found in [6, p. 441]. Part 2 is contained in [32, p. 145]. For Part 3, see [13, p. 374] and Bernstein's Theorem below.

**Proposition 7.3.1.** *1. If $f$ is completely monotone, $g$ nonnegative, and $g'$ completely monotone, then $f \circ g$ is completely monotone.*

*2. If $f$ is absolutely monotone and $g$ completely monotone, then $f \circ g$ is completely monotone.*

*3. If $f$ is completely monotone and $f(0) = 1$, then $f^\alpha$ is completely monotone for all $\alpha \in (0,\infty)$ if and only if $(-\log f)'$ is completely monotone.*

The following theorem, referred to as Bernstein's Theorem, is profound in that it establishes a connection between completely monotone functions on $[0,\infty]$ and Laplace-Stieltjes transforms of distribution functions on $[0,\infty]$, see [6, pp. 439].

**Theorem 7.3.1 ([3]).** *A continuous function $\psi : [0,\infty] \to [0,1]$ is the Laplace-Stieltjes transform of a distribution function $F$ on $[0,\infty]$, denoted by $\psi = \mathscr{LS}[F]$, if and only if $\psi$ is completely monotone on $[0,\infty]$ and $\psi(0) = 1$.*

Note that if $\psi \in \Psi$ is completely monotone with $\psi = \mathscr{LS}[F]$, then $\psi$ is strict and $F(0) = 0$. The following theorem due to [15] precisely characterizes those Archimedean generators, that generate a proper symmetric Archimedean copula in all dimensions $d \geq 2$. The claim of this theorem is already stated in [30], however, without a proof or reference to such. The if part of the proof was originally presented by [15] in the context of t-norms, see, e.g., [2, 16]. However, it can be significantly simplified by using Bernstein's Theorem and the closure of distribution functions under mixtures of products of such.

**Theorem 7.3.2 ([15]).** *Let $\psi \in \Psi$. Then (7.1) is a copula for any $d \geq 2$ if and only if $\psi$ is completely monotone.*

For a fixed dimension $d$, complete monotonicity of $\psi \in \Psi$ is not required for generating a proper $d$-dimensional Archimedean copula. As [21] show, it is precisely the notion of $d$-monotonicity that captures those Archimedean generators that generate a proper Archimedean copula in $d$ dimensions. In this context, the Laplace-Stieltjes transform is replaced by the Williamson $d$-transform, see [21] for more details. In what follows, we will mainly consider completely monotone Archimedean generators due to the connection with Laplace-Stieltjes transforms via Bernstein's Theorem. The subclass of completely monotone Archimedean generators is denoted by $\Psi_\infty$. Many known Archimedean generators belong to $\Psi_\infty$, hence the assumption of complete monotonicity is not very restrictive.

The construction of nested Archimedean copulas is already addressed in [13, p. 88] and implicitly contained in the more general framework of [14]. A more detailed insight into the argument is given in [19]. It is probably best described by considering the trivariate case (7.3) and it becomes clear from the argument that it applies to the general case as well. For this, let $\psi_0, \psi_1 \in \Psi_\infty$ with $\psi_0 = \mathscr{LS}[F_0]$ and $\psi_1 = \mathscr{LS}[F_1]$. For $x \in (0,\infty)$, let $G_0(u;x) := \exp\bigl(-x\psi_0^{-1}(u)\bigr)$, $u \in I$, and

$$\psi_{01}(t;x) := \exp\bigl(-x\psi_0^{-1}(\psi_1(t))\bigr),\ t \in [0,\infty]. \tag{7.5}$$

Further, assume the following *nesting condition* to hold:

$$(\psi_0^{-1} \circ \psi_1)' \text{ completely monotone.} \tag{7.6}$$

By Proposition 7.3.1, Part 1, this condition implies that $\psi_{01} \in \Psi_\infty$ for all $x \in (0,\infty)$. Note that the three-dimensional fully-nested Archimedean copula $C$ as listed in (7.3) can be written as

$$\begin{aligned} C(u) &= \psi_0\bigl(\psi_0^{-1}(u_1) + \psi_0^{-1}\bigl(\psi_1\bigl(\psi_1^{-1}(u_2) + \psi_1^{-1}(u_3)\bigr)\bigr)\bigr) \\ &= \int_0^\infty \exp(-x\psi_0^{-1}(u_1)) \exp\bigl(-x\psi_0^{-1}\bigl(\psi_1\bigl(\psi_1^{-1}(u_2) + \psi_1^{-1}(u_3)\bigr)\bigr)\bigr)\, dF_0(x) \\ &= \int_0^\infty G_0(u_1;x)\,\psi_{01}\bigl(\psi_1^{-1}(u_2) + \psi_1^{-1}(u_3);x\bigr)\, dF_0(x) \\ &= \int_0^\infty G_0(u_1;x)\,\psi_{01}\bigl(\psi_{01}^{-1}(G_0(u_2;x);x) + \psi_{01}^{-1}(G_0(u_3;x);x);x\bigr)\, dF_0(x). \end{aligned} \tag{7.7}$$

From this representation it follows that $C$ is simply a mixture with respect to $F_0$ of the distribution function $G_0$ times the distribution function constructed via the bivariate Archimedean copula with generator $\psi_{01}$ and equal margins $G_0$. Hence, $C$ is a copula. The argument for the general case works similarly by deriving the corresponding mixture representation recursively, see [19]. In essence, for a general nested Archimedean structure to be indeed a proper copula, it is sufficient that all appearing nodes of the form $\psi_i^{-1} \circ \psi_j$ have completely monotone derivatives, i.e. satisfy the nesting condition (7.6), see [19].

Due to the functional forms of the copulas (7.3) and (7.4), we refer to $F_0 = \mathscr{LS}^{-1}[\psi_0]$ as *outer distribution function* and to $F_{01} = \mathscr{LS}^{-1}[\psi_{01}]$ and $F_{0s} = \mathscr{LS}^{-1}[\psi_{0s}]$, $s \in \{1,\dots,S\}$, as *inner distribution functions*. These functions will play an important role in constructing and sampling nested Archimedean copulas. For notational convenience, we mainly focus on the two functions $F_0 = \mathscr{LS}^{-1}[\psi_0]$ and $F_{01} = \mathscr{LS}^{-1}[\psi_{01}]$ related to the copula (7.3). As mentioned before, the given arguments readily apply to all nodes appearing in a general nested Archimedean structure.

*Remark 7.3.1.* It is clear from representation (7.7) that the nesting condition (7.6) is sufficient but not necessary. To see this, simply note that any generator $\psi_1$ which is not completely monotone but such that $\psi_{01}$ generates a proper two-dimensional Archimedean copula leads to the proper copula representation (7.7), e.g. take $\psi_0(t) = \exp(-t)$, a generator of the independence copula, and $\psi_1(t) = \max\{1-t, 0\}$, a proper 2-monotone Archimedean generator. In generalizing this argument, it follows that at least on the lowest level, nesting conditions such as (7.6) may be weakened. However, if one requires $\psi_{01}(\cdot\,;x)$, $x \in (0,\infty)$, to generate a proper symmetric Archimedean copula in all dimensions, i.e. if one requires $\psi_{01}$ to be in $\Psi_\infty$, then (7.6) is necessary and sufficient. Sufficiency is immediate by Proposition 7.3.1 Part 1 and necessity is clear by the following argument. It follows from (7.5) that $\psi_{01}(\cdot\,;x)$ is completely monotone for all $x \in (0,\infty)$ if and only if $\psi_{01}^x(\cdot\,;1)$ is completely monotone for all $x \in (0,\infty)$, which, by Proposition 7.3.1, Part 3, is equivalent to the nesting condition (7.6).

## 7.4 Construction of Nested Archimedean Copulas

Not every combination of generators $\psi_0, \psi_1 \in \Psi_\infty$ is known to lead to a valid nested Archimedean copula as the sufficient nesting condition does not always hold, see [10]. This section introduces two general construction principles such that the sufficient nesting condition (7.6) always holds and therefore proper nested Archimedean copulas result.

The first construction principle is based on the following simple observation. Assume $\psi_0^\alpha \in \Psi_\infty$ for all $\alpha \in (0,\infty)$. By choosing $\psi_1$ as

$$\psi_1(t) := \psi_0(-\log \psi(t)) \tag{7.8}$$

for some generator $\psi$ such that $\psi^\alpha \in \Psi_\infty$ for all $\alpha \in (0,\infty)$, then $\psi_0$ and $\psi_1$ do always fulfill the sufficient nesting condition (7.6). First, let us check that $\psi_1$ defined by (7.8) is a completely monotone Archimedean generator, i.e. that $\psi_1 \in \Psi_\infty$. By Proposition 7.3.1, Part 1, this holds if $-\log \psi$ has completely monotone derivative. By Part 3 of this proposition, this in turn happens if and only if $\psi_0^\alpha$ is completely monotone for all $\alpha \in (0,\infty)$, which we assumed to hold. Now let us check the nesting condition (7.6). Since $\psi_0^{-1}(\psi_1(t)) = -\log \psi(t)$, the complete monotonicity of $(\psi_0^{-1} \circ \psi_1)'$ and therefore the sufficient nesting condition (7.6) again follows from

Proposition 7.3.1, Part 3, due to the assumption $\psi^{\alpha} \in \Psi_\infty$ for all $\alpha \in (0,\infty)$. Note that this assumption on $\psi$ and $\psi_0$ can be considered as rather weak, e.g. all completely monotone Archimedean generators listed in Table 7.1 share this property.

By writing $\psi_1(t) = \psi_0(-\log \psi(t))$ as

$$\psi_1(t) = \int_0^\infty \psi^x(t)\, dF_0(x), \tag{7.9}$$

where $F_0 = \mathscr{L}\mathscr{S}^{-1}[\psi_0]$, we see that $\psi_1$ is simply the generator $\psi$ with randomized parameter $x$ according to the mixing distribution $F_0$. By a similar reasoning one also sees that $\psi_1(t) := \psi_0(\tilde{\psi}^{-1}(\psi(t)))$ is a proper Archimedean generator for any two generators $\psi, \tilde{\psi} \in \Psi_\infty$ that can be mixed according to the sufficient nesting condition (7.6). Further, $\psi_1$ defined in this way also makes sense for all generators $\psi \in \Psi_\infty$ as long as $\psi_0 \circ \tilde{\psi}^{-1}$ is absolutely monotone on $[0,1]$, see Proposition 7.3.1, Part 2.

There is another intuitive interpretation for the construction (7.8). This choice for $\psi_1$ implies that (7.5) takes the form

$$\psi_{01}(t;V_0) = \psi^{V_0}(t),\ t \in [0,\infty],$$

which means that $\psi_1$ is chosen such that $\psi_{01}(t;V_0)$ is simply a power in $\psi$. Starting with a generator $\psi$ such that $\psi^{\alpha} \in \Psi_\infty$ for all $\alpha \in (0,\infty)$, one may thus build a nested Archimedean copula. This idea generalizes to the following argument. Given $\psi_0$, choose $\psi \in \Psi$ such that $\psi_{01}$ as given in (7.5) defines a proper copula. Due to the mixture representation (7.7), a nested Archimedean copula will result.

The second construction principle is also based on a certain transformation of a given Archimedean generator $\psi \in \Psi_\infty$. The proof of the following theorem follows by definition of complete monotonicity and Proposition 7.3.1, Part 1.

**Theorem 7.4.1.** *Let* $\psi \in \Psi_\infty$.

1. *For all* $\vartheta \in [1,\infty)$ *and* $c \in [0,\infty)$, $\psi((c^{\vartheta}+t)^{1/\vartheta}-c) \in \Psi_\infty$.
2. *If for* $i \in \{0,1\}$, $\psi_i(t) = \psi((c^{\vartheta_i}+t)^{1/\vartheta_i}-c)$, $t \in [0,\infty]$, *with* $c \in [0,\infty)$ *and* $\vartheta_i \in [1,\infty)$, *then* $\psi_0^{-1}(\psi_1(t)) = (c^{\vartheta_1}+t)^{\vartheta_0/\vartheta_1} - c^{\vartheta_0}$, *i.e.* $\psi_0$ *and* $\psi_1$ *fulfill the sufficient nesting condition (7.6) for all* $\vartheta_0 \leq \vartheta_1$.

As we will see in the following section, on nesting two Archimedean generators belonging to the same parametric family the construction principle addressed in Theorem 7.4.1 often appears. The resulting generator $\psi_{01}$ is given by

$$\psi_{01}(t;V_0) = \exp(-V_0((c^{\vartheta_1}+t)^{\vartheta_0/\vartheta_1} - c^{\vartheta_0}))$$

By choosing $c = 0$, note that *outer* (or *exterior*) *power families*, see [22, p. 142], follow from this construction principle and hence can be used to build nested Archimedean copulas.

## 7.5 Sampling Nested Archimedean Copulas

In this section, we present sampling algorithms for Archimedean and nested Archimedean copulas, some of the latter are built with the two construction principles addressed in Sect. 7.4. We also provide two tables as references for sampling these copulas.

First, note that the symmetric Archimedean copula (7.1) can be written as

$$C(u) = \int_0^\infty \prod_{j=1}^d \exp(-x\psi^{-1}(u_j))\,dF(x),\ u \in I^d.$$

This mixture representation justifies the following sampling algorithm for the symmetric Archimedean copula $C$, see [18]. First, a random variate $V$ is drawn from the mixture distribution $F$, then exponential random variates are drawn to build the vector of random variates from $C$.

**Algorithm 7.5.1 ([18]).**

1. *Sample* $V \sim F = \mathscr{L}\mathscr{S}^{-1}[\psi]$.
2. *Sample i.i.d.* $X_j \sim U[0,1]$, $j \in \{1,\ldots,d\}$, *independent of* $V$.
3. *Return* $U$, *where* $U_j = \psi((-\log X_j)/V)$, $j \in \{1,\ldots,d\}$.

A sampling algorithm for the fully-nested Archimedean copula of type (7.2) can also be derived from the corresponding mixture representation, see [19]. It is given as follows.

**Algorithm 7.5.2 ([19]).**

1. *Sample* $V_0 \sim F_0 = \mathscr{L}\mathscr{S}^{-1}[\psi_0]$.
2. *Sample* $X_1 \sim U[0,1]$, *independent of* $V_0$.
3. *Sample* $(X_2,\ldots,X_d)^T \sim C(u_2,\ldots,u_d;\psi_{01}(\cdot;V_0),\ldots,\psi_{0d-2}(\cdot;V_0))$, *independent of* $X_1$.
4. *Return* $U$, *where* $U_j = \psi_0((-\log X_j)/V_0)$, $j \in \{1,\ldots,d\}$.

The same argument applies to a partially-nested Archimedean copula of type (7.4). The corresponding sampling algorithm is given as follows.

**Algorithm 7.5.3 ([19]).**

1. *Sample* $V_0 \sim F_0 = \mathscr{L}\mathscr{S}^{-1}[\psi_0]$.
2. *For* $s \in \{1,\ldots,S\}$, *sample* $(X_{s1},\ldots,X_{sd_s})^T \sim C(u_{s1},\ldots,u_{sd_s};\psi_{0s}(\cdot;V_0))$ *independent of each other with the algorithm of [18].*
3. *Return* $U = (U_{11},\ldots,U_{Sd_S})^T$, *where* $U_{sj} = \psi_0(-\log(X_{sj})/V_0)$, $s \in \{1,\ldots,S\}$, $j \in \{1,\ldots,d_s\}$.

For studying these algorithms, we exemplarily consider the three-dimensional fully-nested Archimedean copula of type (7.3). This copula involves a node of the form $\psi_0^{-1} \circ \psi_1$ and $\psi_{01}$ is given by (7.5). As before, the presented arguments generalize to a more complicated nested Archimedean structure by considering all nodes involved.

First assume we have built the three-dimensional fully-nested Archimedean copula of type (7.3) via our first construction principle addressed in Eq. (7.8). Sampling Step 1 of Algorithm 7.5.2 involves sampling $F_0 = \mathscr{LS}^{-1}[\psi_0]$. For sampling Step 3, the algorithm of [18] can be applied. This can always be done on the lowest nesting level. Here, it involves sampling $F_{01} = \mathscr{LS}^{-1}[\psi_{01}] = \mathscr{LS}^{-1}[\psi^{V_0}]$, see (7.9). Hence, if sampling algorithms for the distributions corresponding to $\psi_0$ and $\psi^{\alpha}$, $\alpha \in (0,\infty)$, are known, one can sample the nested Archimedean copula constructed in this way. Note that by choosing $\psi$ among the generators listed in Table 7.1, this is easily achieved in many cases.

Now assume we have built the three-dimensional fully-nested Archimedean copula of type (7.3) via the transformation presented in Theorem 7.4.1. Sampling such a copula is quite involved, indeed, large parts of [10, 11] are devoted to solving this problem. For describing a sampling strategy, we first need to consider a certain five-parametric distribution, an exponentially-tilted Stable distribution.

Let $\mathrm{S}(\alpha,\beta,\gamma,\delta;1)$ denote a *Stable distribution* with *characteristic exponent* $\alpha \in (0,2]$, *skewness* $\beta \in [-1,1]$, *scale* $\gamma \in [0,\infty)$, and *location* parameter $\delta \in \mathbb{R}$, given by the characteristic function

$$\phi(t) = \exp\big(i\delta t - \gamma^{\alpha}|t|^{\alpha}(1 - i\beta\,\mathrm{sgn}(t)w(t,\alpha))\big),\ t \in \mathbb{R}, \tag{7.10}$$

with $\mathrm{sgn}(t) = \mathbb{I}_{[0,\infty)}(t) - \mathbb{I}_{(-\infty,0]}(t)$, $t \in \mathbb{R}$, and

$$w(t,\alpha) = \begin{cases} \tan(\alpha\pi/2), & \alpha \neq 1, \\ -2\log(|t|)/\pi, & \alpha = 1, \end{cases}$$

see [23, p. 8]. A Stable distribution with $\gamma = 0$ is understood to be the unit jump at $\delta$. If, additionally, $\delta \in (0,\infty)$, the Laplace-Stieltjes transform trivially exists. Further, if $\alpha \in (0,1)$, $\beta = 1$, and $\delta = 0$, then the support of the Stable distribution is $[0,\infty)$, see [23, p. 12]. The corresponding Laplace-Stieltjes transform may be derived from (7.10).

Now note that for an Archimedean generator $\psi \in \Psi_\infty$ and all $h \in [0,\infty)$, the function $\tilde{\psi}(t) = \psi(t+h)/\psi(h)$ also defines an Archimedean generator in $\Psi_\infty$. If $\psi$ is the Laplace-Stieltjes transform of a Stable distribution, the distribution $\tilde{F} = \mathscr{LS}^{-1}[\tilde{\psi}]$ is called an *exponentially-tilted Stable distribution* with *tilt* $h \in [0,\infty)$, denoted by $\tilde{\mathrm{S}}(\alpha,\beta,\gamma,\delta,h;1)$. Concerning the Laplace-Stieltjes transform of an exponentially-tilted Stable distribution, one can show that for all $\alpha \in (0,1]$, $h \in [0,\infty)$, and $x \in (0,\infty)$, as well as all $t \in [0,\infty]$,

$$\mathscr{LS}[\tilde{\mathrm{S}}(\alpha,1,(\cos(\pi\alpha/2)x)^{1/\alpha},x\mathbb{I}_{\{\alpha=1\}},h\mathbb{I}_{\{\alpha\neq1\}};1)](t) = e^{-x((h+t)^{\alpha}-h^{\alpha})}. \tag{7.11}$$

Laplace-Stieltjes transforms of this type frequently appear on sampling nested Archimedean copulas. To be more precise, $\psi_{01}(\cdot\,;V_0)$ is often of the form (7.11) with $x = V_0$. For sampling the corresponding distribution, [11] presented an efficient version of the rejection algorithm. Having such a tool at hand, let us go back to our problem of sampling a nested Archimedean copula constructed via the principle addressed in Theorem 7.4.1. The following result provides a solution, see [11].

**Theorem 7.5.4.** *Let* $\psi \in \Psi_\infty$.

1. *If* $\tilde{\psi}(t) = \psi((c^{\vartheta} + t)^{1/\vartheta} - c)$, $t \in [0,\infty]$, *with* $\vartheta \in [1,\infty)$ *and* $c \in [0,\infty)$, *then*
$$\tilde{V} = \tilde{S}V^{\vartheta} \sim \tilde{F} = \mathscr{L}\mathscr{S}^{-1}[\tilde{\psi}],$$
*where* $V \sim F = \mathscr{L}\mathscr{S}^{-1}[\psi]$ *and*
$$\tilde{S} \sim \tilde{\mathrm{S}}(1/\vartheta, 1, \cos^{\vartheta}(\pi/(2\vartheta)), \mathbb{I}_{\{\vartheta=1\}}, (cV)^{\vartheta}\mathbb{I}_{\{\vartheta\neq 1\}}; 1).$$
2. *For* $i \in \{0,1\}$, *let* $\psi_i(t) = \psi((c^{\vartheta_i} + t)^{1/\vartheta_i} - c)$, $t \in [0,\infty]$, *with* $c \in [0,\infty)$ *and* $\vartheta_i \in [1,\infty)$. *Then* $\psi_{01}(t;V_0)$ *is of type (7.11) with* $\alpha = \vartheta_0/\vartheta_1$ *and* $h = c^{\vartheta_1}$, *hence,* $F_{01} = \mathscr{L}\mathscr{S}^{-1}[\psi_{01}(\cdot;V_0)]$ *is an exponentially-tilted Stable distribution.*

Theorem 7.4.1 provides a large source for constructing nested Archimedean copulas. By Theorem 7.5.4, these can be efficiently sampled, as long as $F = \mathscr{L}\mathscr{S}^{-1}[\psi]$ is easy to sample. This is only one appealing property of this generator transformation. Sampling algorithms for various nested Archimedean copulas can be derived from Theorem 7.5.4, e.g. a nested Clayton copula, by taking $c = 1$, and nested outer power Archimedean copulas, for $c = 0$.

Table 7.1 lists commonly applied one-parametric Archimedean generators with corresponding known inverse Laplace-Stieltjes transforms, including the families of Ali-Mikhail-Haq (A), Clayton (C), Frank (F), Gumbel (G), and Joe (J). The list consists of all completely monotone Archimedean generators of [22, pp. 116], with a slightly simplified generator for Clayton's family. The numbers are the ones given in this reference. For the Archimedean family of Ali-Mikhail-Haq, $F$ is a Geometric distribution on $\mathbb{N}$, Clayton's family corresponds to a Gamma distribution, Frank's family to a Logarithmic distribution, and Joe's family to a discrete distribution which puts mass $y_k$ at $x_k$, for $(x_k, y_k)_{k\in\mathbb{N}}$ given in Table 7.1. For the families numbered 12 and 14, the given stochastic representation is easily derived from Theorem 7.5.4, Part 1, where $\mathrm{Exp}(1)$ denotes a standard exponentially-distributed random variate. Note that the family numbered 12 is simply an outer power family based on Clayton's family. Further, the one numbered 13 is related to Gumbel's family via exponential tilting. Table 7.1 may serve as a reference for constructing and sampling Archimedean copulas. For example, one can pick one of these Archimedean families to build a nested Archimedean copula based on the construction principles addressed in Theorems 7.4.1 and 7.5.4.

For constructing a nested Archimedean copula based on generators belonging to the same Archimedean family as given in Table 7.1, Table 7.2 lists parameter restrictions on the generators such that nested Archimedean copulas result. Further, either stochastic representations of $V_{01} \sim F_{01} = \mathscr{L}\mathscr{S}^{-1}[\psi_{01}]$ or the distributions $F_{01}$ are given. For the family of Clayton, as well as for the families numbered 13 and 19, Theorem 7.5.4, Part 2 readily leads to the stated results. The stochastic representations of $V_{01}$ for the families of Ali-Mikhail-Haq, Frank, and Joe are convolutions of independent and identically distributed random variables according to the given distribution. For these and further results, including run times of the presented algorithms, details about the implementation, and examples, see [11]. For the family

**Table 7.1** Known one-parametric Archimedean generators with corresponding inverse Laplace-Stieltjes transforms.

| Family | $\vartheta$ | $\psi(t)$ | $V \sim F = \mathcal{LS}^{-1}[\psi]$ |
|---|---|---|---|
| A | $[0,1)$ | $(1-\vartheta)/(\exp(t)-\vartheta)$ | $\mathrm{Geo}(1-\vartheta)$ |
| C | $(0,\infty)$ | $(1+t)^{-1/\vartheta}$ | $\Gamma(1/\vartheta,1)$ |
| F | $(0,\infty)$ | $-\vartheta^{-1}\log(1-(1-e^{-\vartheta})\exp(-t))$ | $\mathrm{Log}(1-\exp(-\vartheta))$ |
| G | $[1,\infty)$ | $\exp(-t^{1/\vartheta})$ | $\mathrm{S}(1/\vartheta,1,\cos^{\vartheta}(\pi/(2\vartheta)),\mathbb{I}_{\{\vartheta=1\}};1)$ |
| J | $[1,\infty)$ | $1-(1-\exp(-t))^{1/\vartheta}$ | $\left(k,\binom{1/\vartheta}{k}(-1)^{k-1}\right)_{k\in\mathbb{N}}$ |
| 12 | $[1,\infty)$ | $(1+t^{1/\vartheta})^{-1}$ | $\mathrm{S}(1/\vartheta,1,\cos^{\vartheta}(\pi/(2\vartheta)),\mathbb{I}_{\{\vartheta=1\}};1)\,\mathrm{Exp}(1)^{\vartheta}$ |
| 13 | $[1,\infty)$ | $\exp(1-(1+t)^{1/\vartheta})$ | $\tilde{\mathrm{S}}(1/\vartheta,1,\cos^{\vartheta}(\pi/(2\vartheta)),\mathbb{I}_{\{\vartheta=1\}},\mathbb{I}_{\{\vartheta\neq1\}};1)$ |
| 14 | $[1,\infty)$ | $(1+t^{1/\vartheta})^{-\vartheta}$ | $\mathrm{S}(1/\vartheta,1,\cos^{\vartheta}(\pi/(2\vartheta)),\mathbb{I}_{\{\vartheta=1\}};1)\Gamma(\vartheta,1)^{\vartheta}$ |
| 19 | $(0,\infty)$ | $\vartheta/\log(t+e^{\vartheta})$ | $\Gamma(\Gamma(1,1)/\vartheta,e^{\vartheta})$ |
| 20 | $(0,\infty)$ | $\log^{-1/\vartheta}(t+e)$ | $\Gamma(\Gamma(1/\vartheta,1),e)$ |

numbered 14, there are no conditions on the underlying parameters known such that the sufficient nesting condition holds. Finally, let us note that the construction principle addressed in Eq. (7.8) plays an important role in constructing and sampling nested Archimedean copulas based on generators belonging to different Archimedean families, see [11] for more details.

**Table 7.2** Parameter ranges such that nested Archimedean copulas based on the given Archimedean family result; inverse Laplace-Stieltjes transforms corresponding to $\psi_{01}(\cdot;V_0)$.

| Family | Nesting | $V_{01} \sim F_{01} = \mathcal{LS}^{-1}[\psi_{01}(\cdot;V_0)]$ |
|---|---|---|
| A | $\vartheta_0 \le \vartheta_1$ | $\sum_{i=1}^{V_0} V_{01,i},\ V_{01,i} \sim \mathrm{Geo}((1-\vartheta_1)/(1-\vartheta_0))$ |
| C | $\vartheta_0 \le \vartheta_1$ | $\tilde{\mathrm{S}}(\vartheta_0/\vartheta_1,1,(\cos(\pi\vartheta_0/(2\vartheta_1))V_0)^{\vartheta_1/\vartheta_0},V_0\mathbb{I}_{\{\vartheta_0=\vartheta_1\}},\mathbb{I}_{\{\vartheta_0\neq\vartheta_1\}};1)$ |
| F | $\vartheta_0 \le \vartheta_1$ | $\sum_{i=1}^{V_0} V_{01,i},\ V_{01,i} \sim \left(k,\binom{\vartheta_0/\vartheta_1}{k}(-1)^{k-1}(1-e^{-\vartheta_1})^k/(1-e^{-\vartheta_0})\right)_{k\in\mathbb{N}}$ |
| G | $\vartheta_0 \le \vartheta_1$ | $\mathrm{S}(\vartheta_0/\vartheta_1,1,(\cos(\pi\vartheta_0/(2\vartheta_1))V_0)^{\vartheta_1/\vartheta_0},\mathbb{I}_{\{\vartheta_0=\vartheta_1\}};1)$ |
| J | $\vartheta_0 \le \vartheta_1$ | $\sum_{i=1}^{V_0} V_{01,i},\ V_{01,i} \sim \left(k,\binom{\vartheta_0/\vartheta_1}{k}(-1)^{k-1}\right)_{k\in\mathbb{N}}$ |
| 12 | $\vartheta_0 \le \vartheta_1$ | $\mathrm{S}(\vartheta_0/\vartheta_1,1,(\cos(\pi\vartheta_0/(2\vartheta_1))V_0)^{\vartheta_1/\vartheta_0},\mathbb{I}_{\{\vartheta_0=\vartheta_1\}};1)$ |
| 13 | $\vartheta_0 \le \vartheta_1$ | $\tilde{\mathrm{S}}(\vartheta_0/\vartheta_1,1,(\cos(\pi\vartheta_0/(2\vartheta_1))V_0)^{\vartheta_1/\vartheta_0},V_0\mathbb{I}_{\{\vartheta_0=\vartheta_1\}},\mathbb{I}_{\{\vartheta_0\neq\vartheta_1\}};1)$ |
| 14 | - | - |
| 19 | $\vartheta_0 \le \vartheta_1$ | $\tilde{\mathrm{S}}(\vartheta_0/\vartheta_1,1,(\cos(\pi\vartheta_0/(2\vartheta_1))V_0)^{\vartheta_1/\vartheta_0},V_0\mathbb{I}_{\{\vartheta_0=\vartheta_1\}},e^{\vartheta_1}\mathbb{I}_{\{\vartheta_0\neq\vartheta_1\}};1)$ |
| 20 | $\vartheta_0 \le \vartheta_1$ | - |

## 7.6 Conclusion

This article presented an introduction to nested Archimedean copulas. Focus was put on construction and sampling algorithms. The two concepts are related insofar as a construction principle that fulfills the sufficient nesting condition given by [13, p. 88] and [19] directly leads to a sampling algorithm. This is due to a mixture representation of a nested Archimedean copula based on completely monotone Archimedean generators. We presented two general construction principles for these copulas. Further, the resulting copulas are easy to sample in all dimensions. Applying these ideas for construction and sampling of nested Archimedean copulas to commonly known Archimedean generators leads to a rich repertoire of nested Archimedean copulas that can easily be sampled in all dimensions and therefore provide an alternative to elliptical copulas, which are often appreciated for their simple sampling algorithms but are restricted to radial symmetry.

## References

1. Aas, K., Czado, C., Frigessi, A., Bakken, H.: Pair-copula constructions of multiple dependence. Insur. Math. Econ. **44**(2), 182–198 (2009)
2. Alsina, C., Frank, M.J., Schweizer, B.: Associative Functions: Triangular Norms and Copulas. World Scientific Publishing Company, Singapore (2006)
3. Bernstein, S.N.: Sur les fonctions absolument monotones. Acta Mathematica **52**, 1–66 (1928)
4. Choe, G.H., Jang, H.J.: Efficient algorithms for basket default swap pricing with multivariate Archimedean copulas. http://ssrn.com/abstract=1414111 (2009).
5. Choroś, B., Härdle, W., Okhrin, O.: CDO pricing with copulae. Discussion paper 2009–013, SFB 649, Humboldt Universität zu Berlin, Berlin (2009). Available at http://sfb649.wiwi.hu-berlin.de/papers/pdf/SFB649DP2009–013.pdf
6. Feller, W.: An Introduction to Probability Theory and Its Applications, vol. 2, 2nd edn. Wiley, New York, NY (1971)
7. Fischer, M., Köck, C., Schlüter, S., Weigert, F.: An empirical analysis of multivariate copula models. Quant. Finance **9**(7), 839–854 (2009)
8. Genest, C., MacKay, R.J.: Copules archimédiennes et familles de lois bidimensionnelles dont les marges sont données. Can. J. Stat. **14**, 145–159 (1986)
9. Höcht, S., Zagst, R.: Pricing Distressed CDOs with Stochastic Recovery. Forthcoming in Review of Derivatives Research (2009)
10. Hofert, M.: Sampling Archimedean copulas. Comput. Stat. Data Anal. **52**(12), 5163–5174 (2008)
11. Hofert, M.: Efficiently sampling nested Archimedean copulas. Forthcoming in Computational Statistics and Data Analysis (2010)
12. Hofert, M., Scherer, M.: CDO pricing with nested Archimedean copulas. Quant. Finance (2010). In press. Available at http://www.mathematik.uni-ulm.de/numerik/preprints/2008/CDOpricingAC.pdf
13. Joe, H.: Multivariate Models and Dependence Concepts. Chapman & Hall/CRC, Boca Raton, FL (1997)
14. Joe, H., Hu, T.: Multivariate distributions from mixtures of max-infinitely divisible distributions. J. Multivar. Anal. **57**, 240–265 (1996)
15. Kimberling, C.H.: A probabilistic interpretation of complete monotonicity. Aequationes Mathematicae **10**, 152–164 (1974)

16. Klement, E.P., Mesiar, R., Pap, E.: Triangular Norms. Kluwer, Dordrecht (2000)
17. Kurowicka, D., Cooke, R.: Uncertainty Analysis with High Dimensional Dependence Modelling. Wiley, New York, NY (2006)
18. Marshall, A.W., Olkin, I.: Families of multivariate distributions. J. Am. Stat. Assoc. **83**, 834–841 (1988)
19. McNeil, A.J.: Sampling nested Archimedean copulas. J. Stat. Comput. Simul. **78**, 567–581 (2008)
20. McNeil, A.J., Frey, R., Embrechts, P.: Quantitative Risk Management: Concepts, Techniques, and Tools. Princeton University Press, Princeton, NJ (2005)
21. McNeil, A.J., Nešlehová, J.: Multivariate Archimedean copulas, $d$-monotone functions and $l_1$-norm symmetric distributions. Ann. Stat. **37**(5b), 3059–3097 (2009)
22. Nelsen, R.B.: An Introduction to Copulas. Springer, New York, NY (2006)
23. Nolan, J.P.: Stable Distributions—Models for Heavy Tailed Data. Available at http://academic2.american.edu/∼ jpnolan/stable/chap1.pdf (2009)
24. Okhrin, O.: Hierarchical Archimedean copulas: structure determination, properties, applications. Ph.D. thesis (2007)
25. Okhrin, O., Okhrin, Y., Schmid, W.: Properties of hierarchical Archimedean copulas. Discussion paper 2009–014, SFB 649, Humboldt Universität zu Berlin, Berlin (2009). Available at http://sfb649.wiwi.hu-berlin.de/papers/pdf/SFB649DP2009–014.pdf
26. Ridout, M.: Generating random numbers from a distribution specified by its Laplace transform. Statistics and Computing **19**, 439–450 (2009)
27. Savu, C., Trede, M.: Hierarchical Archimedean copulas. Preprint (2006). Available at http://www.uni-konstanz.de/micfinma/conference/Files/papers/Savu_Trede.pdf
28. Schweizer, B., Sklar, A.: Associative functions and statistical triangle inequalities. Publicationes Mathematicae Debrecen **8**, 169–186 (1961)
29. Schweizer, B., Sklar, A.: Associative functions and abstract semigroups. Publicationes Mathematicae Debrecen **10**, 69–81 (1963)
30. Sklar, A.: Random variables, joint distribution functions, and copulas. Kybernetika **9**, 449–460 (1973)
31. Whelan, N.: Sampling from Archimedean copulas. Quant. Finance **4**(3), 339–352 (2004)
32. Widder, D.V.: The Laplace Transform. Princeton University Press, Princeton, NJ (1946)

# Chapter 8
# Tail Behaviour of Copulas

Piotr Jaworski

**Abstract** The study and modeling of interdependencies between extreme events is crucial for many applications of probability theory and statistics. Thanks to Sklar's Theorem such tasks decompose into the study of the tail behaviour of the marginal univariate distributions and of the tail (i.e. corner) behaviour of the corresponding copulas. In this chapter we will deal with the second "subproblem". There are several approaches known in the literature. We shall deal with the one based on the tail expansion of copulas near the vertex $(0,\ldots,0)$ of the unit multicube. We present the notions related to the tail expansion – leading parts, tail dependence functions and limiting invariant measures. We briefly discuss their properties and characterizations and provide several examples of the tail behaviour of copulas. Next we show relations between the tail expansion method and other approaches to the tail behaviour of copulas like the ones based on conditional copulas or associated extreme value copulas. At the end we present possible applications of the notion of tail expansions to quantitative finance, especially to risk measurement.

## 8.1 Introduction

We consider a one-parameter family of homeomorphisms

$$\Phi_t : [0,+\infty)^n \longrightarrow [0,+\infty)^n, \quad \Phi_t(x_1,\ldots,x_n) = (tx_1,\ldots,tx_n), \quad t \in (0,+\infty).$$

If $C$ is an $n$-dimensional copula and $\mu_C$ the associated measure supported on the unit rectangle then for any $d \in (0,+\infty)$,

$$\mu_{d,t}(A) = \frac{1}{t^d}\mu_C(\Phi_t(A))$$

---

Piotr Jaworski
Institute of Mathematics, University of Warsaw, Warszawa, Poland
e-mail: P.Jaworski@mimuw.edu.pl

P. Jaworski et al. (eds.), *Copula Theory and Its Applications*, Lecture Notes in Statistics 198,
DOI 10.1007/978-3-642-12465-5_8, 

is a Borel measure with support $[0,t^{-1}]^n$.

The distribution function of $\mu_{d,t}$ is given by the formula

$$C_t(u) = \frac{1}{t^d}\widetilde{C}(\Phi_t(u)) = \frac{\widetilde{C}(tu_1,\dots,tu_n)}{t^d}, \quad \widetilde{C}(v) = C(\min(1,v_1)^+,\dots,\min(1,v_n)^+).$$

**Problem 1.** Does the limit of $\mu_{d,t}$ exist when $t \to 0$?
**Problem 2.** Does the limit of $C_t$ exist when $t \to 0$?
**Problem 3.** How to describe the set of possible limits?

If the answer to the second question is affirmative then the function

$$L(u) = \lim_{t\to 0^+} \frac{C(tu)}{t^d}, \quad u \in [0,+\infty)^n,$$

is called the *tail dependence function* (see [18, 20, 23]). Note that, for $d = 1$ and $n = 2$, the tail dependence function determines the coefficient of lower tail dependence

$$\lambda_L = \lim_{t\to 0^+} \frac{C(t,t)}{t} = L(1,1).$$

The more analytical point of view can be summarized as follows:

**Problem 4.** Is it possible to approximate a given copula $C$ near the point $(0,\dots,0)$ by a function which is invariant with respect to $\Phi_t$ with multiplicator $t^d$?

Note that a function $F$ invariant with respect to multiplication of coordinates by $t$ with multiplicator $t^d$ is called *homogeneous* of degree $d$:

$$F(\Phi_t(x)) = F(tx_1,\dots,tx_n) = t^d F(x).$$

This leads to the notion of lower tail expansion. We say that a copula has a lower tail expansion if near the origin it can be approximated by a homogeneous function (compare [10, 11, 13]). In more detail:

**Definition 8.1.1.** A copula $C : [0,1]^n \longrightarrow [0,1]$ has a *lower tail expansion* of *degree* $d$ if there exist a homogeneous function $L : [0,+\infty)^n \longrightarrow \mathbb{R}$ of degree $d$,

$$\forall t \geq 0 \;\; L(tu) = t^d L(u),$$

and a function $R : [0,1]^n \longrightarrow \mathbb{R}$ with

$$\lim_{t\to 0} R(tu) = 0,$$

such that

$$\forall u \in [0,1]^n \quad C(u) = L(u) + R(u)(u_1 + \cdots + u_n)^d.$$

Furthermore, we say that the expansion is *uniform* when $R$ is bounded and

$$\lim_{u\to 0} R(u) = 0.$$

The function $L$ will be called *the leading part* of the expansion. When $L \equiv 0$ we shall say that the expansion is *trivial*.

The above definition gives rise to a new question.

**Problem 5.** What are the possible leading parts?

Note that nontrivial expansions are possible only for $d \geq 1$. Indeed, for all $u \in [0,1]^n$ we have $C(u) \leq \min(u_1, \ldots, u_n)$ (the Fréchet-Hoeffding inequality), therefore for every $d < 1$ we get

$$\frac{C(tu)}{(tu_1 + \cdots + tu_n)^d} \leq \frac{\min(tu_1, \ldots, tu_n)}{(tu_1 + \cdots + tu_n)^d} \leq t^{1-d} \longrightarrow 0 \quad \text{as} \quad t \to 0.$$

Hence every copula has a trivial expansion of degree smaller than 1. But there are copulas which do not admit a tail expansion of any degree greater than or equal to 1 (see Corollary 8.3.2).

Furthermore, if $C$ admits a tail expansion of degree $d$, then it admits trivial expansions of all degrees smaller than $d$, and if $C$ admits a nontrivial expansion of degree $d$, then it does not admit an expansion of degree greater than $d$.

## 8.2 Tail Expansions of Copulas

We choose the convergence in the first problem to be the wide (vague) convergence (see [3], Sect. III.1.9, [25], Sect. IV.7, [24, Sect. 3.3.5]). Then Problem 1 is closely related to Problems 2 and 4. So related are also Problems 3 and 5.

We recall that a one-parameter family of measures $\mu_t, t > 0$, converges to the measure $\mu$ in the wide sense as $t$ tends to 0 if for every continuous function $f$ with compact support,

$$\lim_{t\to 0} \int f d\mu_t = \int f d\mu.$$

**Theorem 8.2.1.** *Let $C(u)$ be an n-dimensional copula, $L(u)$ a real valued function on $[0,+\infty)^n$ and $d$ a positive constant. Then the following conditions are equivalent:*
*1. $L$ is the leading part of a lower tail expansion of degree $d$ of the copula $C$;*
*2. $\mu_{d,t}$ converges in the wide sense as $t \to 0$ and $L$ is the Kolmogorov style distribution function of the limit measure $\mu$:*

$$L(u_1, \ldots, u_n) = \mu([0,u_1) \times \cdots \times [0,u_n));$$

3. *for every* $u \in [0,+\infty)^n$ *the ray-like limit* $\lim_{t\to 0^+} \frac{C(tu)}{t^d}$ *exists and equals* $L(u)$.

In the case $d = 1$ the above result can be strengthened – see [13], Theorem 1. The tail expansion is uniform and $L$ is the uniform limit of $\frac{C(tu)}{t}$. Since for $d > 1$ the leading part may not be continuous (see Corollary 8.3.1), the expansions may not be uniform.

To prove Theorem 8.2.1 we will need the basic properties of the tail dependence function.

**Lemma 8.2.1.** *The tail dependence function induced by a copula C,*

$$L(u) = \lim_{t\to 0^+} \frac{C(tu)}{t^d}, \quad u \in [0,+\infty)^n,$$

*is:*
*1) nonnegative,*
*2) grounded (*$L(x) = 0$ *when any coordinate of x is 0 i.e.* $L(\ldots,0,\ldots) = 0$*),*
*3) nondecreasing with respect to every variable,*
*4) n-nondecreasing,*
*5) homogeneous of degree d,*
*6) continuous on* $(0,+\infty)^n$*,*
*7) left-continuous on the whole domain,*
*and satisfies:*
*8) for every* $u \in [0,+\infty)^n$*,*

$$L(1,\ldots,1)\min(u_1^d,\ldots,u_n^d) \le L(u) \le L(1,\ldots,1)\max(u_1^d,\ldots,u_n^d).$$

*Proof.* The first four properties are valid for every copula $C$ and are preserved under the limit.
The fifth property is straightforward. Indeed, for any $s \in (0,+\infty)$ we have

$$L(su) = \lim_{t\to 0} s^d \frac{C(tsu)}{t^d s^d} = s^d \lim_{t\to 0} \frac{C(tu)}{t^d} = s^d L(u).$$

The sixth and eighth ones follow from the homogeneity. Namely, for any two points $u \in [0,+\infty)^n$ and $u^* \in (0,+\infty)^n$ we have, for all $i$,

$$t_1 u_i^* \le u_i \le t_2 u_i^*,$$

where

$$t_1 = \min\left(\frac{u_1}{u_1^*},\ldots,\frac{u_n}{u_n^*}\right), \quad t_2 = \max\left(\frac{u_1}{u_1^*},\ldots,\frac{u_n}{u_n^*}\right).$$

Hence

$$t_1^d L(u^*) \le L(u) \le t_2^d L(u^*).$$

Since $t_1, t_2 \to 1$ when $u \to u^*$, we get

$$\lim_{u \to u^*} L(u) = L(u^*).$$

To get the eighth property one has to put $u^* = (1, \ldots, 1)$.

The left-continuity at the border points $([0,+\infty)^n \setminus (0,+\infty)^n)$ follows from the fact that $L$ is grounded, i.e. equals 0 on the border. □

Now we sketch the proof of Theorem 8.2.1.

*Proof.* The equivalence of 1 and 3 is straightforward. Indeed:
$1 \Rightarrow 3$

$$\frac{C(tu)}{t^d} = \frac{L(tu)}{t^d} + \frac{R(tu)(tu_1 + \cdots + tu_n)^d}{t^d}$$
$$= L(u) + R(tu)(u_1 + \cdots + u_n)^d \to L(u) \quad \text{as} \quad t \to 0.$$

$3 \Rightarrow 1$

$$R(tu) = \frac{C(tu) - L(tu)}{(tu_1 + \cdots + tu_n)^d} = \frac{1}{(u_1 + \cdots + u_n)^d} \left( \frac{C(tu)}{t^d} - L(u) \right) \to 0 \text{ as } t \to 0.$$

The proof of the equivalence of 2 and 3 is more complicated.
$2 \Rightarrow 3$
If a point $u$ has at least one coordinate 0 then $C(tu) = 0$ for all $t$ and the limit $\lim C(tu)t^{-d}$ always exists and is equal to 0.
Now fix $u \in (0,\infty)^n$. For any three real numbers $s_2 > s > s_1 > 0$ there exist two nonnegative, continuous functions $f_1$ and $f_2$ with compact supports such that

$$\mathbb{1}_{s_2 u}(v) \geq f_2(v) \geq \mathbb{1}_{su}(v) \geq f_1(v) \geq \mathbb{1}_{s_1 u}(v), \quad v \in [0,+\infty)^n,$$

where $\mathbb{1}_u$ denotes the indicator (characteristic function) of the rectangle $[0,u_1) \times \cdots \times [0,u_n)$. After integration we get

$$L(s_2 u) \geq \int f_2 d\mu = \lim_{t \to 0} \int f_2 d\mu_{d,t} \geq \limsup_{t \to 0} \frac{C(tsu)}{t^d}$$

$$\geq \liminf_{t \to 0} \frac{C(tsu)}{t^d} \geq \lim_{t \to 0} \int f_1 d\mu_{d,t} = \int f_1 d\mu \geq L(s_1 u).$$

Since the function $F(s) = L(su)$, $s \in [0,+\infty)$, is nondecreasing, it is continuous on a dense subset of its domain. Let $s \in (s_1, s_2)$ be any point of continuity. Then

$$\lim_{s_2 \to s} L(s_2 u) = L(su) = \lim_{s_1 \to s} L(s_1 u)$$

and we get

$$\lim_{t \to 0} \frac{C(tsu)}{t^d} = L(su).$$

Now

$$\lim_{t \to 0} \frac{C(tu)}{t^d} = \lim_{t \to 0} \frac{C(stu)}{(st)^d} = \frac{1}{s^d} \lim_{t \to 0} \frac{C(tsu)}{t^d}.$$

Hence for every $u$ the limit $\lim_{t\to 0}\frac{C(tu)}{t^d}$ exists. Furthermore, since the homogeneous mapping $s\mapsto \lim_{t\to 0}\frac{C(tsu)}{t^d}$ coincides with the nondecreasing mapping $s\mapsto L(su)$ on a dense subset of the positive half-line, they are equal at every point of the positive half-line, in particular at $s=1$. Hence we get

$$\lim_{t\to 0}\frac{C(tu)}{t^d}=L(u).$$

$3\Rightarrow 2$ The tail dependence function $L$ is grounded, $n$-nondecreasing and left-continuous. Therefore there exists a $\sigma$-finite Borel measure $\mu$ such that

$$\mu((-\infty,u_1)\times\cdots\times(-\infty,u_n))=\begin{cases} L(u_1,\dots,u_n) & \text{for } u_1,\dots,u_n\geq 0,\\ 0 & \text{otherwise}\ ,\end{cases}$$

i.e. $L$ (extended by 0) is a Kolmogorov style distribution function of $\mu$. Let $f:\mathbb{R}^n\to\mathbb{R}$ be a continuous function with a compact support $S_f$. Let $M=\max\{|f(x)|:x\in\mathbb{R}^n\}$ and $u=1+\max\{v_i: v\in S_f,\ i=1,\dots,n\}$. If the tail dependence function $L$ equals 0 everywhere, then

$$|\int f d\mu_{d,t}|\leq M\frac{\widetilde{C}(tu,\dots,tu)}{t^d}\to 0 \quad\text{as}\quad t\to 0,$$

the limit measure exists and is equal to 0.
Otherwise, when $L$ does not equal 0 everywhere we have $L(u,\dots,u)=u^dL(1,\dots,1)>0$ (see Lemma 8.2.1, item 8). Let $P_t$ be the probability measure obtained by restricting of $\mu_{d,t}$ to the rectangle $R=[0,u)^n$ and normalizating

$$P_t(A)=\frac{\mu_{d,t}(A\cap R)}{\mu_{d,t}(R)}.$$

For $v\in R$ the distribution function of $P_t$ is equal to

$$F_t(v)=\frac{\widetilde{C}(tv)}{\widetilde{C}(tu)}.$$

Since

$$\lim_{t\to 0}\frac{\widetilde{C}(tv)}{\widetilde{C}(tu)}=\lim_{t\to 0}\frac{C(tv)}{C(tu)}=\frac{L(v)}{L(u)},$$

$P_t$ weakly converges to the measure obtained by restriction of $\mu$ to $R$ and normalization. Therefore ([2], Theorem 29.1)

$$\int f d\mu_{d,t}=\mu_{d,t}(R)\int f dP_t=\frac{\widetilde{C}(tu)}{t^d}\int f dP_t\to L(u)\frac{\int f d\mu}{L(u)}=\int f d\mu.$$

Since the above can be repeated for any continuous function with compact support, $\mu_{d,t}$ widely converges to $\mu$. □

### 8.2.1 Characterization and Properties of Leading Parts

The general properties of the leading part are obvious: $L$ is nonnegative, grounded, nondecreasing in each variable, $n$-nondecreasing and bounded on compacts (see Lemma 8.2.1).

When $d = 1$ we get more interesting properties ([13], Sect. 6).

**Proposition 8.2.1.** *Let $L$ be a leading form of degree 1. Then for all $u, v \in [0, +\infty)^n$ $L$ satisfies the following:*
*1. $L(u)$ is nonnegative and bounded by the smallest coordinate of $u$:*

$$0 \leq L(u) \leq \min(u_1, \ldots, u_n).$$

*2. $L$ is Lipschitz with Lipschitz constant 1:*

$$|L(v) - L(u)| \leq \sum_i |v_i - u_i|.$$

*3. $L$ is superadditive:*

$$L(u+v) \geq L(u) + L(v).$$

*4. $L$ is concave:*

$$\forall \lambda_1, \lambda_2 \geq 0, \lambda_1 + \lambda_2 = 1 \quad L(\lambda_1 u + \lambda_2 v) \geq \lambda_1 L(u) + \lambda_2 L(v).$$

Furthermore, in the case $d = 1$ the characterization of possible $L$ is given by the following (see [13]):

**Theorem 8.2.2.** *For a function $L : [0, +\infty)^n \to \mathbb{R}$, homogeneous of degree 1, the following conditions are equivalent:*
*1. $L$ is the leading part of the lower tail of some copula $C$.*
*2. $L$ is $n$-nondecreasing and*

$$0 \leq L(u) \leq \min(u_1, \ldots, u_n) \quad \text{for} \quad u \in [0, +\infty)^n.$$

*3. There exists a copula $C_L$ whose lower tail equals $L$:*

$$\exists \delta > 0 \ \ \forall u \in [0, \delta]^n \quad C_L(u) = L(u).$$

The case $d > 1$ differs very much from the case $d = 1$. First, $L$ need not be continuous (see Corollary 8.3.1 below). Second, nonzero $L$'s are not concave. Indeed, $L(su) = s^d L(u)$ is convex in $s$. Third if $L(u)$ is a leading part of some copula so is $tL(u)$ for every $t > 0$. Hence there is no upper bound for $L$'s.

It is not known which homogeneous functions of degree $d$ are leading parts. The partial results are given by the following:

**Theorem 8.2.3.** *For a function $L: [0,+\infty)^n \to \mathbb{R}$, homogeneous of degree greater than 1, the following conditions are equivalent:*
*1. L is locally Lipschitz, grounded, n-nondecreasing;*
*2. there exists a copula C whose lower tail equals L:*

$$\exists\delta \;\; \forall u \in [0,\delta]^n \;\; C(u) = L(u).$$

*Proof.*
The case $n = 2$ is done in [5], Proposition 3.4. The general case can be proved using an approach similar to that for [13], Proposition 6.

## 8.2.2 Relatively Invariant Measures on $[0,\infty)^n$

Let us assume that conditions 1, 2 and 3 of Theorem 8.2.1 hold. The limit measure $\mu$ is relatively invariant with respect to $\Phi_t$: for every Borel set $A$ we have

$$\mu(\Phi_t(A)) = t^d \mu(A).$$

Therefore $\mu$ can be represented as the product of a measure $m_d$ on the positive half-line and a measure $\mu_\Delta$ on the unit simplex $\Delta$,

$$\Delta = \left\{ x \in [0,+\infty)^n : \sum_{i=1}^{n} x_i = 1 \right\}.$$

$m_d$ is an absolutely continuous measure with density $dx^{d-1}$. If $d = 1$ it is the standard Lebesgue measure. The measure $\mu_\Delta$ is induced by the projection

$$\left\{ x \in [0,+\infty)^n : 0 < \sum_{i=1}^{n} x_i \leq 1 \right\} \longrightarrow \Delta, \quad x \mapsto \frac{1}{x_1 + \cdots + x_n} x.$$

On a Borel subset $A$ of $\Delta$ it is given by

$$\mu_\Delta(A) = \mu(\{t\xi : \xi \in A, \; t \in [0,1]\}).$$

**Proposition 8.2.2.** *For any bounded Borel set $A \subset \langle 0,+\infty)^n$,*

$$\mu(A) = \int_\Delta m_d(A \cap \{t\xi : t \in [0,+\infty)\}) d\mu_\Delta(\xi).$$

When $d = 1$ one can also characterize all the measures $\mu_\Delta$ corresponding to tail expansions.

**Proposition 8.2.3.** *A measure $\mu$ relatively invariant with respect to $\Phi_t$ with multiplicator t is the limit measure of a copula if and only if the associated measure $\mu_\Delta$ satisfies*

$$\int_{\Delta} \frac{1}{q_i} d\mu_{\Delta}(q) \leq 1 \quad for \quad i = 1, \ldots, n.$$

When $d > 1$ the characterization of all the measures $\mu$ corresponding to tail expansions is an open problem.

## 8.3 Examples of Tail Expansions

### 8.3.1 Homogeneous Copulas

We start with examples of homogeneous copulas. In such cases $L = C$ and $R = 0$.

*Example 8.3.1.* The copula of independent random variables

$$C(u) = \prod_i u_i$$

has a nontrivial expansion of degree $n$.

*Example 8.3.2.* The copula of comonotonic random variables

$$C(u) = \min(u_1, \ldots, u_n)$$

has a nontrivial expansion of degree 1.

*Example 8.3.3.* The "exterior" product of a comonotonic copula and an independent copula

$$C(u,v) = \min(u_1, \ldots, u_n) v_1 \cdot \cdots \cdot v_m, \quad n,m = 1,2,\ldots,$$

has a nontrivial expansion of degree $m+1$.

*Example 8.3.4.* The Cuadras-Augé copulas, i.e. "interior" weighted products of a comonotonic copula and an independent copula,

$$C(u_1,u_2) = \min(u_1,u_2)^{\alpha}(u_1 u_2)^{1-\alpha}, \quad \alpha \in (0,1),$$

have an expansion of degree $2 - \alpha$.

### 8.3.2 Diagonal Copulas

We recall that the function

$$\delta(t) = C(t, \ldots, t), \quad t \in [0,1]$$

is called the *diagonal section* (or *diagonal* for short) of the copula $C$. The limit behaviour of $\delta$ at 0 is closely related to the existence of a tail expansion of the copula $C$. Namely, if $C$ admits a tail expansion of degree $d$ with leading part $L$ then

$$\lim_{t\to 0}\frac{\delta(t)}{t^d} = L(1,\ldots,1).$$

Diagonals satisfy the following conditions:

a. $\forall t\in[0,1]\quad \delta(t)\le t,$
b. $\delta(1)=1,\quad \delta(0)=0,$
c. $\forall s,t\in[0,1]\quad s\le t \;\Rightarrow\; 0\le \delta(t)-\delta(s)\le n(t-s).$

It is well known that every function $\delta:[0,1]\to[0,1]$ which satisfies a,b and c is a diagonal section of some $n$-variate copula. For example, $\delta$ is a diagonal section of the so called *diagonal copula* (see [16], Proposition 2.1)

$$C_\delta(u)=\frac{1}{n}\sum_{i=1}^{n}\min(f(u_{\tau^i(1)}),\ldots,f(u_{\tau^i(n-1)}),\delta(u_{\tau^i(n)})),$$

where

$$f(t)=\frac{nt-\delta(t)}{n-1}\quad\text{and}\quad \tau(k)=\begin{cases}k+1 & \text{for } k=1,\ldots,n-1,\\ 1 & \text{for } \quad k=n.\end{cases}$$

For $n=2$ the above formula simplifies to (compare [22], Sect. 3.2.6)

$$C_\delta(u_1,u_2)=\min\left(u_1,u_2,\frac{1}{2}(\delta(u_1)+\delta(u_2))\right).$$

The tail expansion of $C_\delta$ is completely determined by $\delta$.

**Proposition 8.3.1.** *The diagonal copula $C_\delta$ has a tail expansion of degree d if and only if $\frac{\delta(t)}{t^d}$ has a limit at 0. Furthermore, if*

$$\lim_{t\to 0}\frac{\delta(t)}{t}=\lambda,\quad \textit{then}\quad L(u)=\frac{1}{n}\sum_{i=1}^{n}\min(\frac{n-\lambda}{n-1}u_1,\ldots,\lambda u_i,\ldots,\frac{n-\lambda}{n-1}u_n),$$

*and if $d>1$ and*

$$\lim_{t\to 0}\frac{\delta(t)}{t^d}=\lambda,\quad \textit{then}\quad L(u)=\begin{cases}\frac{\lambda}{n}(u_1^d+\cdots+u_n^d) & \textit{if all } u_i>0,\\ 0 & \textit{if at least one } u_i=0,\end{cases}$$

**Corollary 8.3.1.** *For every $d\ge 1$ there exists an n-variate copula with a nontrivial tail expansion of degree d. Furthermore, for $d>1$ the leading part of such an expansion may not be continuous.*

*Proof.* It is easy to check that the function

$$\delta(t) = \max\left(\frac{t^d}{d}, nt - (n-1)\right)$$

satisfies the conditions a, b and c. Hence it is a diagonal. Furthermore, since

$$\lim_{t \to 0} \frac{\delta(t)}{t^d} = \frac{1}{d} > 0,$$

the associated diagonal copula $C_\delta$ has a nontrivial tail expansion with a leading part which is not continuous for $d > 1$. □

**Corollary 8.3.2.** *There exists an n-variate copula which does not admit a tail expansion of any degree $d \geq 1$.*

*Proof.* It is easy to check that the function

$$\delta(t) = \begin{cases} \max\left(2(t - 2^{-i}), 2^{-i}\right) & \text{for} \quad t \in (2^{-i}, 2^{1-i}], \quad i = 1, 2, \ldots, \\ 0 & \text{for} \quad t = 0, \end{cases}$$

satisfies the conditions a, b and c. Hence it is a diagonal. Furthermore, since the limit at 0 of $\frac{\delta(t)}{t^d}$ does not exist for any $d \geq 1$, the associated diagonal copula $C_\delta$ does not admit a tail expansion of degree $d \geq 1$. □

### 8.3.3 Absolutely Continuous Copulas

Let $C(u)$ be an absolutely continuous copula with density $c(u)$. The tail behaviour of $C$ depends heavily on the tail behaviour of $c$.

**Proposition 8.3.2.** *Let $d$ be a nonnegative constant and $l(u)$ a homogeneous function of degree $d - n$. If*

$$\lim_{u \to 0} \frac{c(u)}{l(u)} = c_0$$

*then $C$ has an expansion of degree $d$ and*

$$L(u) = c_0 \int_0^{u_1} \cdots \int_0^{u_n} l(u) dv_1 \ldots dv_n.$$

*Proof.*

$$\frac{C(tu)}{t^d} = \frac{1}{t^d} \int_0^{tu_1} \cdots \int_0^{tu_n} c(v) dv_1 \ldots dv_n = t^{n-d} \int_0^{u_1} \cdots \int_0^{u_n} c(tv) dv_1 \ldots dv_n$$

$$= \int_0^{u_1} \cdots \int_0^{u_n} \frac{c(tv)}{l(tv)} l(v) dv_1 \ldots dv_n \xrightarrow{t \to 0} \int_0^{u_1} \cdots \int_0^{u_n} c_0 l(v) dv_1 \ldots dv_n.$$

□

The most important example of an absolutely continuous copula is the Gaussian one.

**Proposition 8.3.3.** *The Gaussian copula*

$$C_N(u) = F_{\mathcal{N}}(F^{-1}(u_1), \ldots, F^{-1}(u_n)),$$

*where $F_{\mathcal{N}}$ is the distribution function of an n-dimensional normal distribution ($N(0, (\sigma_{ij})_{i,j=1,\ldots,n})$) with standardized margins (i.e. $\sigma_{ii} = 1$) and $F$ is the distribution function of the standard normal distribution ($N(0,1)$), has a trivial expansion of degree d, for all d satisfying*

$$d < \frac{n^2}{\sum_{i,j=1}^{n} \sigma_{ij}}.$$

The proof of the triviality of the expansion of degree 1 is given in [13] (Proposition 8). The general case can be proved in a similar way.

Another important example is the Student $t$-copula.

**Proposition 8.3.4 (see [23], Theorem 2.3).** *The Student t-copula*

$$C_t(u) = F_{\mathcal{T}}(F^{-1}(u_1), \ldots, F^{-1}(u_n)),$$

*where $F_{\mathcal{T}}$ is the distribution function of an n-dimensional Student t-distribution ($T(0, \Sigma, \nu)$) and $F$ is the distribution function of its one-dimensional margins, has a nontrivial expansion of degree 1.*

A more general family of absolutely continuous copulas consists of the copulas which are $n$-times continuously differentiable on $(0,1)^n$. Their tails were studied in [18] (in connection with Vine Copulas) and [10] (in connection with higher degree expansions).

*Remark 8.3.1.* Note that the leading part $L$ of an absolutely continuous copula $C$ need not be absolutely continuous, even if $C$ is $C^\infty((0,1)^n)$ – see Example 8.3.7 below.

### *8.3.4 Archimedean Copulas*

We recall that a copula $C : [0,1]^n \longrightarrow [0,1]$ is called *Archimedean* if it can be expressed in the form

$$C(u) = \psi\left(\sum_{i=1}^{n} \varphi(u_i)\right),$$

where $\psi : [0,+\infty] \longrightarrow [0,1]$ and $\varphi : [0,1] \longrightarrow [0,+\infty]$ are continuous convex nonincreasing functions such that

$$\psi \circ \varphi = id \quad \text{and} \quad \varphi(1) = 0.$$

Depending on the source, either $\psi$ [6, 21] or $\varphi$ [22] is called the generator. Furthermore, it is known that if $\psi$ and $\varphi$ are as above and moreover $\psi$ is $n-2$ times differentiable and $(-1)^n \psi^{(n-2)}$ is convex (i.e. $\psi$ is an $n$-monotone function) then $\psi(\sum_{i=1}^n \varphi(u_i))$ is a copula ([13], Lemma 3 or [21], Theorem 2.2).

The lower tail expansion of an Archimedean copula depends on the value of its generator $\varphi(x)$ at 0 and on the limit elasticity of $\varphi$ at 0:

$$\varphi_0 = \varphi(0) = \lim_{x\to 0} \varphi(x), \quad \mathscr{E}_x(0) = \lim_{x\to 0^+} \frac{x\varphi'(x)}{\varphi(x)},$$

and also on the index $\kappa$ of regular variation of the function $-\frac{\psi(t)}{\psi'(t)}$ at infinity. Note that $\psi$ and $\varphi$, being convex, are differentiable except at most in a countable set of points.

If $\varphi_0$ is finite then $C_\varphi(u)$ vanishes in some neighbourhood of the origin, so $L(u) = 0$.

**Proposition 8.3.5.** *If $\varphi_0 = +\infty$ and the limit $\mathscr{E}_x(0)$ exists then the Archimedean copula $C_\varphi$ has a uniform lower tail expansion of degree 1.*
*Moreover, if $\mathscr{E}_x(0) = -d$, $0 < d < +\infty$, then*

$$L(u) = \frac{1}{\sqrt[d]{u_1^{-d} + \cdots + u_n^{-d}}};$$

*if $\mathscr{E}_x(0) = -\infty$, then $L(u) = \min(u_1,\ldots,u_n)$; and if $\mathscr{E}_x(0) = 0$, then $L(u) = 0$.*

For the proof see [13], Proposition 10 and [11], Theorem 6 or [4], Theorem 3.1.

In the case of a trivial expansion of degree 1, i.e. when $\mathscr{E}_x(0) = 0$, we get as a direct corollary from [4, Theorem 3.3] the following characterization of expansions of higher degree.

**Proposition 8.3.6.** *If $\varphi_0 = +\infty$, $\mathscr{E}_x(0) = 0$ and the function $-\frac{\psi(t)}{\psi'(t)}$ is regularly varying at infinity of finite index $\kappa$ then $\kappa \le 1$ and the Archimedean copula $C_\varphi$ has trivial tail expansions of degrees smaller than $n^{1-\kappa}$ and does not admit an expansion of degree greater than $n^{1-\kappa}$. Furthermore, if*

$$\lim_{t\to 0} \frac{\delta_n(t)}{t^{n^{1-\kappa}}} = a > 0$$

*then $C_\varphi$ has a tail expansion of degree $n^{1-\kappa}$ and*

$$L(u) = a\prod_{i=1}^{n} u_i^{n-\kappa}.$$

We recall that the *survival copula* of a copula $C$ is given by the following formula:

$$\widehat{C_\varphi}(u) = \mu_C(\times_{i=1}^{n}[1-u_i,1]) = V_C([1-u_1,1]\times\cdots\times[1-u_n,1]),$$

where $V_C$ denotes the induced volume, i.e. the signed sum of the values of $C$ at the vertices of the given rectangle. The lower tail expansion of the survival copula of an Archimedean copula (i.e. the upper tail expansion of the underlying Archimedean copula) depends on the limit elasticity of its generator $\varphi$ at 1,

$$\mathscr{E}_x(1) = \lim_{x\to 0^+}\frac{-x\varphi'(1-x)}{\varphi(1-x)}.$$

**Proposition 8.3.7.** *If the limit $\mathscr{E}_x(1)$ exists then the survival copula of an Archimedean copula $\widehat{C_\varphi}$ has a uniform tail expansion of degree 1.*
*Moreover, if $\mathscr{E}_x(1) = d,\ 1 < d < +\infty$, then*

$$L(u) = (-1)^{n-1}V_{L^*}(0,u), \quad L^*(u) = \sqrt[d]{u_1^d+\cdots+u_n^d};$$

*if $\mathscr{E}_x(1) = +\infty$, then $L(u) = \min(u_1,\ldots,u_n)$; and if $\mathscr{E}_x(1) = 1$, then $L(u) = 0$.*

For the proof see [13], Proposition 9 and [11], Theorem 5 or [4], Theorem 4.1.

When the tail expansion of degree 1 is trivial, i.e. $\mathscr{E}_x(1) = 1$, a further characterization is obtainable when $\psi$ is $n$-times continuously differentiable. Then $C_\varphi$ and $\widehat{C_\varphi}$ are absolutely continuous. The density of $C_\varphi$ is given by the formula

$$c_\varphi(u) = \psi^{(n)}(\varphi(u_1)+\cdots+\varphi(u_n))\prod_{i=1}^{n}\varphi'(u_i)$$

and the density of $\widehat{C_\varphi}$ by

$$\widehat{c_\varphi}(u) = c_\varphi(1-u_1,\ldots,1-u_n).$$

If the limit at 0 of $\psi^{(n)}(t)$ is finite, then also the limit at 1 of $\varphi'(t)$ is finite and

$$c_0 = \lim_{u\to 0}\widehat{c_\varphi}(u) = \lim_{t\to 0}\psi^{(n)}(t)(\lim_{t\to 1}\varphi'(t))^n$$

is finite. We obtain the following corollary from Proposition 8.3.2 (compare [4], Theorem 4.3):

**Corollary 8.3.3.** *If $\mathscr{E}_x(1) = 1$ and the limit at 0 of $\psi^{(n)}(t)$ is finite, then the survival copula of an Archimedean copula $\widehat{C_\varphi}$ has a tail expansion of degree n. Furthermore,*

$$L(u) = \lim_{t \to 0} \psi^{(n)}(t) (\lim_{t \to 1} \varphi'(t))^n \prod_{i=1}^{n} u_i.$$

*Example 8.3.5.* The Clayton copula

$$C_{Cl}(u) = (u_1^{-\theta} + \cdots + u_n^{-\theta} - n + 1)^{-\frac{1}{\theta}}, \ \ \theta > 0,$$

is an Archimedean copula with generators

$$\varphi(x) = \frac{1}{\theta}(x^{-\theta} - 1), \ \ \psi(y) = (1 + \theta y)^{-\frac{1}{\theta}}.$$

We get $\mathscr{E}_x(1) = 1$ and $\mathscr{E}_x(0) = -\theta$. Therefore $C_{Cl}$ has a nontrivial tail expansion of degree 1 with leading part

$$L(u) = (u_1^{-\theta} + \cdots + u_n^{-\theta})^{-\frac{1}{\theta}}.$$

Furthermore, $\widehat{C_{Cl}}$ has a nontrivial expansion of degree $n$ with leading part

$$L(u) = (1 + \theta) \cdot \cdots \cdot (1 + (n-1)\theta) u_1 \ldots u_n.$$

In Fig. 8.1 we show how the order of the tail expansion is reflected by the *corner* distribution of the sample points drawn from the Clayton copula.

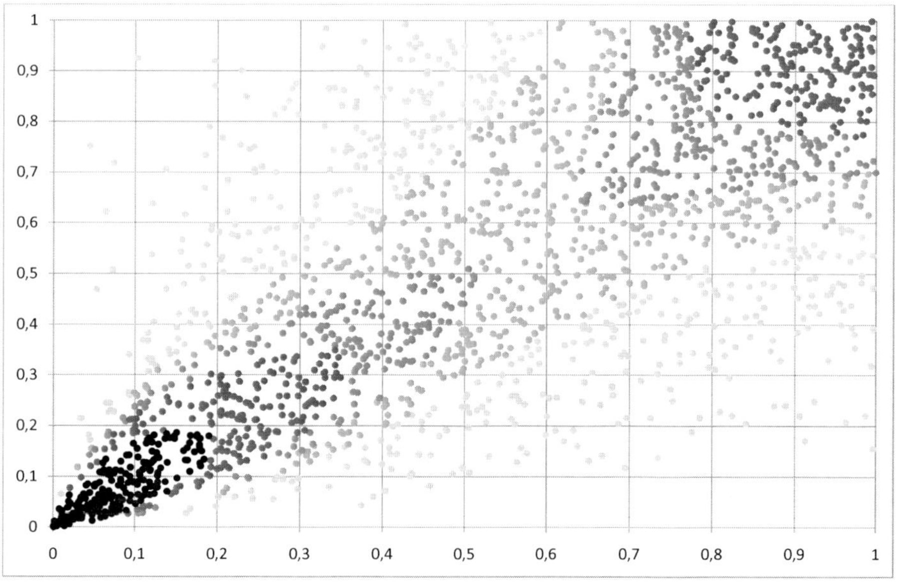


**Fig. 8.1** The scattergraph of the Clayton copula, $\theta = 2$.

*Example 8.3.6.* The Gumbel copula

$$C_G(u) = \exp\left(-((-\ln u_1)^\theta + \cdots + (-\ln u_n)^\theta)^{\frac{1}{\theta}}\right), \quad \theta > 1,$$

is an Archimedean copula with generators

$$\varphi(x) = (-\ln x)^\theta, \quad \psi(y) = \exp\left(-y^{-\frac{1}{\theta}}\right).$$

We get $\mathscr{E}_x(1) = \theta$ and $\mathscr{E}_x(0) = 0$. Therefore $\widehat{C_G}$ has a nontrivial expansion of degree 1 with leading part

$$L(u) = (-1)^{n-1} V_{L^*}(0,u), \quad L^*(u) = \left(u_1^\theta + \cdots + u_n^\theta\right)^{\frac{1}{\theta}}.$$

Furthermore, $C_G$ has a nontrivial tail expansion of degree $n^{1/\theta}$ with leading part

$$L(u) = u_1^{n^{-1+1/\theta}} \ldots \ u_n^{n^{-1+1/\theta}}.$$

In Fig. 8.2 we show how the order of the tail expansion is reflected by the *corner* distribution of the sample points drawn from the Gumbel copula.

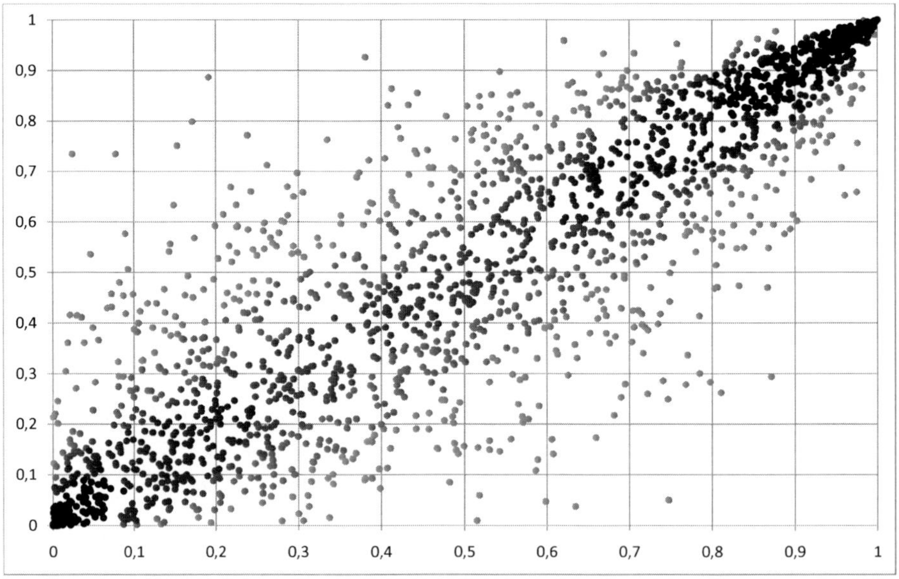


**Fig. 8.2** The scattergraph of the Gumbel copula, $\theta = 3$.

*Example 8.3.7.* The Archimedean copula

$$C(u) = \frac{\theta}{\ln\left(e^{\theta/u_1} + \dots e^{\theta/u_n} - (n-1)e^{\theta}\right)}, \qquad \theta > 0,$$

has generators

$$\varphi(x) = e^{\frac{\theta}{x}} - e^{\theta}, \quad \psi(y) = \frac{\theta}{\ln(e^{\theta}+y)}.$$

We get $\mathscr{E}_x(1) = 1$ and $\mathscr{E}_x(0) = \infty$. Therefore $C$ has a nontrivial tail expansion of degree 1 with leading part

$$L(u) = \min(u_1, \dots, u_n).$$

Furthermore, $\widehat{C}$ has a nontrivial expansion of degree $n$ with leading part

$$L(u) = \left\{ (n-1)! \sum_{i=0}^{n-1} (i+1)!\, \sigma_i \theta^{n-i} \right\} \cdot u_1 \dots u_n,$$

where $\sigma_i$ are symmetric polynomials of $\frac{1}{1}, \frac{1}{2}, \dots, \frac{1}{n-1}$, i.e.

$$\forall x \quad \prod_{i=1}^{n-1}\left(x + \frac{1}{i}\right) = \sum_{i=0}^{n-1} \sigma_i x^{n-i-1}.$$

### 8.3.5 Multivariate Extreme Value Copulas

In modeling a multivariate extreme value distribution many authors based on the following family of copulas (see [9, 19] and [22], Sect. 3.3.4):

$$C_a(u_1, \dots, u_n) = \begin{cases} \exp(-a(-\ln(u_1), \dots, -\ln(u_n)) & \text{if } \prod u_i > 0, \\ 0 & \text{if } \prod u_i = 0, \end{cases}$$

$$a(z_1, \dots, z_n) = \int_{\Delta} \max(w_1 z_1, \dots, w_n z_n) dH(w),$$

where $\Delta$ is the unit simplex in $\mathbb{R}^n$, $w_1 + \cdots + w_n = 1$, and $H$ is a positive dependence measure (spectral measure) subject to

$$\int_{\Delta} w_j dH(w) = 1 \quad j = 1, \dots, n.$$

*Remark 8.3.2.* The function $a$ is called *tail dependence function* (the same as the tail dependence function $L$ introduced in Sect. 8.1!). Being homogeneous of degree 1 it is determined by its restriction to the simplex $\Delta$

$$a(u) = A\left(\frac{u}{u_1 + \cdots + u_n}\right), \quad \text{where } A = a_{|\Delta}.$$

$A$ is known under the name of *Pickands dependence function* (see [9]).

**Proposition 8.3.8 (see [13], Proposition 12).** *Every copula $\widehat{C_a}$ has a nontrivial tail expansion of degree 1 with leading term*

$$L(u) = (-1)^{n+1} V_a(0,u) = \int_{\Delta} \min(w_1 u_1, \ldots, w_n u_n) dH(w).$$

**Proposition 8.3.9.** *If $a(1,\ldots,1) > d$ then the copula $C_a$ has a trivial lower tail expansion of degree $d$.*

The case $d = 1$ was proved in [13] (Proposition 13). The proof of the case $d > 1$ can be done in the same way.

*Remark 8.3.3.* The exceptional case $a(1,\ldots,1) = 1$ occurs only when the measure $H$ is concentrated in the one point (the center of the simplex). In this case $a(z) = \max(z_1,\ldots,z_n)$ and

$$C_a(u) = \exp(-\max(-\ln(u_1),\ldots,-\ln(u_n))) = \min(u_1,\ldots,u_n),$$

hence the lower tail expansion is uniform.

## 8.4 Applications

### 8.4.1 Tail Conditional Copulas

Let $v \in (0,1]^n$. We consider an increasing linear function

$$\psi : [0,1] \longrightarrow [0,1]^n, \quad \psi(t) = \Phi_t(v) = tv = (tv_1,\ldots,tv_n).$$

If $\mu_C$ is a measure on $[0,1]^n$ associated to a copula $C$ such that $C(\psi(t)) > 0$ for $t > 0$, then

$$\upsilon_t(A) = \frac{\mu_C(\Phi_t(A))}{C(\psi(t))}$$

is a probability measure on $\times_{i=1}^n [0,v_i]$. Let $C_t^*$ be the copula associated to $\upsilon_t$.

Let $F_t$ be the distribution function of $\upsilon_t$. Then

$$F_t(x) = \frac{C(\Phi_t(x))}{C(\psi(t))} = P(U_1 \le tx_1,\ldots,U_n \le tx_n \mid U_1 \le tv_1,\ldots,U_n \le tv_n),$$

where $U = (U_1,\ldots,U_n)$ is an $n$-variate random variable with distribution function $C$.

*Remark 8.4.1.* $F_t$ is a rescaled conditional distribution function and $C_t^*$ is a threshold (conditional) copula with thresholds $\psi_i(t) = tv_i,\ i = 1,\dots,n$.

**Proposition 8.4.1.** *If $C$ has a nontrivial degree $d$ continuous leading part $L$ then, as $t$ tends to 0, $C_t^*$ converges to*

$$C_0(u) = \frac{L(L_1^{-1}(u_1),\dots,L_n^{-1}(u_n))}{L(v)},$$

*where*

$$L_i(s) = \frac{L(v_1,\dots,v_{i-1},s,v_{i+1},\dots,v_n)}{L(v)}.$$

*Furthermore if all right-side partial derivatives $\partial_i L(v_1,\dots,v_{i-1},0^+,v_{i+1}\dots,v_n)$ exist and are positive then $C_0$ has a leading part $L_0$ of the same degree $d$,*

$$L_0(u) = L(v)^{d-1} L\left(\frac{u_1}{c_1},\dots,\frac{u_n}{c_n}\right), \quad c_i = \partial_i L(v_1,\dots,v_{i-1},0^+,v_{i+1}\dots,v_n).$$

*Proof.* Since the measures $\mu_{d,t}$, $\mu_{d,t}(A) = t^{-d}\mu_C(\Phi_t(A))$, converge to a measure $\mu$ with a nontrivial distribution function $L$, the probability measures $\upsilon_t$ converge to a measure $\upsilon$ with the distribution function

$$F_0(x) = \frac{L(x)}{L(v)}, \quad x \in \times_{i=1}^n [0,v_i].$$

Since $L$ is continuous the corresponding copula $C_0$ is uniquely determined and therefore it is the limit of $C_t^*$. Since the one-dimensional marginal distributions of $\upsilon$ are

$$L_i(s) = \frac{L(v_1,\dots,v_{i-1},s,v_{i+1},\dots,v_n)}{L(v)},$$

the copula $C_0$ equals

$$C_0(u) = \frac{L(L_1^{-1}(u_1),\dots,L_n^{-1}(u_n))}{L(v)}.$$

To prove the second part of the proposition let us observe that

$$\lim_{s\to 0}\frac{L_i(s)}{s} = \frac{1}{L(v)}\partial_i L(v_1,\dots,v_{i-1},0^+,v_{i+1},\dots,v_n) = \frac{c_i}{L(v)} > 0$$

and

$$\lim_{t\to 0}\frac{L_i^{-1}(t)}{t} = \lim_{s\to 0}\frac{s}{L_i(s)} = \frac{L(v)}{c_i}.$$

Therefore for $u \in (0,+\infty)^n$,

$$\lim_{t\to 0}\frac{C_0(tu)}{t^d} = \lim_{t\to 0}\frac{1}{L(v)}L\left(\frac{u_1L_1^{-1}(tu_1)}{tu_1},\ldots,\frac{u_nL_n^{-1}(tu_n)}{tu_n}\right)$$

$$= \frac{1}{L(v)}L\left(\frac{L(v)u_1}{c_1},\ldots,\frac{L(v)u_n}{c_n}\right) = L(v)^{d-1}L\left(\frac{u_1}{c_1},\ldots,\frac{u_n}{c_n}\right).$$

□

*Remark 8.4.2.* In the case $d = 1$ the leading part $L$ is always continuous and the $c_i$ are not greater than 1. When furthermore all $c_i$ are equal to 1 then both copulas $C$ and $C_0$ have the same leading part $L$.

## 8.4.2 Extreme Value Copulas of a Given Copula

Another approach to describing the tail behaviour of a copula $C$ is to study the limits of $C^k(u_1^{1/k},\ldots,u_n^{1/k})$ and $\widehat{C}^k(u_1^{1/k},\ldots,u_n^{1/k})$ when $k$ tends to infinity. If these limits exist they are called the *upper extreme value* and *lower extreme value copulas* of $C$ [18, 23]

$$C^{LEV} = \lim_{k\to+\infty}\widehat{C}^k(u_1^{1/k},\ldots,u_n^{1/k}),\quad C^{UEV} = \lim_{k\to+\infty}C^k(u_1^{1/k},\ldots,u_n^{1/k}).$$

In [18, Theorem 3.2] it is shown that the extreme value copulas and the corresponding tail dependence functions $a$ can be expressed in terms of the leading parts of the copula $C$ and its marginal copulas. Namely, for any nonempty subset $S = \{i_1,\ldots,i_{|S|}\}$ of the set of indices $I = \{1,\ldots,n\}$ we denote by $\pi_S$ the projection from $\mathbb{R}^n$ to $\mathbb{R}^{|S|}$,

$$\pi_S((u_1,\ldots,u_n)) = (u_{i_1},\ldots,u_{i_{|S|}}),$$

and by $\tau_S$ the right inverse of $\pi_S$ given by the following formula:

$$\tau_S((v_1,\ldots,v_{|S|})) = (u_1,\ldots,u_n),\quad u_i = 1 \text{ if } i\notin S \text{ and } u_i = v_j \text{ if } i = i_j.$$

Furthermore, we denote by $C_S$ the marginal copula induced by $S$,

$$C_S(v) = C(\tau_S(v)).$$

For consistency we put $C_I = C$ and $C_{\{i\}}(v) = v = L_{\{i\}}(v)$.

**Proposition 8.4.2.** *If a copula $C$ and all its marginal copulas $C_S$ admit nontrivial tail expansions of degree 1 then the lower extreme value copula of $C$ exists. Furthermore, it can be expressed as*

$$C^{LEV}(u) = \exp(-a(-\ln(u_1),\ldots,-\ln(u_n))),\quad a(w) = \sum_{\emptyset\neq S\subset I}(-1)^{|S|-1}L_S(\pi_S(w)),$$

*where $L_S$ is the leading part of $C_S$.*

**Proposition 8.4.3.** *Let $C$ be an $n$-dimensional copula. If the survival copula $\widehat{C}$ and all its marginal copulas $\widehat{C}_S$ admit nontrivial tail expansions of degree 1 then the upper extreme value copula of $C$ exists. Furthermore, it can be expressed as*

$$C^{UEV}(u) = \exp(-a(-\ln(u_1),\ldots,-\ln(u_n))), \quad a(w) = \sum_{\emptyset \neq S \subset I} (-1)^{|S|-1} L_S(\pi_S(w)),$$

*where $L_S$ is the leading part of $\widehat{C}_S$.*

### 8.4.3 Regularly Varying Random Vectors with a Given Copula

We recall (see [24], Sect. 2.2) that a random variable $Y$ has a distribution with *regularly varying upper tail* with *index* $\alpha \in \mathbb{R}$ if for every $x > 0$,

$$\lim_{t\to\infty} \frac{1-F_Y(tx)}{1-F_Y(t)} = x^{\alpha}.$$

$\alpha$ is called the *exponent* (*index*) *of variation*.

This notion has the following multivariate generalization (see [24], Chap. 6). The random vector $X$ has a distribution with *regularly varying upper tail with limit function* $\lambda$ if for every $x \in [0,+\infty)^n \setminus \{(0,\ldots,0)\}$,

$$\lim_{t\to\infty} \frac{1-F_X(tx_1,\ldots,tx_n)}{1-F_X(t,\ldots,t)} = \lambda(x).$$

Note that the limit function is homogeneous. Its degree is a generalization of the exponent of variation.

**Proposition 8.4.4.** *Let $X = (X_1,\ldots,X_n)$ be a random vector with distribution function $F_X$ and copula $C_X$. Assume that the components $X_i$ are positive and have continuous distribution functions $F_i$. If the upper tails of the distributions of all the components $X_i$, $i = 1,\ldots,n$, are regularly varying with the same exponent $\alpha \geq 0$ and are tail equivalent:*

$$\lim_{t\to\infty} \frac{1-F_i(t)}{1-F_j(t)} = \frac{r_i}{r_j}, \quad i,j = 1,\ldots,n,$$

*where $r_1,\ldots,r_n \in (0,+\infty)$, then the following conditions are equivalent:*
*i. The survival copula $\widehat{C}$ and all its marginal copulas $\widehat{C}_S$ admit nontrivial tail expansions of degree 1.*
*ii. The random vector $X$ and all the projected vectors $\pi_S(X)$ have distributions with regularly varying upper tails.*
*Furthermore, when the above conditions hold, then*

$$\lim_{t\to\infty}\frac{1-F_X(tx_1,\dots,tx_n)}{1-F_X(t,\dots,t)}=\frac{a(r_1x_1^{\alpha},\dots,r_nx_n^{\alpha})}{a(r_1,\dots,r_n)},\quad a(w)=\sum_{\emptyset\neq S\subset I}(-1)^{|S|-1}L_S(\pi_S(w)),$$

*where $L_S$ is the leading part of $\widehat{C}_S$.*

*Proof.* The implication $ii \Rightarrow i$ follows from [20], Theorem 2.4.
The converse is a consequence of the equality

$$1-C(q_1,\dots,q_n)=\sum_{\emptyset\neq S\subset I}(-1)^{|S|-1}\widehat{C}_S(\pi_S(1-q_1,\dots,1-q_n)).$$

Indeed, due to the uniform convergence we can replace $\widehat{C}_S$ by the leading form $L_S$ and get

$$\lim_{t\to\infty}\frac{1-F_X(tx_1,\dots,tx_n)}{1-F_X(t,\dots,t)}$$

$$=\lim_{t\to\infty}\frac{\sum_{\phi\neq S\subset I}(-1)^{|S|-1}\widehat{C}_S(\pi_S(1-F_1(tx_1),\dots,1-F_n(tx_n)))}{\sum_{\phi\neq S\subset I}(-1)^{|S|-1}\widehat{C}_S(\pi_S(1-F_1(t),\dots,1-F_n(t)))}$$

$$=\lim_{t\to\infty}\frac{\sum_{\phi\neq S\subset I}(-1)^{|S|-1}L_S(\pi_S(1-F_1(tx_1),\dots,1-F_n(tx_n)))}{\sum_{\phi\neq S\subset I}(-1)^{|S|-1}L_S(\pi_S(1-F_1(t),\dots,1-F_n(t)))}$$

$$=\frac{\sum_{\phi\neq S\subset I}(-1)^{|S|-1}L_S(\pi_S(r_1x_1^{\alpha},\dots,r_ntx_n^{\alpha}))}{\sum_{\phi\neq S\subset I}(-1)^{|S|-1}L_S(\pi_S(r_1,\dots,r_n)}=\frac{a(r_1x_1^{\alpha},\dots,r_nx_n^{\alpha})}{a(r_1,\dots,r_n)}.$$

Hence $X$ has a regularly varying upper tail. In the same way we show that the projected vectors $\pi_S(X)=(X_{i_1},\dots,X_{i_{|S|}})$ have this property. □

*Remark 8.4.3.* For a more general study of the relations between multivariate regular variation and the copula tail dependence function the reader is referred to [20].

### 8.4.4 Value at Risk

Decisions in finance are made under uncertainty. The outcome of present decisions depends on quantities (like future stock prices or exchange rates) which are yet unknown. The usual approach is to represent such quantities by random variables. As a consequence, the outcome of a decision (e.g. the future value of an investment) is also a random variable. The randomness adds the *risk dimension* to the problem. A natural question is how to measure the risk. In this section we deal with Value at Risk, nowadays one of the most popular risk measures (for more information about risk measurement the reader is referred to [8] and about risk aggregation to [7]).

We shall consider the following simple case of a portfolio consisting of long positions. An investor has several highly dependent assets in his portfolio. Let $s_i$, $i=1,\dots,n$, be the quotients of the prices of these assets at the end and at the beginning of the investment period. Let $w_i$ be the part of the capital invested in the $i$-th asset,

$\sum w_i = 1$, $w_i > 0$. So the final value of the investment equals

$$W_1(w) = (w_1 s_1 + \cdots + w_n s_n) \cdot W_0,$$

where $W_0 > 0$ is the initial value. "Value at Risk" (VaR), at the confidence level $1-\alpha$, is defined by the formula

$$\mathrm{VaR}_{1-\alpha}(w) = \sup\{V : P(W_0 - W_1(w) \leq V) < 1-\alpha\}.$$

If we denote by $Q_\alpha$ the quantile of the distribution of $\frac{W_1(w)}{W_0}$, then

$$\mathrm{VaR}_{1-\alpha} = W_0(1-Q_\alpha).$$

Roughly speaking, the idea is to determine the biggest amount one can lose at a certain confidence level $1-\alpha$.

We assume that the distribution functions $F_i$ of $s_i$ are continuous. This implies that the copula $C$ is the joint distribution function of the random variables $F_i(s_i)$.

We apply the "quantile transformation" $s_i \mapsto F_i(s_i)$ and reduce our problem to the $\mu_C$-measurement of subsets of the unit multicube $[0,1]^n$.

The set of events such that the final value of the portfolio is less than $r$ ($r > 0$) is transformed into

$$V_r = \left\{ q : \sum_{i=1}^{n} w_i F_i^{-1}(q_i) \leq r \right\}.$$

Note that

$$\mathrm{VaR}_{1-\beta}(w) = (1-r)W_0, \text{ where } \beta = \mu_C(V_r).$$

Since for $r$ close to 0, $V_r$ is contained in a small neighbourhood of 0, we see that the asymptotic behaviour of $\mathrm{VaR}_{1-\alpha}$, when $\alpha \to 0$, depends only on the tail properties of the distributions of $s_i$ and the lower tail of the copula $C$. As an illustration let us consider the following case:

**Model assumptions 1.**
We assume that there exists a positive constant $\bar{x}$ such that:
• $\ln(s_i)$ have continuous cumulative distributions with power-like lower tails with the same index $\gamma > 2$. For $x < -\bar{x}$,

$$F_i(x) = P(\ln(s_i) \leq x) = a_i(-x)^{-\gamma}.$$

• The lower tail part of the copula of $s_i$'s is a nonzero function $L$ which is homogeneous of degree 1. For $q = (q_1, \ldots, q_n)$ such that $0 \leq q_i \leq a_i \bar{x}^{-\gamma}$, $C(q) = L(q)$.
• The measure $\mu$ associated to $L$ is absolutely continuous with respect to the Lebesgue measure and its density is continuous off the origin.

Under these assumptions we get:

**Proposition 8.4.5 (see [12], Theorem 1.1).** *For $\alpha$ close enough to 0,*

$$\mathrm{VaR}_{1-\alpha} = W_0 - W_0 \left(\prod_{i=1}^{n} w_i^{g_i}\right) \exp\left(-\left(\frac{L(a)}{\alpha}\right)^{\frac{1}{\gamma}}\right)(1+O(\alpha^{\frac{1}{\gamma}})),$$

*where the $g_i$ are equal to the elasticities of $L$*

$$g_i = \frac{a_i}{L(a)} \frac{\partial L}{\partial a_i}(a).$$

If we only want to check whether the diversification of the portfolio decreases the risk, i.e. to estimate $\mathrm{VaR}_{1-\alpha}$ of the portfolio in terms of VaR's of its components, we need weaker assumptions on the marginal distributions.

**Model assumptions 2.**
We assume that there is a constant $\delta \in (0,1)$ such that:
**A1.** The joint probability distribution of $s_i$'s is continuous with respect to Lebesgue measure and is determined by a copula $C$ which is a weighted mean of copulas $C_j$, $j=0,\dots,m$, such that for $q=(q_1,\dots,q_n)$ and $0 \le q_i \le \delta$, $C_j(q) = L_j(q)$, where $L_j$ is some nonzero positive homogeneous function of degree $k_j$, $1 = k_0 < k_1 < \dots < k_m$.
**A2.** For $t > 0$ the distribution functions of $s_i$ – $F_i(t)$ are positive and the functions $G_i(t) = \frac{1}{F_i(t)}$ restricted to $t \in F_i^{-1}((0,\delta])$ are convex.
**A3.** For $i = 1,\dots,n$,

$$\forall z>0 \;\; \exists \alpha_0 \;\; \forall 0<\alpha\le\alpha_0 \;\; F_i(z\cdot(F_1^{-1}(\alpha)+\dots+F_n^{-1}(\alpha))) \le \begin{cases} \delta\alpha^{\frac{1}{k_1}} & \text{if } m\ge 1, \\ \delta & \text{if } m=0. \end{cases}$$

**Proposition 8.4.6 (see [14], Theorem 2).** *For $\alpha \in (0,1)$ such that*

$$\sum_{i=1}^{n} w_i F_i^{-1}(\alpha) \le \min\{w_j F_j^{-1}(\delta^*) : j=1,\dots,n\}, \quad \delta^* = \begin{cases} \delta\alpha^{\frac{1}{k_1}} & \text{if } m\ge 1, \\ \delta & \text{if } m=0, \end{cases}$$

*the following inequality holds:*

$$\mathrm{VaR}_{1-\alpha}(w) \le (w_1 \mathrm{VaR}_{1-\alpha}(s_1) + \dots + w_n \mathrm{VaR}_{1-\alpha}(s_n))W_0.$$

The case $m=0$ was considered also in [15].

The leading parts are also applied to study the portfolio consisting of positions which may generate infinitely big losses (see for example [1]). The basic requirement is that the copula $C$ admits a nontrivial expansion of degree 1 and for all $i$ the leading part $L$ satisfies

$$\partial_i L(1,\dots,1,0_i^+,1,\dots,1) = \lim_{t\to 0} \frac{L(1,\dots,1,t_i,1,\dots,1)}{t}$$

$$= \lim_{x\to\infty} L(x,\dots,x,1_i,x,\dots,x) = 1$$

i.e. the threshold probability mass of $\mu_C$ fully concentrates in the corner

$$\lim_{t\to 0}\lim_{s\to 0} P(U_1 \le s, \dots, U_n \le s | U_i \le ts) = 1, \quad i = 1, \dots, n,$$

where $U = (U_1, \dots, U_n)$ is a random variable with distribution function $C$.

*Remark 8.4.4.* In this section we have discussed the case when the loss cannot exceed the invested capital. Hence it is bounded. A more general approach is presented for example in [1, 17].

## References

1. Alink, S., Löwe, M., Wüthrich, M.: Diversification for general copula dependence. Statistica Neerlandica **61**, 446–465 (2007)
2. Billingsley, P.: Probability and Measure. Wiley, New York, NY (1979)
3. Bourbaki, N.: Éléments de mathématique. Intégration. Hermann, Paris (1959, 1963)
4. Charpentier, A., Segers, J.: Tails of multivariate Archimedean copulas. J. Multivar. Anal. **100**, 1521–1537 (2009)
5. Durante, F., Jaworski, P.: A new characterization of bivariate copulas. Commun. Stat. Theory Methods To appear (2010)
6. Durante, F., Sempi, C.: Copula theory: an introduction. In: Jaworski, P., Durante, F., Härdle, W., Rychlik, T. (eds.) Copula Theory and Its Applications, Proceedings of the Workshop, Warsaw, 25–26 Sept 2009. Springer, Dordrecht (2010)
7. Embrechts, P., Puccetti, G.: Risk aggregation. In: Jaworski, P., Durante, F., Härdle, W., Rychlik, T. (eds.) Copula Theory and Its Applications, Proceedings of the Workshop, Warsaw, 25–26 Sept 2009. Springer, Dordrecht (2010)
8. Föllmer, H., Schied, A.: Stochastic Finance. An Introduction in Discrete Time. de Gruyter, Berlin (2004)
9. Gudendorf, G., Segers, J.: Extreme-value copulas. In: Jaworski, P., Durante, F., Härdle, W., Rychlik, T. (eds.) Copula Theory and Its Applications, Proceedings of the Workshop, Warsaw, 25–26 Sept 2009. Springer, Dordrecht (2010)
10. Iwaniec, E.: Characterization of asymptotic expansions of copulas by the use of homogeneous functions. Demonstratio Mathematica **42**, 567–579 (2009)
11. Jaworski, P.: On uniform tail expansions of bivariate copulas. Appl. Math. (Warsaw) **31**, 397–415 (2004)
12. Jaworski, P.: Value at risk in the presence of the power laws. Acta Physica Polonica B **36**, 2575–2587 (2005)
13. Jaworski, P.: On uniform tail expansions of multivariate copulas and wide convergence of measures. Appl. Math. (Warsaw) **33**, 159–184 (2006)
14. Jaworski, P.: Bounds for value at risk for multiasset portfolios. Acta Physica Polonica A **114**, 619–627 (2008)
15. Jaworski, P.: Bounds for value at risk – the approach based on copulas with homogeneous tails. MathWare Soft Comput. **15**, 113–124 (2008)
16. Jaworski, P.: On copulas and their diagonals. Inf. Sci. **179**, 2863–2871 (2009)
17. Joe, H., Li, H.: Tail Risk of Multivariate Regular Variation. Preprint (2009)
18. Joe, H., Li, H., Nikoloulopoulos, A.: Tail dependence functions and vine copulas. J. Multivar. Anal. **101**, 252–270 (2010)
19. Ledford, A., Tawn, J.: Statistics for near independence in multivariate extreme values. Biometrika **83**, 169–187 (1996)
20. Li, H., Sun, Y.: Tail dependence for heavy-tailed scale mixtures of multivariate distributions. J. Appl. Probab. **46**, 925–937 (2009)

21. McNeil, A.J., Nešlehová, J.: Multivariate Archimedean copulas, d-monotone functions and $\lambda_1$-norm symmetric distributions. Ann. Statist. **37**(5B), 3059–3097 (2009)
22. Nelsen, R.B.: An Introduction to Copulas. Springer, New York, NY (2006)
23. Nikoloulopoulos, A., Joe, H., Li, H.: Extreme value properties of multivariate *t* copulas. Extremes **12,** 129–148 (2009)
24. Resnick, S.: Heavy-Tail Phenomena, Probabilistic and Statistical Modeling. Springer, New York, NY (2007)
25. Schwartz, L.: Analyse Mathématique. Hermann, Paris (1967)

# Chapter 9
# Copulae in Reliability Theory (Order Statistics, Coherent Systems)

Tomasz Rychlik

**Abstract** We discuss useful representations of lifetime distributions of coherent systems by means of convex combinations of marginal distributions of order statistics based on the lifetimes of exchangeable components. The representations are applied for characterizing distributions of system lifetimes composed of exchangeable units with given marginal distribution and joint absolutely continuous copula. The characterizations are used for calculating sharp bounds on the expectations and variances of system lifetimes by means of respective parameters of single unit lifetime distribution.

## 9.1 Coherent Systems

In the engineering practice, some sophisticated devices composed of simple units are mathematically modelled by means of coherent systems. Throughout the paper, we denote the number of components by $n$. The coherent system operates iff some fixed combinations of its components work. For instance, the system whose components are connected in a series (the so called series system) works iff all its components do so. The parallel system works iff at least one of its components works. There are popular more general systems which operate iff at least $k$ of its $n$ components, $1 \le k \le n$, do so. They are called $k$-out-of-$n$ systems. Dependence between the working status of the system and those of its components is described by means of its structure function $\varphi : \{0,1\}^n \mapsto \{0,1\}$. By convention, 1 means that the system or the component works, and 0 denotes the respective failure. By definition, we say that a system is coherent if its structure function satisfies two conditions:

- for all $x_i,\ y_i \in \{0,1\}$ such that $x_i \le y_i$, $i = 1,\dots,n$, we have

Tomasz Rychlik
Institute of Mathematics, Polish Academy of Sciences, Toruń, Poland
e-mail: trychlik@impan.gov.pl

P. Jaworski et al. (eds.), *Copula Theory and Its Applications*, Lecture Notes in Statistics 198,
DOI 10.1007/978-3-642-12465-5_9, 

$$\varphi(x_1,\ldots,x_n) \le \varphi(y_1,\ldots,y_n),$$

- For every $1 \le i \le n$ there exist $x_1,\ldots x_{i-1},x_{i+1},\ldots,x_n \in \{0,1\}$ such that

$$\varphi(x_1,\ldots,x_{i-1},0,x_{i+1},\ldots,x_n) < \varphi(x_1,\ldots,x_{i-1},1,x_{i+1},\ldots,x_n).$$

The first condition means that the failure of one or more units cannot result in recovering of the system which is out of order. Certainly, failures of some components may either cause the failure of the working system or they do not change the working status of the system. The latter assumption implies that each component is relevant i.e., there are some combinations of failures of other units such that the failure of the distinguished component results in the break-down of the whole system. The definition implies $\varphi(0,\ldots,0) = 0$ and $\varphi(1,\ldots,1) = 1$ in particular. For instance, we easily check that the structure functions of $k$-out-of-$n$ systems

$$\varphi_k(x_1,\ldots,x_n) = \begin{cases} 1, \text{ if } \sum_{i=1}^n x_i \ge k, \\ 0, \text{ if } \sum_{i=1}^n x_i < k, \end{cases} \qquad 1 \le k \le n,$$

are coherent.

It is assumed that the lifetimes (working times) of system components are random. Let $X_1,\ldots,X_n$ be non-negative random variables representing the lifetimes of consecutive components of the system. Then $W_i(t) = \mathbf{1}_{[t,+\infty)}(X_i)$ describes the working status of the $i$th unit at time $t$, and $W(t) = \varphi(W_1(t),\ldots,W_n(t))$ is the respective working status of the system at the moment. It follows that the system lifetime $T = \int_0^\infty W(t)\,dt$ is a random variable which is a deterministic function of the component lifetimes $X_1,\ldots,X_n$. Moreover, the value of $T$ coincides with either of $X_1,\ldots X_n$ which means that that system fails immediately at the moment of failure of a component. Clearly, the fact that a failure of a given unit causes the derangement of the system, depends on the system structure as well.

The reliability theory is a branch of applied probability theory which is devoted to evaluations of system lifetimes based on analysis of their structures and the distributions of lifetimes of its components. The theory is rapidly developing since early 60s of the former century when the foundations were laid by Birnbaum et al. [6]. The classic references are Barlow and Proschan [4, 5]. We also refer to more recent monographs of Aven and Jensen [3], Gertsbakh [18], Rausand and Høyland [47] and Samaniego [58]. We do not pretend to presenting a comprehensive review of reliability theory here. We focus on a particular problem of analyzing how dependencies between identical component lifetimes affect the lifetime of a system. To this end, we use the notions of order statistics and signatures whose definitions are presented below.

## 9.2 Signatures

We show a useful representation of lifetime distributions of general coherent systems as convex combinations of lifetime distributions of $k$-out-of-$n$ systems which is valid when the joint distribution of component lifetimes is exchangeable. The vector of the combination coefficients depends only on the system structure, and is called the system signature. The distributions of $k$-out-of-$n$ systems depend on the joint lifetime distribution of the components. We also discuss more general mixed systems which are randomly chosen among the systems of fixed size with given choice distribution.

### *9.2.1 Components with i.i.d. Lifetimes*

For simplicity of presentation, we first assume that the unit lifetimes $X_1,\ldots,X_n$ are independent and identically continuously distributed. Let $F$ stand for the common distribution function. Identical distributions mean that the system is composed of identical units, and arbitrary rearrangements of them do not affect the system efficiency. The continuity assumption is naturally preserved if we continuously observe the system. The independence condition is usually more controversial, and a special care should be taken if this can be applied in practical analysis of real technical systems. Nevertheless, a significant part of the system analysis was based on the classic i.i.d. assumption, and the conclusions are useful in practice.

As we already noted, the failure time of the system coincides with the moment of failure of a unit. Moreover, if we know the system structure and the order of failures of its components, we are able to indicate the component whose failure decided on the definite failure of the system. We can also say which consecutive failure of components resulted in that for the system. Shortly, we can check for which $i$ and $j$, $1 \le i, j \le n$, we have $T = X_i$ and $X_{j:n}$, where $X_{j:n}$ denotes the $j$th smallest value among $X_1,\ldots,X_n$, and is called the $j$th order statistic. Samaniego [57] (see also [58, Chap. 3]) presented a useful representation of the system lifetime distribution

$$G(t) = \mathbb{P}(T \le t) = \sum_{i=1}^{n} \mathbb{P}(T = X_{i:n} \le t) = \sum_{i=1}^{n} \mathbb{P}(T = X_{i:n})\mathbb{P}(X_{i:n} \le t) \qquad (9.1)$$

under the assumption that $X_i$, $i = 1,\ldots,n$, are i.i.d. and its common distribution is continuous. Then clearly

$$F_{i:n}(t) = \mathbb{P}(X_{i:n} \le t) = \sum_{k=i}^{n} \binom{n}{k} F^k(t)[1-F(t)]^{n-k}, \quad i = 1,\ldots,n, \qquad (9.2)$$

(cf. e.g., [11, p. 9]) are independent of the system structure. Also,

$$s_i = \mathbb{P}(T = X_{i:n}) = \frac{\#\{\pi \in \Pi : X_{\pi(1)} < \ldots < X_{\pi(n)}, T = X_{\pi(i)}\}}{n!}, \qquad i = 1, \ldots, n, \tag{9.3}$$

is the proportion of these permutations $\pi : \{1,\ldots,n\} \mapsto \{1,\ldots,n\}$ of unit indices corresponding to the consecutive failures of respective units for which the system failure coincides with the $i$th subsequent failure of elements to the total number $n!$ of all rearrangements $\Pi$. Under the assumptions, the probabilities of events $\{X_{\pi(1)} < \ldots < X_{\pi(n)}\}$, $\pi \in \Pi$, are equal and sum up to 1. Pairs $\{T = X_{i:n}\}$ and $\{X_{i:n} \le t\}$, $i = 1,\ldots,n$, $t > 0$, are independent. The Samaniego representation (9.1) is convenient in view of the fact that it is independent of the way of numbering identical system units. Vector $\mathbf{s} = (s_1,\ldots,s_n)$ of non-negative rationals adding up to 1, and representing the system structure is called the Samaniego signature of the system. In particular, for the $k$-out-of-$n$ system, we have $T = X_{n+1-k:n}$, the $k$th greatest order statistic, and so $s_{n+1-k} = 1$, $s_i = 0$, $i \neq n+1-k$.

It is a challenging problem of establishing all the coherent systems of a fixed size $n$. There are only 2 systems of size 2: parallel and series ones. There are 5 and 20 essentially different systems of sizes 3 and 4, which were presented in Kochar et al. [24], and Shaked and Suarez-Llorens [59], respectively. The essential difference means that the systems which differ only by numbering of units are identified. Recently Navarro an Rubio [33] presented an algorithm allowing to determine all the systems of arbitrary size. They established that there are 180 and 16,145 essentially different systems composed of 5 and 6 components. In the paper, they listed all the 5-component systems, and calculated the respective signatures. It is worth pointing out that there are essentially different systems whose signature and the lifetime distributions, in consequence, are equal. For instance, the systems with 4 components, represented by their lifetime functions as $T_1 = \max\{\min\{X_1,X_2\},\min\{X_1,X_3\},\min\{X_2,X_3,X_4\}\}$ and $T_2 = \max\{\min\{X_1,X_2\},\min\{X_3,X_4\}\}$ have the signature $(0,2/3,1/3,0)$ (cf. [34]).

### 9.2.2 Mixed Systems

Representation (9.1) with (9.2) and (9.3) shows that the lifetime distribution of a coherent system is identical with the lifetime of randomly chosen $k$-out-of-$n$ system with probability $s_{n+1-k}$, $k = 1,\ldots,n$. The distribution of the randomization rule coincides with the Samaniego signature $\mathbf{s} = (s_1,\ldots,s_n)$ representing the system structure function. The family of signatures is a finite subset of the simplex $\mathbb{S}^n = \{(s_1,\ldots,s_n) : s_i \ge 0, \sum_{i=1}^n s_i = 1\}$. It certainly contains the vertices, but in general it is difficult to determine.

Boland and Samaniego [8] introduced the notion of mixed systems with $n$ units. By definition, it is the system chosen at random from the family of systems of size $n$ with arbitrary probabilities. By the Samaniego representation, it is equivalent to the system which is randomly chosen from the set of $k$-out-of-$n$ systems, $k = 1,\ldots,n$, with arbitrary probabilities $(s_n,\dots,s_1)\in\mathbb{S}^n$. The mixed system has the lifetime distribution

$$G_{\mathbf{s}}(t)=\sum_{i=1}^{n}s_iF_{i:n}(t)=\sum_{i=1}^{n}s_i\sum_{k=i}^{n}\binom{n}{k}F^k(t)[1-F(t)]^{n-k}$$
$$=\sum_{i=1}^{n}\left(\sum_{k=1}^{i}s_k\right)\binom{n}{k}F^k(t)[1-F(t)]^{n-k}. \tag{9.4}$$

In some aspects, analysis of mixed systems is easier than for the proper coherent systems. Especially for the systems of large sizes, optimization problems can be easier solved analytically for the mixed systems represented by the signatures from simple convex sets $\mathbb{S}^n$ than for the coherent systems with discrete subsets of the simplices. Moreover, the family of all mixed systems of size $n$ is actually identical with the family of mixed systems of size not greater than $n$. It follows from the fact that each order statistic from the samples of size $n-1$ is equal in law to a random mixture of order statistics of size $n$, because

$$F_{i:n-1}=\left(1-\frac{i}{n}\right)F_{i:n}+\frac{i}{n}F_{i+1:n},\qquad 1\le i\le n-1. \tag{9.5}$$

In particular, we can check that the uniform mixture of $n$ $k$-out-of-$n$ systems with i.i.d. components is equivalent to the system with one component.

Note that the distribution function of (9.4) is the composition of the polynomial

$$H_{\mathbf{s}}(x)=\sum_{i=1}^{n}s_iH_{i:n}(x)=\sum_{i=1}^{n}s_i\sum_{k=i}^{n}\binom{n}{k}x^k(1-x)^{n-k}$$
$$=\sum_{i=1}^{n}\left(\sum_{k=1}^{i}s_k\right)\binom{n}{k}x^k(1-x)^{n-k},\qquad 0\le x\le 1, \tag{9.6}$$

of degree $n$ at most, and the component lifetime distribution function $F$. Evidently, (9.6) is strictly increasing on $[0,1]$, and $H_{\mathbf{s}}(0)=0$ and $H_{\mathbf{s}}(1)=1$. So this is the distribution function supported on the unit interval. It has the representation $H_{\mathbf{s}}(x)=\sum_{i=1}^{n}b_ix^i$ with possibly negative coefficients such that $\sum_{i=1}^{n}b_i=H_{\mathbf{s}}(1)=1$. The respective reliability function $\overline{H}_{\mathbf{s}}=1-H_{\mathbf{s}}$ can be written as $\overline{H}_{\mathbf{s}}(x)=\sum_{i=1}^{n}a_i(1-x)^i$, where non-necessarily non-negative coefficients sum up to $\sum_{i=1}^{n}a_i=\overline{H}_{\mathbf{s}}(0)=1$. Due to representations

$$F_{\mathbf{s}}(t)=\sum_{i=1}^{n}b_iF_{i:i}(t), \tag{9.7}$$

$$\overline{F}_{\mathbf{s}}(t)=\sum_{i=1}^{n}a_i\overline{F}_{1:i}(t), \tag{9.8}$$

vectors $\mathbf{b}=(b_1,\dots,b_n)$ and $\mathbf{a}=(a_1,\dots,a_n)$ are called the system maximal and minimal signatures, respectively. They were introduced by Navarro et al. [38], and applied for analysis of asymptotic properties of the system lifetimes. Behavior of the system lifetime for large arguments depends on the first non-zero element of the minimal signature and the right-hand tail behavior of the parent reliability function $\overline{F}$. The system lifetime at the left-end of its support depends on the first non-zero coefficient of the maximal signature and properties of $F$ at the left-end. We finally mention the notion of dynamic minimal signatures introduced by Navarro et al. [39] which allows to represent the distribution of the residual lifetime of the system $(X-t|T>t)$ as a linear combination of series system distributions composed of units with residual lifetimes $(X-t|X>t)$.

### 9.2.3 Components with Exchangeable Lifetimes

The assumption that identical components have independent lifetimes seems to be inadequate in various practical applications. It often happens that the failure of one or several units increases the burden on the others, and that results in shortening their operating times. A natural relaxation of the i.i.d. model consists in assuming that the lifetimes are exchangeable. It means that $X_1,\dots,X_n$ are possibly dependent, and their joint distribution is identical with the distribution of arbitrary permutation of $X_1,\dots,X_n$. In particular, it implies that all the one-dimensional marginals are identical. Practically, the exchangeability assumption means that the system is built of $n$ identical units, and replacing the roles of two ot more units in the system structure does not affect its reliability. Navarro and Rychlik [34, Lemma 1] formally proved that for the coherent system whose components have the exchangeable lifetimes $X_1,\dots,X_n$ satisfying

$$\mathbb{P}(X_i = X_j) = 0, \qquad i \neq j, \tag{9.9}$$

the Samaniego representation (9.1) is possible. This was known and applied earlier by several authors (see, e.g., [7, 37]). Assumption (9.9) implies that $\mathbb{P}(X_{1:n} < \ldots < X_{n:n}) = 1$, and is satisfied when the joint distribution is absolutely continuous. The Samaniego signature can be calculated by means of (9.3) as in the i.i.d. case, but the distribution functions of order statistics differ from (9.2) when the lifetimes are actually dependent. Navarro et al. [40] showed that the representation

$$\mathbb{P}(T \leq t) = \sum_{i=1}^{n} s_i \mathbb{P}(X_{i:n} \leq t) \tag{9.10}$$

still holds if we get rid of (9.9). However, then the interpretation $s_i = \mathbb{P}(T = X_{i:n})$, $i = 1,\dots,n$, is not valid since the probabilities do not sum up to one. The coefficients of the Samaniego vector are calculated with use of the last expression in (9.3) as if the continuity assumption excluding the knots were valid. These results allow to introduce the notion of mixed systems in the exchangeable model as well. The mixed system is a randomly chosen coherent system composed of $n$ units. It is equivalent to a mixed system composed of $n$ $k$-out-of-$n$ systems with arbitrary probabilities $s_n,\dots,s_1$. The distribution of each order statistic $X_{i:n}$ can be represented by means of linear combinations of marginal distribution or reliability functions of dimensions not exceeding $n$ at a given common level (cf. [11, p. 99]). These represent the distribution and reliability functions of the parallel and series systems, respectively, of smaller sizes. In the exchangeable case, we have the formulae

$$\mathbb{P}(X_{i:n} \le t) = \sum_{j=i}^{n} (-1)^{j-i} \binom{j-1}{i-1} \binom{n}{j} F_{j:j}(t),$$

$$\mathbb{P}(X_{i:n} \ge t) = \sum_{j=n+1-i}^{n} (-1)^{j-n-1+i} \binom{j-1}{n-i} \binom{n}{j} \overline{F}_{1:j}(t),$$

(cf. [11, p. 46]). Accordingly, there exist maximal and minimal signature representation (9.7) and (9.8), respectively, of every mixed system with $n$ elements. As in the i.i.d. case, they may contain negative elements, and both add up to 1. They were defined and studied by Navarro et al. [37] (see also [35]).

We now come back to analysis of marginal distribution functions of order statistics in the exchangeable case which in the Samaniego representation (9.10) play the role as important as the signature. Rychlik [49] proved that the vector of distribution functions $(F_1,\dots,F_n)$ of consecutive order statistics is characterized by two relations

$$F_1 \ge \dots \ge F_n, \tag{9.11}$$

$$\textstyle\sum_{i=1}^{n} F_i = nF, \tag{9.12}$$

where $F$ is the common marginal distribution function of each $X_i$. The former condition is evident, and the latter is the consequence of taking expectations of both sums of the trivial identity $\sum_{i=1}^{n} \mathbf{1}_{(-\infty,t]}(X_{i:n}) = \sum_{i=1}^{n} \mathbf{1}_{(-\infty,t]}(X_i)$. It asserts that the uniform combination of all $k$-out-of-$n$ systems works as a single component. Moreover, relation (9.5) is still valid in the exchangeable case which allows to represent all the coherent and mixed systems of sizes up to $n$ as mixed systems of the maximal size.

A construction of an exchangeable sequence with marginal $F$ whose order statistics satisfy (9.11) and (9.12) is quite simple. We first determine the ordered variables taking the quantile transformations $X_{i:n} = F_i^{-1}(U)$, $i = 1,\dots,n$, of a single standard random variable $U$. Then we define $X_1,\dots,X_n$ as an arbitrarily chosen random permutation of $X_{1:n},\dots,X_{n:n}$. In this construction, the copula of ordered variables coincides with the Fréchet-Hoeffding upper bound copula $M(x_1,\dots,x_n) = \min\{x_1,\dots,x_n\}$. The support of the exchangeable copula of $X_1,\dots,X_n$ consists of $n!$ curves $x \mapsto (F_{\pi(1)}(F^{-1}(x)),\dots,F_{\pi(n)}(F^{-1}(x)))$, $0 \le x \le 1$, $\pi \in \Pi$, the probabilities of all curves are identical, and the probability of a curve section is the length of its projection onto the $x$-axis multiplied by the scaling factor $1/n!$.

We finally note that (9.11) and (9.12) characterize the distributions of order statistics for unnecessarily exchangeable possibly dependent random variables with a common marginal $F$. For instance, we define non-exchangeable random variables with given distributions of order statistics when we replace all the permutations in the above construction by its subset $\{\pi_i : i = 0, \dots, n-1\} \subset \Pi$ with $\pi_i(j) = i + j \pmod n$, $j = 1, \dots, n$. This example shows that in the context of order statistics many problems for exchangeable and identically distributed samples have identical solutions. If we construct a vector of identically distributed non-exchangeable random variables with the distribution of order statistics satisfying a desired property, we derive the same distribution of order statistics when we rearrange numbering of the original variables. Taking the uniform mixture over all $n!$ rearrangements, we obtain an exchangeable joint distribution preserving the distribution of order statistics with the desired property.

## 9.3 Bounds for Exchangeable Lifetime Components

Applying the characterizations (9.11) and (9.12), we determine sharp lower and upper bounds on the system lifetime distributions, dependent on the system signature and common marginal lifetime distribution of exchangeable units. We also establish bounds on the system lifetime expectations, dependent on the distribution function, mean, and mean and variance of the component lifetime.

### *9.3.1 Distribution Bounds*

Suppose that the components of a system have exchangeable lifetimes $X_1, \dots, X_n$ with a common known marginal $F$. Without loss of generality, we can relax the exchangeability conditions and assume that $X_1, \dots, X_n$ are merely identically distributed. Let $\mathbf{s} = (s_1, \dots, s_n) \in \mathbb{S}^n$ be the Samaniego signature of the system. We admit that $\mathbf{s}$ is arbitrarily fixed, and corresponds to either proper coherent or mixed system. We present sharp lower and upper bounds on the lifetime distribution function of the system, dependent on the parent marginal $F$ and signature $\mathbf{s}$. These are based on a construction of Rychlik [49] applied for evaluating expectations of linear combinations of order statistics from identically distributed dependent samples. The construction is based on the following notions. Define $\underline{S} : [0,1] \mapsto [0,1]$ as the greatest convex function satisfying $\underline{S}(\frac{j}{n}) \le \sum_{i=1}^{j} s_i$, $j = 0, \dots, n$. This is called the greatest convex minorant of points $(\frac{j}{n}, \sum_{i=1}^{j} s_i)$, $j = 0, \dots, n$. Let $0 = j_0 < \dots < j_M = n$ for some $1 \le M \le n$ denote all the integers satisfying $\underline{S}(\frac{j_m}{n}) = \sum_{i=1}^{j_m} s_i$, $m = 1, \dots, M$. Note that $\underline{S}$ is a piecewise linear, non-decreasing, and convex function satisfying $\underline{S}(0) = 0$ and $\underline{S}(1) = 1$. It has a non-decreasing derivative almost everywhere. Similarly, we define $\overline{S} : [0,1] \mapsto [0,1]$ as the smallest concave majorant of points $(\frac{j}{n}, \sum_{i=1}^{j} s_i)$, $j = 0, \dots, n$, and determine integers $0 = k_0 < \dots < k_L = n$ for some $1 \le L \le n$ such that $\overline{S}(\frac{k_l}{n}) = \sum_{i=1}^{k_l} s_i$, $l = 1, \dots L$. Under the above assumptions and notation, we have

$$\underline{S}(F(t)) \le G(t) = \mathbb{P}(T \le t) \le \overline{S}(F(t)) \tag{9.13}$$

for all real $t$. The lower and upper bounds are attained at every argument $t$ by the joint distribution functions of component lifetimes satisfying

$$\mathbb{P}\left(F^{-1}\left(\frac{j_{m-1}}{n}\right) \le X_{j_{m-1}+1:n} = \ldots = X_{j_m:n} \le F^{-1}\left(\frac{j_m}{n}\right)\right) = 1,\ m = 1,\ldots,M, \tag{9.14}$$

$$\mathbb{P}\left(F^{-1}\left(\frac{k_{l-1}}{n}\right) \le X_{k_{l-1}+1:n} = \ldots = X_{k_l:n} \le F^{-1}\left(\frac{k_l}{n}\right)\right) = 1,\ l = 1,\ldots,L, \tag{9.15}$$

respectively, together with (9.11) and (9.12). Bounds (9.13) are obtained by minimizing and maximizing the linear combination $\sum_{i=1}^{n} s_i F_i(t)$ under the constraints (9.11) and (9.12) for arbitrarily fixed values of $F(t) \in [0,1]$. These are linear programming problems. It is worth pointing out that there exist distributions which attain the bounds uniformly for all values of $F(t) \in [0,1]$.

We present a construction of the copula of order statistics satisfying conditions (9.11), (9.12) and (9.14), which leads to the lower distribution bounds. We simply take $U_{j_{m-1}+1:n} = \ldots = U_{j_m:n} = V_m$, $m = 1,\ldots,M$, where $V_m$ is uniformly distributed on $\left[\frac{j_{m-1}}{n}, \frac{j_m}{n}\right]$. A random permutation of $U_{1:n},\ldots,U_{n:n}$ has an exchangeable joint distribution and standard uniform marginals. Taking the quantile transformation $F^{-1}$ of the permuted variables we complete the construction. There are no special assumptions on the joint distributions of $V_1,\ldots,V_M$. For instance, they can be independent or functionally dependent, e.g., $V_m = \frac{j_{m-1}}{n} + \frac{j_m - j_{m-1}}{j_1} V_1$, $m = 2,\ldots,M$. The construction for the upper bound is analogous. Navarro and Rychlik [34] proved that the evaluations (9.13) are also optimal for components with absolutely continuous joint distributions. The main obstacles are here (9.14) and (9.15) which force some lifetimes to be identical almost surely. One can overcome them by use of ordinal sum of copulae (see [28]). We take $V_1,\ldots,V_M$ independent, and partition each $\left[\frac{j_{m-1}}{n}, \frac{j_m}{n}\right]$ onto a large number of small intervals. If $V_m$ falls into some interval, we take $j_m - j_{m-1}$ independent random variables uniformly distributed on the interval. Completing the above construction results in exchangeable standard uniform random variables $U_1,\ldots,U_n$ with a joint density. Increasing the number of partitions so that the interval lengths tend to zero, we approach the lower bound in (9.13) uniformly. Of course, we get the same result once we replace the product distributions used for constructing $V_1,\ldots,V_M$ and the random variables supported on small subintervals by other absolute continuous copulae.

We note that in the case that each $j_m/n$, $m = 1,\ldots,M-1$, is the breaking point of the broken line $\underline{S}$ (i.e. $\underline{S}'(j_m -) < \underline{S}'(j_m/n+)$), then conditions (9.11), (9.12) and (9.14) uniquely determine the marginal distributions of order statistics. Otherwise condition (9.14) may be weakened: the indicated order statistics should be still mutually equal, but their support can be extended. An analogous comment concerns the necessity condition for the upper bound. In the special case of order statistics that correspond to the lifetimes of $k$-out-of-$n$ systems we have

$$\max\left\{\frac{nF(t)-n+k}{k},0\right\} \le G(t) = \mathbb{P}(X_{n+1-k:n} \le t) \le \min\left\{\frac{nF(t)}{n+1-k},1\right\}, \quad (9.16)$$

and the necessary conditions for the lower and upper bounds attainability are

$$\mathbb{P}\left(X_{n-k:n} \le F^{-1}\left(1-\frac{k}{n}\right) \le X_{n+1-k:n} = \ldots = X_{n:n}\right) = 1, \quad (9.17)$$

$$\mathbb{P}\left(X_{1:n} = \ldots = X_{n+1-k:n} \le F^{-1}\left(1-\frac{k-1}{n}\right) \le X_{n+2-k:n}\right) = 1, \quad (9.18)$$

respectively. The evaluations were first presented in Caraux and Gascuel [10] and Rychlik [48]. For the parallel systems in particular, the bounds

$$\max\{nF(t)-n+1,0\} \le \mathbb{P}(X_{n:n} \le t) \le F(t) \quad (9.19)$$

correspond to the special case of the Fréchet-Hoeffding bounds (see, e.g., [42, p. 47]) with the specific diagonal argument $(x,\ldots,x) \in [0,1]^n$. In contrast to the general argument case, the lower bounds are here uniformly attained by the same distributions. The upper bound is trivially attained by identical random variables with marginal $F$ (cf. (9.18)). Mallows [27] constructed an absolutely continuous copula providing the equality in the upper bound of (9.19). Other constructions based on transformations of a single random variable were presented in Lai and Robbins [25, 26]. They also presented sequences of random variables attaining the lower bound in (9.19) for each $n$, and established their asymptotic distributions. Generalizations of (9.16) and (9.19) to the samples with non-identical marginal distributions consist in replacing $nF$ by the sum of marginals. Lai and Robbins [25] verified attainability of the respective lower bound for the sample maximum, and Tchen [61] determined sequences satisfying the property for all $n$. Rychlik [52] described the conditions under which the corresponding lower and upper bounds for arbitrary order statistics are attainable.

### *9.3.2 Expectation Bounds*

For the mixed system with signature **s** composed of exchangeable units with common lifetime distribution $F$, (9.13) implies the following sharp bounds for the lifetime expectation

$$\int_0^1 F^{-1}(t)\overline{S}'(t)dt \le \mathbb{E}T \le \int_0^1 F^{-1}(t)\underline{S}'(t)dt. \quad (9.20)$$

Since the above derivatives are constant on every interval $[\frac{i-1}{n},\frac{i}{n}]$, $i=1,\ldots,n$, the bounds simplify to linear combinations of the integrals of the quantile function on the intervals. We denote the respective constant levels by $\overline{s}_1 = \ldots \overline{s}_{k_1} \ge \ldots \ge \overline{s}_{k_{L-1}+1} = \ldots = \overline{s}_n$, and $\underline{s}_1 = \ldots \underline{s}_{j_1} \le \ldots \le \underline{s}_{j_{M-1}+1} = \ldots = \underline{s}_n$, respectively. The attainability conditions (9.11), (9.12) with either (9.14) or (9.15), respectively, for the upper and lower inequalities are identical with those for the lower and upper distribution bounds, respectively. Evaluations (9.20) are identical with those for the expectations of convex combinations of order statistics $\sum_{i=1}^{n} s_i X_{i:n}$ in the exchangeable model. Rychlik [49] proved that they hold for arbitrary linear combinations. In the special case of $k$-out-of-$n$ system lifetimes, (9.20) rewrites as

$$\frac{n}{n+1-k}\int_0^{1-\frac{k-1}{n}} F^{-1}(t)dt \leq \mathbb{E}X_{n+1-k:n} \leq \frac{n}{k}\int_{1-\frac{k}{n}}^{1} F^{-1}(t)dt.$$

It often happens that we do not know precisely the marginal distribution of the component lifetime, but only some its parameters like mean and variance. Formulae (9.20) make it possible to evaluate the system lifetime in terms of the mean component lifetime as follows

$$\mathbb{E}T \leq \int_0^1 F^{-1}(t)\underline{S}'(t)dt \leq \operatorname{ess\,sup}_{0\leq t\leq 1}\underline{S}'(t)\int_0^1 F^{-1}(t)dt = \underline{S}'(1-)\mu = \underline{s}_n\mu. \tag{9.21}$$

The equality in the latter inequality appears if $F^{-1}$ is positive on a left neighborhood of 1 where $\underline{S}'$ attains its maximum, and vanishes elsewhere. This means that the component lifetime distribution $F$ has an atom at 0 with a mass not less than the argument of the last breaking point of $\underline{S}$. Conditions (9.11), (9.12) and (9.14) on the joint distribution assert the equality in the first inequality. Similarly, we get the lower bound $\mathbb{E}T \geq \overline{s}_n\mu$ with the respective attainability conditions.

More subtle bounds, expressed in terms of both mean and variance of exchangeable components are established by use of the Schwarz inequality

$$\mathbb{E}T - \mu \leq \int_0^1 [F^{-1}(t)-\mu]\underline{S}'(t)dt = \int_0^1 [F^{-1}(t)-\mu][\underline{S}'(t)-1]dt$$

$$\leq \left[\int_0^1 [F^{-1}(t)-\mu]^2dt\int_0^1 [\underline{S}'(t)-1]^2dt\right]^{1/2} = \sigma\left[\frac{1}{n}\sum_{i=1}^{n}\underline{s}_i^2 - 1\right]^{1/2} = \sigma B_{\mathbf{s}}, \tag{9.22}$$

say (cf. [50]). The equality in the Schwarz inequality is attained by the marginal distribution satisfying

$$\frac{F^{-1}(x)-\mu}{\sigma} = \frac{\underline{S}'(x)-1}{B_{\mathbf{s}}} = \frac{\sum_{i=1}^{n}\underline{s}_i\mathbf{1}_{[(i-1)/n,i/n)}(x)-1}{B_{\mathbf{s}}}, \tag{9.23}$$

which assign probabilities $\frac{j_m - j_{m-1}}{n}$ to the points $\mu+\sigma(\underline{s}_{j_m}-1)/B_{\mathbf{s}}$, $m = 1,\ldots,M$. Adding the conditions for the joint distribution, implies deterministic distributions of order statistics $X_{i:n} = x_{i:n} = \mu+\sigma(\underline{s}_i - 1)/B_{\mathbf{s}}$, $i = 1,\ldots,n$. The respective sampling model is the exclusive drawing from the $n$-valued population $x_{1:n},\ldots,x_{n:n}$. Since the common marginal is discrete, there is not a unique copula representing the dependence structure. The corresponding lower bound is

$$\mathbb{E}T-\mu \geq -\sigma\left[\frac{1}{n}\sum_{i=1}^{n}\bar{s}_i^2-1\right]^{1/2} = -\sigma b_{\mathbf{s}}, \tag{9.24}$$

attained in the exhaustive drawing without replacement model from the set $\mu + \sigma(\bar{s}_i - 1)/b_{\mathbf{s}}$, $i=1,\dots,n$. In the special case of $k$-out-of-$n$ systems, we obtain

$$0 \leq \frac{\mathbb{E}X_{n+1-k:n}}{\mu} \leq \frac{n}{k}, \tag{9.25}$$

$$\sqrt{\frac{k-1}{n+1-k}} \leq \frac{\mathbb{E}X_{n+1-k:n}-\mu}{\sigma} \leq \sqrt{\frac{n-k}{k}}, \quad k=1,\dots,n, \tag{9.26}$$

(in the latter case cf. [17], and [1] for the special case $k=1$). Note that the upper bound in (9.22) may vanish. This happens for the systems with signatures satisfying $\sum_{i=1}^{j} s_i \geq \frac{j}{n}$, $j=1,\dots,n-1$, which holds for the series system in particular. Then $\underline{S}(x)=x$, $0\leq x\leq 1$, and $\underline{s}_1 = \dots = \underline{s}_n = 1$, and $B_{\mathbf{s}}=0$. Formula (9.23) cannot be written and serve as the attainability condition. However, Goroncy [19] proved that for such systems the zero mean-variance are still sharp, and are attained in the limit by some sequences of product lifetime distributions. Similar remarks can be formulated for the lower bound (9.24).

There are known sharp mean-variance bounds on the lifetimes of $k$-out-of-$n$ systems with exchangeable components whose marginal lifetime distributions belong to restricted families of distributions. For instance, families of DFR, IFR, DFRA nad IFRA distributions were studied in Rychlik [54]. A general idea of such evaluations for various non-parametric families of marginals and some further examples was presented in Rychlik [55, Chap. 5]. Other refinements are based on additional assumptions about the dependence structure of unit lifetimes. Some achievements in this direction were presented by Kemperman [23]. We note that the bounds on the expected lifetime of systems linearly depend on location and scale transformations of the lifetime distributions. For instance, in bound (9.22) they are represented by the mean and standard deviation of the unit lifetimes. In (9.21), the mean is the scale parameter, whereas the left-end support point 0 is the location. Other pairs of location and scale were also studied in literature (see, e.g., [2, 50, 53]). We finally mention that the counterparts of (9.25) and (9.26) in the case of i.i.d. component lifetimes were established by Papadatos [46] and Moriguti [30], respectively.

## 9.4 Characterizations of *k*-Out-of-*n* System Lifetime Distributions

We characterize the distributions of $k$-out-of-$n$ systems whose components have a joint distribution with arbitrary exchangeable copula, and with absolutely continuous exchangeable copula. Then we use the characterization for calculating precise bounds on the system lifetime variance, expressed in terms of the marginal distribution of the exchangeable units, and respective variance.

### *9.4.1 General Copula Joint Distribution*

It is a natural and important problem of characterizing all possible system lifetime distributions of mixed systems whose units have exchangeable lifetimes with a fixed marginal distribution $F$. This is equivalent with describing all the convex combinations $\sum_{i=1}^{n} s_i F_i$ of distribution functions satisfying (9.11) and (9.12) with fixed coefficients representing the signature of the system. This is still a challenging open problem, and we know its solution only in the special cases of $k$-out-of-$n$ systems. Rychlik [51] proved that these distributions are fully determined by conditions (9.16) together with

$$0 \le G(t) - G(s) = \mathbb{P}(s < X_{n+1-k:n} \le t) \le n[F(t) - F(s)], \quad 0 \le s < t. \tag{9.27}$$

The first asserts that the distribution function should run between two extreme elements of the class, determined in Sect. 9.3.1. The other says that $G$ is absolutely continuous with respect to $F$, and the density is bounded by the number of units $n$. We immediately obtain it from the relations

$$0 \le \mathbb{P}(s < X_{n+1-k:n} \le t) \le \sum_{i=1}^{n} \mathbb{P}(s < X_{i:n} \le t) = n[F(t) - F(s)],$$

following from (9.12). The proof of the fact that the simple conditions (9.16) and (9.27) are sufficient consists in constructing $n-1$ distribution functions $F_1, \dots, F_{n-k}$, $F_{n+2-k}, \dots F_n$ which together with $G = F_{n+1-k}$ satisfy conditions (9.11) and (9.12). It is easy to check that

$$F_1 = \dots = F_{n-k} = \min\left\{\frac{nF - G}{n-k}, 1\right\}, \qquad \text{if } k < n, \tag{9.28}$$

$$F_{n+2-k} = \dots = F_n = \max\left\{\frac{nF - G - n + k}{k-1}, 0\right\}, \quad \text{if } k > 1, \tag{9.29}$$

are the distribution functions that actually satisfy the conditions. Further construction of exchangeable random variables with marginal $F$ and distribution functions of order statistics $F_1, \dots, F_n$ is analogous to that presented in Sect. 3.1. Obviously, the respective copulae of order statistics and exchangeable components are supported on one-dimensional curves. Moreover, constructions (9.28) and (9.29) imply that $n$-component random vector $X_1, \dots, X_n$ takes only 2 values in the cases of parallel and series systems, and 3 values otherwise.

Since $F_2 = 2F - F_1$ in the case $n = 2$, we easily conclude the characterization of lifetime distributions of two-component mixed systems with signature $(s_1, s_2) = (s_1, 1 - s_1)$, $0 \le s_1 \le 1$, by the pairs of relations

$$\max\{2s_1F(t), 2s_2[F(t)-1]+1\} \le G(t) \le F(t),$$
$$2s_1[F(t)-F(s)] \le G(t)-G(s) \le 2s_2[F(t)-F(s)],$$

when $s_1 \le 1/2 \le s_2$, and

$$F(t) \le G(t) \le \min\{2s_1F(t), 2s_2[F(t)-1]+1\}$$
$$2s_2[F(t)-F(s)] \le G(t)-G(s) \le 2s_1[F(t)-F(s)],$$

when $s_1 \ge 1/2 \ge s_2$. It follows that in the boundary case $s_1 = s_2 = 1/2$, each of the relations characterizes uniquely the distribution function $G = F$. It can be easily seen that for $n \ge 3$, we have similar conditions (9.16) together with

$$n \min_{1\le i\le n} s_i[F(t)-F(s)] \le G(t)-G(s) \le n \max_{1\le i\le n} s_i[F(t)-F(s)],$$

but it seems that they are not sufficient for characterizing the system lifetime distributions.

Observe that the problem of characterizing the lifetime distribution of the parallel system with unit lifetimes uniformly distributed on $[0,1]$ is equivalent with characterizing the diagonal sections of copulae. In the case $n = 2$, a characterization equivalent to (9.19) and (9.27), and an exemplary construction of bivariate copula with arbitrarily fixed diagonal were presented in Fredericks and Nelsen [16] (see also [43] and [42, pp. 84–85]) Other constructions of copulae with given marginal sections were described, e.g., in Durante et al. [13, 14] and Nelsen et al. [44].

### *9.4.2 Absolute Continuous Copula Joint Distribution*

Constructions of copulae with given diagonal sections cited above provided singular distributions. The joint component lifetime distributions with given distributions of order statistics have same properties: all the components fail at 3 different moments at most. It does not reflects practical situations. Usually, a failure of a component may cause further failures in a short time, but common failures of several units just at the same time are unlikely. In an adequate mathematical model, it is assumed that each unit lifetime has a common density function as well as the dependence structure is represented by an exchangeable absolutely continuous copula. We focus on the latter problem. Durante and Jaworski [12] presented bivariate absolute continuous copulae with fixed diagonal sections. The construction was generalized by Jaworski [21] to higher dimensions. Another approach to the problem was presented in Erdely and Gonzáles-Barrios [15]. Jaworski and Rychlik [22] modified the methods of Durante and Jaworski for constructing absolute continuous copulae with given distributions of $k$-out-of-$n$ system, which satisfy relations (9.16) and (9.27) with $F(t)$ replaced by $t \in [0,1]$. Below we outline basic ideas.

We note that it is impossible to construct an absolutely continuous copula for every distribution satisfying the characterization conditions. For instance, the distributions attaining the lower and upper bounds in (9.16) with $k > 1$ and $k < n$, respectively, satisfy conditions (9.17) and (9.18), respectively, and so are necessarily singular. Moreover, if $F_{n+1-k}(t)$ coincides with the lower $\frac{nt-n+k}{k}$ or upper $\frac{nt}{n+1-k}$ bounds on an interval, conditions (9.11) and (9.12) imply that there $F_{n+1-k}(t) = \ldots = F_n(t)$ and $F_1(t) = \ldots = F_{n+1-k}(t)$, respectively, and this contradicts existence of the joint density. It occurs that the joint density may exist if we exclude the cases of such type.

We first solve an auxiliary problem of constructing exchangeable standard uniform jointly absolutely continuous random variables with given marginal distributions of order statistics $G_1, \ldots, G_n$. It is possible under conditions (9.11) and (9.12) (with $F(t) = t$, $0 \le t \le 1$), and an extra condition

$$\lambda\left(\bigcup_{k=1}^{n-1} \Sigma_k\right) = \lambda\left(\bigcup_{k=1}^{n-1} \{0 < t < 1 : \exists\, 0 < s < 1 \;\; t = G_k(s) = G_{k+1}(s)\}\right) = 0, \tag{9.30}$$

with $\lambda$ denoting the Lebesgue measure on the unit interval. The intuitive meaning of the additional condition is clear: if the distribution functions of two consecutive order statistics were identical on an interval of their increase, the order statistics would be identical with a positive probability. The fundamental step of proof that (9.11), (9.12) and (9.30) are sufficient for existing a joint density is based on the following observation. Suppose that $V_1, \ldots V_n$ have a joint distribution being an absolutely continuous copula, and the corresponding order statistics have marginal distribution functions $H_1 \ldots, H_n$, respectively. If

$$H_1^{-1} \circ G_1 \ge \ldots \ge H_n^{-1} \circ G_n, \tag{9.31}$$

then the ordered random variables $Z_1 = G_1^{-1}(H_1(V_{1:n})) \le \ldots \le Z_n = G_n^{-1}(H_n(V_{n:n}))$ have marginal distribution functions $G_1, \ldots, G_n$, and their random permutations are exchangeable, jointly absolutely continuous, and standard uniform. Clearly, their order statistics have desired distributions. So our original problem can be treated as the construction of an absolute continuous copula with distribution functions of order statistics satisfying (9.31). Note that we have (9.11) and $H_1^{-1} \le \ldots \le H_n^{-1}$ (with sharp inequalities a.e. by the absolute continuity assumption), and we need (9.31). On one hand, it means that all $H_i$ $i = 1, \ldots, n$, should be sufficiently close to the identity functions, and on the other, they should be ordered and different almost everywhere. It can be shown that it is sufficient in order to obtain (9.31) that all $H_i$ lie between two lines $L(t) \le U(t)$, $0 \le t \le 1$, which are defined as the inverses of transformations

$$t \mapsto \frac{3}{4}t + \min_{1\le k\le n-1} G_k(G_{k+1}^{-1}(t)),$$

$$t \mapsto \frac{3}{4}t + \max_{1\le k\le n-1} G_{k+1}(G_k^{-1}(t)), \quad 0 \le t \le 1.$$

Moreover, $L(t) < t < U(t)$ for some $t$ iff $t \notin \bigcup_{k=1}^{n-1} \Sigma_k$. Then we can take an ordinal sum of copulae with countably many absolute continuous (e.g., product) components such that the distribution functions of order statistics $H_1, \dots, H_n$ actually run between the bounds $L$ and $U$. Their non-overlapping supports $(a_i, b_i)$, $i = 1, 2, \dots$, are chosen so that $\bigcup_{i=1}^{\infty}(a_i, b_i) \subset [0,1] \setminus \bigcup_{k=1}^{n-1} \Sigma_k$, $\lambda(\bigcup_{i=1}^{\infty}(a_i, b_i)) = 1$, and all the squares $(a_i, b_i)^2$ are contained in the space between the graphs of $L$ and $U$. The properties of ordinal sums imply that the graphs of all $H_i$, $i = 1, \dots, n$, are situated in the sum of squares and on the diagonal points $(s, \dots, s)$ for $s \in [0,1] \setminus \bigcup_{i=1}^{\infty}(a_i, b_i)$. This completes the construction.

Now we are in the position to characterize the distribution functions of $k$-out-of-$n$ system lifetimes in the case that the corresponding joint distribution of unit lifetimes is an absolutely continuous copula. Jaworski and Rychlik [22] proved that distribution functions satisfy the requirements if assumptions (9.16) and (9.27) hold together with

$$\lambda(\Sigma_-) = \lambda(\Sigma_+) = 0 \tag{9.32}$$

for

$$\Sigma_- = \left\{ 0 < s < 1 : \, G(s) = \frac{ns - n + k}{k} \right\}, \qquad k > 1, \tag{9.33}$$

$$\Sigma_+ = \left\{ 0 < s < 1 : \, G(s) = \frac{ns}{n + 1 - k} \right\}, \qquad k < n, \tag{9.34}$$

with $\Sigma_- = \emptyset$ for $k = 1$ and $\Sigma_+ = \emptyset$ for $k = n$. Arguments for the necessity of the last condition were given above. The sufficiency proof consists in modifying functions (9.28) and (9.29) so that they are distinct almost everywhere, and together with $G_{n+1-k} = G$ satisfy conditions (9.11), (9.12) and (9.30). The modifications have the forms

$$G_i^*(s) = \min\left\{ \frac{ns - G(s) + a_i A(ns - G(s))}{n - k}, 1 \right\}, \qquad 1 \le i \le n - k, \tag{9.35}$$

$$G_i^*(s) = \max\left\{ \frac{ns - G(s) - n + k + b_i B(ns - G(s))}{k - 1}, 0 \right\}, \; n + 2 - k \le i \le n, \tag{9.36}$$

with $a_i$, $i = 1, \dots, n-k$, $b_i$, $i = n+1-k, \dots, n$, being two arithmetic sequences decreasing from 1 to $-1$. The crucial point of the proof consists in subtle constructions of non-negative functions $A, B$, depending on $G$, and such that (9.35) and (9.36) are distribution functions which together with $G_{n+1-k} = G$ satisfy (9.11) and (9.12). Moreover, the functions fulfill the implications

$$\begin{aligned}
t = G(s) \notin \Sigma_- \cup \Sigma_+ &\Rightarrow G_1^*(s) > \ldots > G_n^*(s), \\
t = G(s) \in \Sigma_- &\Rightarrow G_1^*(s) = \ldots = G_{n+1-k}^*(s), \\
t = G(s) \in \Sigma_+ &\Rightarrow G_{n+1-k}^*(s) = \ldots = G_n^*(s).
\end{aligned}$$

Due to definitions (9.30), (9.33) and (9.34), it follows that $\Sigma_- \cup \Sigma_+ = \bigcup_{k=1}^{n-1} \Sigma_k$, and so conditions (9.30) and (9.32) are equivalent here. For general $G$, it is not possible to present closed formulae for $A$ and $B$. We omit the details of describing them, and refer the reader to the original paper [22].

### 9.4.3 Variance Bounds

Characterization results of Sect. 4.1 make it possible to establish sharp evaluations of lifetime variances for $k$-out-of-$n$ systems presented in Rychlik [51]. It is obvious that the problems of providing lower and upper bounds of variances over a family of distributions may be treated as a problem of indicating the least and most dispersed distributions, respectively, about their means. More generally, we can look for distributions which are least and most dispersed about arbitrary points of the distribution supports. Evidently, the supports of $k$-out-of-$n$ systems are contained in that of the parent marginal $F$. Suppose that a distribution function $G$ satisfying (9.16) and (9.27) attains the value $0 < \varsigma = G(\tau) < 1$ at some support point $\tau$ of $F$. It is clear that either of the distribution functions

$$F_{\varsigma\tau}^{c1}(t) = \max\left\{0, \min\left\{n[F(t) - F(\tau)] + \varsigma, \frac{nF(t)}{n+1-k}, 1\right\}\right\}, \tag{9.37}$$

if $\varsigma \geq n[G(\tau) - 1] + 1 - k$, and

$$F_{\varsigma\tau}^{c2}(t) = \min\left\{1, \max\left\{0, \frac{nF(t) - n + k}{k}, n[F(t) - F(\tau)] + \varsigma\right\}\right\}, \tag{9.38}$$

when the inequality is reversed, is more concentrated about $\tau$ than $G$, and satisfies both (9.16) and (9.27). Also, distribution function

$$F_{\varsigma}^{d}(t) = \begin{cases} \frac{nF(t)}{n+1-k}, & t \leq F^{-1}\left(\frac{(n+1-k)\varsigma}{n}\right), \\ \varsigma, & F^{-1}\left(\frac{(n+1-k)\varsigma}{n}\right) \leq t < F^{-1}\left(\frac{n-k+k\varsigma}{n}\right), \\ \frac{nF(t)-n+k}{k}, & t \geq F^{-1}\left(\frac{n-k+k\varsigma}{n}\right), \end{cases} \tag{9.39}$$

is more dispersed about $\tau \in \left[F^{-1}\left(\frac{(n+1-k)\varsigma}{n}\right), F^{-1}\left(\frac{n-k+k\varsigma}{n}\right)\right]$ than $G$, and satisfies (9.16) and (9.27) as well. Therefore the variance minimization and maximization problems can be reduced to calculating the extremes among the functions of shapes (9.37), (9.38) and (9.39), respectively. As we see, for fixed $F$, the latter family is parametrized by the single parameter $0 < \varsigma < 1$. It can be also checked that the distributions (9.37) and (9.38) can be also jointly parametrized as follows

$$F_{\theta}^{c}(t) = \begin{cases} \max\left\{0, \min\left\{n[F(t) - \theta], \frac{nF(t)}{n+1-k}, 1\right\}\right\}, & 0 < \theta \leq \frac{n-k}{n}, \\ \min\left\{1, \max\left\{0, \frac{nF(t)-n+k}{k}, n[F(t) - \theta], 1\right\}\right\}, & \frac{n-k}{n} \leq \theta < \frac{n-1}{n}. \end{cases} \tag{9.40}$$

Obviously, the copulae corresponding with the most dispersed and concentrated distributions of $k$-out-of-$n$ systems are singular. For instance, in the dispersed case, by (9.28) and (9.29) events $\left\{X_{n+1-k:n} \leq F^{-1}\left(\frac{(n+1-k)\varsigma}{n}\right)\right\}$ and $\left\{X_{n+1-k:n} \geq F^{-1}\left(\frac{n-k+k\varsigma}{n}\right)\right\}$ imply $X_{1:n} = \ldots = X_{n+1-k:n}$ and $X_{n+1-k:n} = \ldots = X_{n:n}$, respectively. Also, the interval $\left[F^{-1}\left(\frac{(n+1-k)\varsigma}{n}\right), F^{-1}\left(\frac{n-k+k\varsigma}{n}\right)\right]$, where the system does not fail almost surely, can be split at $F^{-1}\left(\frac{n-k+\varsigma}{n}\right)$ into two parts so that $X_{1:n} = \ldots = X_{n+k:n} \leq F^{-1}\left(\frac{n-k+\varsigma}{n}\right) \leq X_{n+2-k:n} = \ldots = X_{n:n}$. However, the singular copulae can be properly approximated by sequences of absolutely continuous ordinal sums.

It easily follows from (9.40) that the minimal variances of all $k$-out-of-$n$ systems composed of units with exchangeable standard uniform lifetimes amount to $1/(3n^2)$. It was first proved by Lai and Robbins [26] in the case of parallel system. More calculations are needed in order to prove that the maximal variances in the uniform case are equal to

$$\max \mathbb{V}arX_{\frac{n+1}{2}:n} = \frac{7n^2+4n+1}{12n^2}$$

for the sample median, and

$$\begin{aligned}\max \mathbb{V}arX_{n+1-k:n} &= 1 + \frac{k^2+(n+1-k)^2}{n^2} \\ &+ \frac{n(n+1)[(|n+1-2k|+2)\sqrt{n+1-2k)^2+4}-4]}{3k(n+1-k)|n+1-2k|}\end{aligned}$$

otherwise. This increases with respect to $\left|\frac{k}{n} - \frac{n+1}{2n}\right|$ from $\frac{7}{12} \approx 0.58333$ at 0 to $\frac{5}{6}(3-\sqrt{5}) \approx 0.63661$ as $\left|\frac{k}{n} - \frac{n+1}{2n}\right| \to \frac{1}{2}$.

Rychlik [56] used (9.39) for establishing sharp non-parametric bounds on the lifetime variances of $k$-out-of-$n$ systems with exchangeable components which are expressed in terms of lifetime variances of the system components. They have the forms

$$0 \leq \frac{\mathbb{V}arX_{n+1-k:n}}{\mathbb{V}arX_1} \leq \max\left\{\frac{n}{k}, \frac{n}{n+1-k}\right\}. \tag{9.41}$$

The lower bounds are trivially attained by the exhaustive drawing without replacement scheme from a numerical population taking on at least two different values. The upper one is approximated by the following model. We choose one of two urn with some probabilities $0 < \varsigma < 1$ and $1-\varsigma$. The first urn contains $n+1-k$ balls with value 0, and $k-1$ balls with value $\tau(\varsigma)$, and we have $n-k$ and $k$ balls with values $\tau(\varsigma)$ and 1, respectively, in the other, where

$$0 < \tau(\varsigma) = \frac{(1-\varsigma)k}{(1-\varsigma)k+\varsigma(n+1-k)} < 1.$$

Then we draw without replacement all the balls from the chosen urn. The outcomes of consecutive drawings $X_i$, $i = 1,\ldots,n$, are exchangeable. They have the marginal distribution

$$\mathbb{P}(X_i = 0) = \varsigma \frac{n+1-k}{n},$$
$$\mathbb{P}(X_i = \tau(\varsigma)) = \frac{n-k}{n} + \varsigma \frac{2k-n-1}{n},$$
$$\mathbb{P}(X_i = 1) = (1-\varsigma)\frac{k}{n},$$

mean $\tau(\varsigma)$, and variance

$$\mathbb{V}ar X_1 = \frac{\varsigma(1-\varsigma)k(n+1-k)}{n[(1-\varsigma)k + \varsigma(n+1-k)]}.$$

The $k$th greatest order statistic has the two-point distribution

$$\mathbb{P}(X_{n+1-k:n} = 0) = \varsigma = 1 - \mathbb{P}(X_{n+1-k:n} = 1),$$

mean $\varsigma$ and variance $\varsigma(1-\varsigma)$. Maximizing the ratio

$$\frac{\mathbb{V}ar X_{n+1-k:n}}{\mathbb{V}ar X_1} = \frac{n[(1-\varsigma)k + \varsigma(n+1-k)]}{k(n+1-k)}$$

with respect to $0 < \varsigma < 1$, we obtain (9.41). Papadatos [45] determined analogous upper bounds in the i.i.d. case. They coincide with (9.41) for the series and parallel systems, and are strictly less in the other cases. The respective lower bounds are strictly positive for the series and parallel systems (see [29]), and amount to 0 otherwise. Jasiński et al. [20] generalized the results of Papadatos [45] to the general coherent and mixed systems.

## 9.5 Final Remarks

We merely mention several results concerning order relations and asymptotics of lifetimes of coherent system composed of exchangeable units. Navarro et al. [37] delivered some comparisons in the stochastic, hazard rate and likelihood orders. Navarro and Shaked [35] analyzed the likelihood ratio order of series systems, and using the minimal signature representation, obtained some asymptotic evaluations for general coherent and mixed systems. Similar ideas we used in Navarro and Hernandez [32] (see also [9]) for analyzing asymptotic properties of the system mean residual life. Navarro [31] studied the implications of the likelihood ratio ordering of the series and parallel systems, and concluded asymptotic relations for general systems. Block et al. [7] analyzed the initial and final behavior of the system failure rate. Navarro et al. [38] provided bounds on the system reliability function based on the notions of hyperminimal and hypermaximal distributions. There is a vast literature concerning the treatment of systems with components whose lifetimes are independent and identically or non-identically distributed. There were some attempts of applying the signature representations for the systems with dependent non-exchangeable components. Navarro et al. [41] introduced notions of average and projected systems (see also [60]), and analyzed their relations with the original systems. Navarro and Spizzichino [36] studied connections of the copulae of order statistics with those of original variables.

## References


1. Arnold, B.C.: Distribution-free bounds on the mean of the maximum of a dependent sample. SIAM J. Appl. Math. **38**, 163–167 (1980)
2. Arnold, B.C.: $p$-Norm bounds on the expectation of the maximum of possibly dependent sample. J. Multivar. Anal. **17**, 316–332 (1985)
3. Aven, T., Jensen, U.: Stochastic Models in Reliability. Applications of Mathematics, vol. 41. Springer, New York, NY (1999)
4. Barlow, R.E., Proschan, F.: Mathematical Theory of Reliability. Wiley, New York, NY (1965), reprinted in Classics in Applied Mathematics, vol. 17. SIAM, Philadelphia, PA (1996)
5. Barlow, R.E., Proschan, F.: Statistical Theory of Reliability and Life Testing. Holt, Rinehart and Winston, New York, NY (1975)
6. Birnbaum, Z.W., Esary, J.D., Saunders, S.C.: Muliticomponent systems and structures, and their reliability. Technometrics **3**, 55–77 (1961)
7. Block, H.W., Li, Y., Savits, T.H.: Initial and final behavior of failure rate functions for mixtures and systems. J. Appl. Probab. **40**, 721–740 (2003)
8. Boland, P.J., Samaniego, F.: The signature of a coherent system and its applications in reliability. In: Soyer, R., Mazzuchi, T., Singpurwalla, N. (eds.) Mathematical Reliability: An Expository Perspective, International Series in Operational Research and Management Science, vol. 67, pp. 1–29. Kluwer, Boston (2004)
9. Bradley, D.M., Gupta, R.C.: Limiting behaviour of the mean residual life. Ann. Inst. Stat. Math. **55**, 217–226 (2003)
10. Caraux, G., Gascuel, O.: Bounds on distribution functions of order statistics for dependent variates. Stat. Probab. Lett. **14**, 103–105 (1992)
11. David, H.A., Nagaraja, H.N.: Order Statistics, 3rd edn. Wiley, Hoboken, NJ (2003)
12. Durante, F., Jaworski, P.: Absolutely continuous copulas with given diagonal sections. Commun. Stat. Theory Methods **37**, 2924–2942 (2009)
13. Durante, F., Kolesárowa, A., Mesiar, R., Sempi, P.: Copulas with given diagonal sections: novel constructions and applications. Int. J. Uncertain. Fuzziness Knowl-Based Syst. **15**, 397–410 (2007)
14. Durante, F., Mesiar, R., Sempi, P.: On a family of copulas constructed from the diagonal section. Soft Comput. **10**, 490–494 (2006)
15. Erdely, A., González-Barrios, J.M.: On the construction of families of absolutely continuous copulas with given restrictions. Commun. Stat. Theory Methods **35**, 649–659 (2006)
16. Fredericks, G.A., Nelsen, R.B.: Copulas constructed from diagonal sections. In: Beneş, V., Štěpán, J. (eds.) Distributions with Given Marginals and Moment Problems, pp. 129–136. Kluwer Academic Publishers, Dordrecht (1997)
17. Gascuel, O., Caraux, G.: Bounds on expectations of order statistics via extremal dependences. Stat. Probab. Lett. **15**, 143–148 (1992)
18. Gertsbakh, I.: Reliability Theory, with Applications to Preventive Maintenance. Springer, Berlin (2000)
19. Goroncy, A.: Lower bounds on positive $L$-statistics. Commun. Stat. Theory Methods **38**, 1989–2002 (2009)
20. Jasiński, K., Navarro, J., Rychlik, T.: Bounds on variances of lifetimes of coherent and mixed systems. J. Appl. Probab. **46**, 894–908 (2009)

21. Jaworski, P.: On copulas and their diagonals. Inform. Sci. **179**, 2863–2871 (2009)
22. Jaworski, P., Rychlik, T.: On distributions of order statistics for absolutely continuous copulas with applications to reliability problems. Kybernetika **44**, 757–776 (2008)
23. Kemperman, J.H.B.: Bounding moments of an order statistic when each $k$-tuple is independent. In: Beneş, V., Štěpán, J. (eds.) Distributions with Given Marginals and Moment Problems, pp. 291–304. Kluwer Academic Publishers, Dordrecht (1997)
24. Kochar, S., Mukerjee, H., Samaniego, F.J.: The "signature" of a coherent system and its application to comparison among systems. Naval Res. Logist. **46**, 507–523 (1999)
25. Lai, T.L., Robbins, H.: Maximally dependent random variables. Proc. Natl. Acad. Sci. USA. **73**, 286–288 (1976)
26. Lai, T.L., Robbins, H.: A class of dependent random variables and their maxima. Z. Wahrsch. Verw. Gebiete **42**, 89–111 (1978)
27. Mallows, C.L.: Extrema of expectations of uniform order statistics. SIAM Rev. **11**, 410–411 (1969)
28. Mesiar, R., Sempi, C.: Ordinal sums and idempotents of copulas. Aequationaes Math. **79**, 39–52 (2010)
29. Moriguti, S.: Extremal properties of extreme value distributions. Ann. Math. Stat. **22**, 523–536 (1951)
30. Moriguti, S.: A modification of Schwarz's inequality with applications to distributions. Ann. Math. Stat. **24**, 107–113 (1953)
31. Navarro, J.: Likelihood ratio ordering of order statistics, mixtures and systems. J. Stat. Plann. Inference **138**, 1242–1257 (2008)
32. Navarro, J., Hernandez, P.J.: Mean residual life functions of finite mixtures, order statistics and coherent systems. Metrika **67**, 277–298 (2008)
33. Navarro, J., Rubio, R.: Computations of signatures of coherent systems with five components. Commun. Stat. Simul. Comput. **39**, 68–84 (2010)
34. Navarro, J., Rychlik, T.: Reliability and expectation bounds for coherent systems with exchangeable components. J. Multivar. Anal. **98**, 102–113 (2007)
35. Navarro, J., Shaked, M.: Hazard rate ordering of order statistics and systems. J. Appl. Probab. **43**, 391–408 (2006)
36. Navarro, J., Spizzichino, F.: On the relationships between copulas of order statistics and marginal distributions. Stat. Probab. Lett. **80**, 473–479 (2010)
37. Navarro, J., Ruiz, J.M., Sandoval, C.J.: A note on comparisons among coherent systems with dependent components using signatures. Stat. Probab. Lett. **72**, 179–185 (2005)
38. Navarro, J., Ruiz, J.M., Sandoval, C.J.: Properties of coherent systems with dependent components. Commun. Stat. Theory Methods **36**, 175–191 (2007)
39. Navarro, J., Balakrishnan, N., Samaniego, F.J.: Mixture representations of residual lifetimes of used systems. J. Appl. Prob. **45**, 1097–112 (2008)
40. Navarro, J., Samaniego, F.J., Balakrishnan, N., Bhattacharya, D.: On the application and extension of system signatures to problems in engineering reliability. Naval Res. Logist. **55**, 313–327 (2008)
41. Navarro, J., Spizzichino, F., Balakrishnan, N.: Applications of average and projected systems to the study of coherent systems. J. Multivariate Anal. **101**, 1471–1482 (2010)
42. Nelsen, R.B.: An Introduction to Copulas. Springer Series in Statistics, 2nd edn. Springer, New York, NY (2006)
43. Nelsen, R.B., Fredericks, G.A.: Diagonal copulas. In: Beneş, V., Štěpán, J. (eds.) Distributions with Given Marginals and Moment Problems, pp. 121–128. Kluwer Academic Publishers, Dordrecht (1997)
44. Nelsen, R.B., Quesada-Molina, J.J., Rodríquez-Lallena, J.A., Úbeda-Flores, M.: On the construction of copulas and quasi-copulas with given diagonal sections. Insur. Math. Econ. **42**, 473–483 (2008)
45. Papadatos, N.: Maximum variance of order statistics. Ann. Inst. Stat. Math. **47**, 185–193 (1995)

46. Papadatos, N.: Exact bounds for the expectations of order statistics from non-negative populations. Ann. Inst. Stat. Math. **49**, 727–736 (1997)
47. Rausand, M., Høyland, A.: System Reliability Theory: Models, Statistical Methods, and Applications. Wiley Series in Probability and Statistics, 2nd edn. Wiley-Interscience, Hoboken, NJ (2004)
48. Rychlik, T.: Stochastically extremal distributions of order statistics for dependent samples. Stat. Probab. Lett. **13**, 337–341 (1992)
49. Rychlik, T.: Bounds for expectation of *L*-estimates for dependent samples. Statistics **24**, 1–7 (1993)
50. Rychlik, T.: Sharp bounds on *L*-estimates and their expectations for dependent samples. Commun. Stat. Theory Methods **22**, 1053–1068 (1993)
51. Rychlik, T.: Distributions and expectations of order statistics for possibly dependent random variables. J. Multivar. Anal. **48**, 31–42 (1994)
52. Rychlik, T.: Bounds for order statistics based on dependent variables with given nonidentical distributions. Stat. Probab. Lett. **23**, 351–358 (1995)
53. Rychlik, T.: Bounds for expectations of *L*-estimates. In: Balakrishnan, N., Rao, C.R. (eds.) Order Statistics: Theory & Methods, Handbook of Statistics, vol. 16, pp. 105–145. North-Holland, Amsterdam (1998)
54. Rychlik, T.: Mean-variance bounds for order statistics from dependent DFR, IFR, DFRA and IFRA samples. J. Stat. Plann. Infer. **92**, 21–38 (2001)
55. Rychlik, T.: Projecting Statistical Functionals. Lecture Notes in Statistics, vol. 160. Springer, New York, NY (2001)
56. Rychlik, T.: Extreme variances of order statistics in dependent samples. Stat. Probab. Lett. **78**, 1577–1582 (2008)
57. Samaniego, F.J.: On closure of the IFR class under formation of coherent systems. IEEE Trans. Reliab. **R-34**, 69–72 (1985)
58. Samaniego, F.J.: (2007). System Signatures and Their Applications in Engineering Reliability. International Series in Operations Research and Management Science, vol. 110. Springer, New York, NY (2007)
59. Shaked, M., Suarez-Llorens, A.: On the comparison of reliability experiments based on the convolution order. J. Am. Stat. Assoc. **98**(463), 693–702 (2003)
60. Spizzichino, F.: The role of symmetrization and signature for systems with non-exchangeable components. In: Bedford, T., Quigley, J., Walls, L., Alkali, B., Daneshkhah, A., Hardman, G. (eds.) Advances in Mathematical Modelling for Reliability, pp. 138–148. IOS Press, Amsterdam (2008)
61. Tchen, A.: Inequalities for distributions with given marginals. Ann. Probab. **8**, 814–827 (1980)

# Chapter 10
# Copula-Based Measures of Multivariate Association

Friedrich Schmid, Rafael Schmidt, Thomas Blumentritt,
Sandra Gaißer and Martin Ruppert

**Abstract** This chapter constitutes a survey on copula-based measures of multivariate association – i.e. association in a $d$-dimensional random vector $\mathbf{X} = (X_1, ..., X_d)$ where $d \geq 2$. Some of the measures discussed are multivariate extensions of well-known bivariate measures such as Spearman's rho, Kendall's tau, Blomqvist's beta or Gini's gamma. Others rely on information theory or are based on $L_p$-distances of copulas. Various measures of multivariate tail dependence are derived by extending the coefficient of bivariate tail dependence. Nonparametric estimation of these measures based on the empirical copula is further addressed.

## 10.1 Introduction and Definitions

The measurement of bivariate association is well established and measures such as Spearman's rho, Kendall's tau, Blomqvist's beta, Gini's gamma, Spearman's

---

Friedrich Schmid
Department of Economic and Social Statistics, University of Cologne, Cologne, Germany
e-mail: schmid@wiso.uni-koeln.de

Rafael Schmidt
Risk Control, Bank for International Settlements, Basel, Switzerland
e-mail: rafael.schmidt@gmx.de

Thomas Blumentritt
Department of Economic and Social Statistics, University of Cologne, Cologne, Germany
e-mail: blumentritt@wiso.uni-koeln.de

Sandra Gaißer
Department of Economic and Social Statistics, University of Cologne, Cologne, Germany
e-mail: sandra.gaisser@uni-koeln.de

Martin Ruppert
Graduate School of Risk Management, University of Cologne, Cologne, Germany
e-mail: ruppert@wiso.uni-koeln.de

P. Jaworski et al. (eds.), *Copula Theory and Its Applications*, Lecture Notes in Statistics 198,
DOI 10.1007/978-3-642-12465-5_10, 

footrule, and some lesser known are widely used in economics and social sciences. All these measures share one important property: For continuous random variables they are invariant with respect to the two marginal distributions, i.e. they can be expressed as a function of their copula. This property is also known as 'scale-invariance'. Note that not all measures of association satisfy this property, e.g. Pearson's linear correlation coefficient (see [26] for related discussions).

It is natural to generalize these bivariate copula-based measures to the multivariate case, i.e. to try to measure the amount of association in a $d$-dimensional random vector $\mathbf{X} = (X_1, ..., X_d)$ where $d > 2$. This is of interest in many fields of application, e.g. in risk management or in the multivariate analysis of financial asset returns. In a multivariate setting, a number of additional problems and questions occur which are not present in the bivariate case. In dimension $d = 3$ e.g., three perfectly negatively associated variables do not exist. This is also expressed by the fact that the lower Fréchet-Hoeffding bound of a copula is not a copula itself for $d \geq 3$, implying that a natural lower bound for the measures does not exist in this case. While desirable analytical properties of a bivariate measure of association are fairly clear and well investigated, this is different for $d \geq 3$. Indeed, there might be differing views concerning the normalization of the multivariate measure or its preferred behaviour regarding the addition, deletion or transformation of one or several components of $\mathbf{X} = (X_1, ..., X_d)$. We do not think that a best measure of multivariate association, satisfying all of the desirable features, has already been found or even exists. We therefore give a survey and a short discussion of some of the measures which have been suggested in the past. There is, however, room for further contributions. Note that we focus on multivariate versions that take into account the multivariate association structure as represented by the $d$-dimensional copula of $\mathbf{X}$. We thus do not consider the type of multivariate measures which is given by the average of pairwise bivariate measures with respect to all distinct bivariate margins of the copula. We further do not address measures of complete or functional dependence (see [50, 62, 68, 101]).

Throughout this chapter, we assume that the $d$-dimensional random vector $\mathbf{X}$ has distribution function $F$ with continuous marginal distribution functions $F_i, i = 1, \ldots, d$. The associated copula $C$ of $\mathbf{X}$ is thus uniquely defined, which allows for the definition of well-defined copula-based measures of multivariate association. Regarding the case of non-continuous marginal distributions, we refer to Vandenhende and Lambert [112], Nešlehová [80, 81], Denuit and Lambert [19], Mesfioui and Tajar [75], Genest and Nešlehová [37] as well as Feidt et al. [28].

We further address the statistical estimation of the multivariate measures, which, in our opinion, has not been sufficiently treated in the literature yet but needs further consideration. To do so, we introduce additional notation and definitions in the following, which are not given in Durante and Sempi [22]. Note that, in order to ease notation, we omit the subscript referring to the dimension in the notation of the copula.

Let $(\mathbf{X}_j)_{j=1,\ldots,n}$ be a random sample of $\mathbf{X}$ and assume that the distribution function $F$, the marginal distribution functions $F_i$, $i = 1,\ldots,d$, and the copula $C$ of $\mathbf{X}$ are completely unknown. The marginal distribution functions $F_i$ are estimated by their empirical counterparts

$$\hat{F}_{i,n}(x) = \frac{1}{n}\sum_{j=1}^{n}\mathbf{1}_{\{X_{ij}\leqslant x\}} \text{ for } i = 1,\ldots,d \text{ and } x \in \mathbb{R}.$$

Further, set $\hat{U}_{ij,n} := \hat{F}_{i,n}(X_{ij})$ for $i = 1,...,d$, $j = 1,...,n$, and $\hat{\mathbf{U}}_{j,n} = \left(\hat{U}_{1j,n},...,\hat{U}_{dj,n}\right)$. Since $\hat{U}_{ij,n} = \frac{1}{n}(\text{rank of } X_{ij} \text{ in } X_{i1},...,X_{in})$, we consider rank order statistics. The copula $C$ is then estimated by the empirical copula which is defined as

$$\hat{C}_n(\mathbf{u}) = \frac{1}{n}\sum_{j=1}^{n}\prod_{i=1}^{d}\mathbf{1}_{\{\hat{U}_{ij,n}\leqslant u_i\}} \quad \text{for } \mathbf{u} = (u_1,...,u_d) \in [0,1]^d. \tag{10.1}$$

Empirical copulas were introduced by Rüschendorf [88] and Deheuvels [18]. The asymptotic statistical theory for the related estimators of the multivariate measures is based on the following proposition concerning the asymptotic behaviour of the empirical copula process $\mathbb{C}_n = \sqrt{n}\left\{\hat{C}_n(\mathbf{u}) - C(\mathbf{u})\right\}$, which has been discussed e.g. by Rüschendorf [88], Gänßler and Stute [33], Fermanian et al. [29], and Tsukahara [110].

**Proposition 10.1.1.** *Let $F$ be a continuous $d$-dimensional distribution function with copula $C$. Under the additional assumption that the $i$th partial derivatives $D_iC(\mathbf{u})$ exist and are continuous for $i = 1,...,d$, we have*

$$\mathbb{C}_n = \sqrt{n}\left\{\hat{C}_n(\mathbf{u}) - C(\mathbf{u})\right\} \xrightarrow{w} \mathbb{G}_C(\mathbf{u}).$$

*Weak convergence takes place in $\ell^\infty\left([0,1]^d\right)$ and*

$$\mathbb{G}_C(\mathbf{u}) = \mathbb{B}_C(\mathbf{u}) - \sum_{i=1}^{d} D_iC(\mathbf{u})\mathbb{B}_C(\mathbf{u}^{(i)}). \tag{10.2}$$

*The vector $\mathbf{u}^{(i)}$ denotes the vector where all coordinates, except the $i$th coordinate of $\mathbf{u}$, are replaced by 1. The process $\mathbb{B}_C$ is a tight centered Gaussian process on $[0,1]^d$ with covariance function*

$$E\left\{\mathbb{B}_C(\mathbf{u})\mathbb{B}_C(\mathbf{v})\right\} = C(\mathbf{u}\wedge\mathbf{v}) - C(\mathbf{u})C(\mathbf{v}),$$

*i.e., $\mathbb{B}_C$ is a $d$-dimensional Brownian bridge.*

A similar result can be obtained for the survival function $\overline{C}$ (cf. [94]). Consider the estimator

$$\hat{\overline{C}}_n(\mathbf{u}) = \frac{1}{n}\sum_{j=1}^{n}\prod_{i=1}^{d}\mathbf{1}_{\{\hat{U}_{ij,n}>u_i\}} \quad \text{for} \quad \mathbf{u} = (u_1,\ldots,u_d) \in [0,1]^d. \tag{10.3}$$

Under the assumptions of Proposition 10.1.1, weak convergence of the process $\overline{C}_n = \sqrt{n}\{\hat{\overline{C}}_n(\mathbf{u}) - \overline{C}(\mathbf{u})\}$ in $\ell^\infty([0,1]^d)$ to the Gaussian process $\mathbb{G}_{\overline{C}}$ can be established, where $\mathbb{G}_{\overline{C}}$ has the form

$$\mathbb{G}_{\overline{C}}(\mathbf{u}) = \mathbb{B}_{\overline{C}}(\mathbf{u}) - \sum_{i=1}^{d} D_i\overline{C}(\mathbf{u})\mathbb{B}_C(\mathbf{u}^{(i)}) \tag{10.4}$$

with $d$-dimensional Brownian bridges $\mathbb{B}_C, \mathbb{B}_{\overline{C}}$.

## 10.2 Aspects of Multivariate Association

The works by Rényi [86], Scarsini [91] as well as Schweizer and Wolff [99] introduce various axioms to characterize bivariate measures of association. However, the derivation of a comparable set of axioms to comprehensively describe multivariate measures of association is not straightforward. We thus concentrate on providing an overview of existing criteria in the literature that are considered to be relevant for distinguishing measures of multivariate association.

A measure of association is a functional

$$\mathscr{M} : \mathscr{C}_d \to D \subseteq \mathbb{R},$$

which we denote by $\mathscr{M}(C)$ or equivalently by $\mathscr{M}(\mathbf{X}) = \mathscr{M}(X_1,...,X_d)$. The following criteria summarize and extend those presented in Wolff [114], Taylor [109], and Dolati and Úbeda-Flores [21]:

**W** **Well-definedness:** The measure $\mathscr{M}$ is well-defined for every random vector $\mathbf{X} = (X_1,...,X_d)$ with continuous marginals and is a function of the copula $C \in \mathscr{C}_d$, i.e. $\mathscr{M}(X_1,...,X_d) = \mathscr{M}(C)$.

A measure $\mathscr{M}$ satisfying **W** is invariant with respect to its marginal distributions; in particular, moment assumptions are not required for $\mathscr{M}(\mathbf{X})$ to be defined.

**P** **Invariance with respect to permutations:** For every permutation $\pi$ we have $\mathscr{M}(X_1,...,X_d) = \mathscr{M}\left(X_{\pi(1)},...,X_{\pi(d)}\right)$.

In general, the measures further vary regarding their range and maximal and minimal arguments. We differentiate the following normalization attributes:

**N** **Normalization:**
**N1** If $\Pi$ is the copula of $\mathbf{X}$ then $\mathscr{M}(\mathbf{X}) = \mathscr{M}(\Pi) = 0$.
**N2** If $\mathscr{M}(\mathbf{X}) = 0$ then $\mathbf{X}$ has copula $\Pi$.
**N3** If $M$ is the copula of $\mathbf{X}$ then $\mathscr{M}(\mathbf{X}) = \mathscr{M}(M) = 1$.
**N4** If $\mathscr{M}(\mathbf{X}) = 1$ then $\mathbf{X}$ has copula $M$ or $W$ in dimension $d = 2$. If $\mathscr{M}(\mathbf{X}) = 1$ then $\mathbf{X}$ has copula $M$ in higher dimension.

**N5** If the joint distribution of $\mathbf{X}$ is multivariate normal and all pairwise correlations $\rho_{ij}$ of $X_i$ and $X_j$ are either nonnegative or nonpositive, then $\mathscr{M}(\mathbf{X})$ is a strictly increasing function of the absolute value of each of the pairwise correlations.

Note that **N4** considers the lower Fréchet-Hoeffding bound W in order to cover those measures that are based on notions of distance to independence. It does not impose a lower bound for the measure's range.

Multivariate measures of association further support different notions of orderings in the set of copulas. Here, we consider the partial order $\precsim$, where $C_1 \precsim C_2$ if and only if $C_1(\mathbf{u}) \leq C_2(\mathbf{u})$ for all $\mathbf{u} \in [0,1]^d$. Further, $C_1$ is smaller than $C_2$ according to the concordance (partial) order, denoted by $C_1 \precsim_C C_2$, if and only if $C_1(\mathbf{u}) \leq C_2(\mathbf{u})$ and $\overline{C}_1(\mathbf{u}) \leq \overline{C}_2(\mathbf{u})$ for all $\mathbf{u} \in [0,1]^d$.

**M Monotonicity and concordance:**

**M1** For $\Pi \precsim C_1 \precsim C_2 \precsim M$ we have $\mathscr{M}(C_1) \leq \mathscr{M}(C_2)$.
**M2** For $C_1 \precsim C_2$ we have $\mathscr{M}(C_1) \leq \mathscr{M}(C_2)$.
**M3** For $C_1 \precsim_C C_2$ we have $\mathscr{M}(C_1) \leq \mathscr{M}(C_2)$.

Note that **M3** implies **M2** which itself implies **M1**. Criteria **M2** and **M3** are equivalent for bivariate measures, cf. Joe [58]. **M1** is relevant for all measures relying on some notion of distance between an arbitrary copula and the independence copula. **M3** is important in the context of measures of concordance which are defined later in this section.

If one or several components of a random vector $\mathbf{X}$ are transformed strictly monotonously, then the copula either stays invariant or changes in a well-known way. The behaviour of multivariate measures of association under strictly monotonous transformations of the random vector can be characterized by:

**T Behaviour under transformations:**

**T1** For strictly increasing and continuous transformations $I_i$ we have $\mathscr{M}(X_1,...,X_d) = \mathscr{M}(I_1(X_1),\dots,I_d(X_d))$.
**T2** For strictly decreasing and continuous transformations $D_i$ of all components we have $\mathscr{M}(X_1,\dots,X_d) = \mathscr{M}(D_1(X_1),\dots,D_d(X_d))$.
**T3** For a strictly decreasing and continuous transformation $D_i$ of one arbitrary component $i$ we have $\mathscr{M}(X_1,\dots,X_d) = \mathscr{M}(X_1,\dots,D_i(X_i),\dots,X_d))$.

Since the copulas of $(X_1,\dots,X_d)$ and $(I_1(X_1),...,I_d(X_d))$ are identical, **T1** follows from **W**. Wolff [114] points out that **T2** is equivalent to

$$\mathscr{M}(X_1,...,X_d) = \mathscr{M}(-X_1,...,-X_d), \tag{10.5}$$

independent of the particular choice of transformations. The literature on concordance measures refers to Eq. (10.5) as the Duality axiom. Note that criterion **T3** implies **T2** whereas the converse does not hold.

The following criterion is technical and allows to consider sequences of random variables:

**C** **Continuity:** If $(\mathbf{X}_n)_{n\in\mathbb{N}}$ is a sequence of random vectors and corresponding copulas $(C_n)_{n\in\mathbb{N}}$ and if $\lim_{n\to\infty} C_n(\mathbf{u}) = C(\mathbf{u})$ for all $\mathbf{u} \in [0,1]^d$ and a copula $C$, then $\lim_{n\to\infty} \mathscr{M}(C_n) = \mathscr{M}(C)$.

To generalize the bivariate axiom

$$\mathscr{M}(X,Y) = -\mathscr{M}(-X,Y) = -\mathscr{M}(X,-Y) = \mathscr{M}(-X,-Y), \qquad (10.6)$$

validity of **T2** as well as an additional symmetry property are required. Here, Taylor [109] considers the following: Assume that $\delta = (\delta_1,\dots,\delta_d)$ is a vector of independent Rademacher variables, i.e. $\delta_i \in \{-1;+1\}$ where probability 0.5 is assigned to each value. Furthermore, the random vector $\mathbf{X}$ and $\delta$ are assumed to be independent. For measures of concordance, Taylor [109] then assumes that $\mathscr{M}(\delta_1 X_1,\dots,\delta_d X_d) = 0$. Calculating the conditional expectation given $\mathbf{X}$ of the left-hand side of the latter equation yields the following criterion:

**R** **Reflection symmetry:** $\sum_{\varepsilon_1\in\{-1;+1\}} \cdots \sum_{\varepsilon_d\in\{-1;+1\}} \mathscr{M}(\varepsilon_1 X_1,\dots,\varepsilon_d X_d) = 0.$

In contrast, Dolati and Úbeda-Flores [21] argue that there is no analogous multivariate generalization of Eq. (10.6) and thus do not consider **R** .

The following criterion relates $(d-1)$- and $d$-dimensional measures of association in order to quantify changes in the measure that are solely caused by the transition to a higher dimension:

**TP** **Transition property:** For every $\mathbf{X} = (X_1,\dots,X_d)$ a sequence $(r_d)_{d\geq 3}$ exists, such that $r_{d-1}\mathscr{M}(X_2,\dots,X_d) = \mathscr{M}(X_1,X_2,\dots,X_d) + \mathscr{M}(-X_1,X_2,\dots,X_d)$.

A measure satisfying the afore listed properties except **N2, N4, N5** and **T3** is called a measure of concordance. Whether or not **R** is required to hold depends on the respective definition of Taylor [109] or Dolati and Úbeda-Flores [21]. For further discussions on multivariate measures of concordance, see Joe [58] and Nelsen [78].

The behaviour of multivariate measures of association may differ if an independent component is added to the random vector $\mathbf{X}$ (cf. [30]). This might be of interest in portfolio analysis, when an additional independent asset is incorporated into an existing portfolio.

**A** **Addition of an independent component:**
**A1** $\mathscr{M}(X_1,...,X_d) \geq \mathscr{M}(X_1,...,X_d,X_{d+1})$ if $X_{d+1}$ is independent of $(X_1,...,X_d)$.
**A2** $\mathscr{M}(X_1,...,X_d) = \mathscr{M}(X_1,...,X_d,X_{d+1})$ if $X_{d+1}$ is independent of $(X_1,...,X_d)$.

In order to justify the use of sophisticated multivariate measures of association, we need to investigate whether they can be expressed as a function of lower dimensional measures:

**I** **Irreducibility:** For every dimension $d$ and every copula $C$ the measure $\mathcal{M}(C)$ cannot be written as a function of lower dimensional measures $\{\mathcal{M}(C')\}_{C'\in\mathcal{F}}$, where $\mathcal{F}$ denotes the set of all marginal copulas $C'$ of $C$.

Note that even if **I** applies, there can be exceptions in particular cases, e.g. for radially symmetric copulas (cf. [94, 114]).

## 10.3 Multivariate Generalizations of Spearman's Rho, Kendall's Tau, Blomqvist's Beta, and Gini's Gamma

This section describes how the well-known measures of bivariate association Spearman's rho, Kendall's tau, Blomqvist's beta, and Gini's gamma can be generalized to the multivariate case. In the bivariate case, these measures are often referred to as measures of concordance since they fulfill the set of axioms given by Scarsini [91] (cf. Sect. 10.2). As shown below, all multivariate versions can solely be expressed in terms of the copula $C$ of the random vector $\mathbf{X}$ and satisfy properties **W**, **P**, **T1**, **C**, and **I**; further properties are stated separately next. For similar discussions regarding the measure of association Spearman's footrule we refer to Genest et al. [39] and references therein.

### *10.3.1 Spearman's Rho*

Spearman's rank correlation coefficient (or Spearman's rho) represents one of the best-known measures to quantify the degree of association between two random variables and was first studied by Spearman [106]. For the two random variables $X_1$ and $X_2$ with bivariate distribution function $F$ and continuous univariate margins $F_1$, $F_2$, Spearman's rho is defined as

$$\rho(X_1,X_2) = \frac{Cov(F_1(X_1),F_2(X_2))}{\sqrt{Var(F_1(X_1))}\sqrt{Var(F_2(X_2))}}.$$

Assuming that $X_1$ and $X_2$ have copula $C$, this is equivalent to

$$\begin{aligned}\rho(C) &= \frac{\int_0^1\int_0^1 u_1u_2dC(u_1,u_2)-\left(\frac{1}{2}\right)^2}{\left(\frac{1}{12}\right)} = 12\int_0^1\int_0^1 C(u_1,u_2)\,du_1du_2-3\\ &= \frac{\int_{[0,1]^2} C(u_1,u_2)\,du_1du_2-\int_{[0,1]^2}\Pi(u_1,u_2)\,du_1du_2}{\int_{[0,1]^2} M(u_1,u_2)\,du_1du_2-\int_{[0,1]^2}\Pi(u_1,u_2)\,du_1du_2}\end{aligned} \tag{10.7}$$

because of $\int_{[0,1]^2} M(u_1,u_2)\,du_1du_2 = 1/3$ and $\int_{[0,1]^2}\Pi(u_1,u_2)\,du_1du_2 = 1/4$. Thus, $\rho$ can be interpreted as the normalized average difference between the copula $C$ and the independence copula $\Pi$. Several multivariate extensions of Spearman's rho and their estimation have been discussed in the literature, we mention Ruymgaart and van Zuijlen [89], Wolff [114], Joe [57], Nelsen [76], Stepanova [107], and Schmid and Schmidt [94]. Further, Schmid and Schmidt [92] suggest a related class of multivariate measures of tail dependence, cf. Sect. 10.6. Based on Eq. (10.7), the following $d$-dimensional extension of $\rho$ is straightforward

$$\rho_1(C) = \frac{\int_{[0,1]^d} C(\mathbf{u})d\mathbf{u} - \int_{[0,1]^d} \Pi(\mathbf{u})d\mathbf{u}}{\int_{[0,1]^d} M(\mathbf{u})d\mathbf{u} - \int_{[0,1]^d} \Pi(\mathbf{u})d\mathbf{u}} = h_\rho(d)\left\{2^d \int_{[0,1]^d} C(\mathbf{u})d\mathbf{u} - 1\right\},$$

with $h_\rho(d) = (d+1)/\{2^d - (d+1)\}$. In a similar way, another multivariate version of Spearman's rho can be derived, which is given by

$$\rho_2(C) = h_\rho(d)\left\{2^d \int_{[0,1]^d} \Pi(\mathbf{u})dC(\mathbf{u}) - 1\right\}.$$

Nelsen [76] further considers the average of the two versions, i.e. $\rho_3 = (\rho_1 + \rho_2)/2$. All three measures satisfy **N1**, **N3**, **N4**, **M3**, **R**, **TP**, and **A1**. In addition, **T2** can be verified for $\rho_3$, which, thus represents a multivariate measure of concordance according to Taylor [109]. For $d = 2$, the three versions coincide and reduce to Spearman's rho as given in (10.7). For $d = 3$, Nelsen [76] points out that $\rho_3$ is equal to the average of the pairwise Spearman's rho coefficients, which is, for example, discussed in Kendall [61]. A lower bound for $\rho_i$, $i \in \{1,2,3\}$ is given by

$$\frac{2^d - (d+1)!}{d!\{2^d - (d+1)\}}, \qquad d \geq 2,$$

see Nelsen [76]. However, to our knowledge, there exist no literature on the best-possible lower bound for $\rho_i$ (see e.g. Úbeda-Flores [111]). Consider further an index set $I \subset \{1,\ldots,d\}$ with cardinality $2 \leq |I| \leq d$ and denote by $C_I$ the $|I|$-dimensional marginal copula of $C$ corresponding to those components $X_i$ of $\mathbf{X}$ where $i \in I$. Then, the following relationship between $\rho_1$ and $\rho_2$ holds (cf. Schmid and Schmidt [94]):

$$\rho_2(C) = \sum_{k=2}^{d} (-1)^k \frac{h_\rho(d)}{h_\rho(k)} \frac{2^d}{2^k} \sum_{\substack{I \subset \{1,\ldots,d\} \\ |I| = k}} \rho_1(C_I).$$

It immediately follows from this relationship that $\rho_1$ and $\rho_2$ coincide in case the copula $C$ is radially symmetric.

Statistical inference for $\rho_i$, $i = 1,2$, based on the empirical copula is discussed in Schmid and Schmidt [94]. By replacing the copula $C$ with its empirical counterpart $\hat{C}_n$, we obtain the following nonparametric estimators for $\rho_i$, $i = 1,2$ :

$$\rho_1(\hat{C}_n) = h_\rho(d)\left\{2^d \int_{[0,1]^d} \hat{C}_n(\mathbf{u})d\mathbf{u} - 1\right\} = h_\rho(d)\left\{\frac{2^d}{n}\sum_{j=1}^{n}\prod_{i=1}^{d}(1 - \hat{U}_{ij,n}) - 1\right\},$$

$$\rho_2(\hat{C}_n) = h_\rho(d)\left\{2^d \int_{[0,1]^d} \Pi(\mathbf{u})\, d\hat{C}_n(\mathbf{u}) - 1\right\} = h_\rho(d)\left\{\frac{2^d}{n}\sum_{j=1}^{n}\prod_{i=1}^{d}\hat{U}_{ij,n} - 1\right\}.$$

Under the assumptions of the Proposition 10.1.1 (cf. Sect. 10.1), it can be shown that

$$\sqrt{n}\left(\rho_i(\hat{C}_n) - \rho_i(C)\right) \xrightarrow{w} Z_i \sim N(0, \sigma_i^2), \quad n \to \infty, \quad i = 1,2.$$

The variances are given by

$$\sigma_1^2 = 2^{2d} h_\rho(d)^2 \int_{[0,1]^d} \int_{[0,1]^d} E\left\{\mathbb{G}_C(\mathbf{u})\mathbb{G}_C(\mathbf{v})\right\} d\mathbf{u}d\mathbf{v},$$

$$\sigma_2^2 = 2^{2d} h_\rho(d)^2 \int_{[0,1]^d} \int_{[0,1]^d} E\left\{\mathbb{G}_{\overline{C}}(\mathbf{u})\mathbb{G}_{\overline{C}}(\mathbf{v})\right\} d\mathbf{u}d\mathbf{v},$$

with the tight Gaussian processes $\mathbb{G}_C$ and $\mathbb{G}_{\overline{C}}$ as stated in Eqs. (10.2) and (10.4). Asymptotic normality of $\rho_3$ can analogously be established based on the weak convergence of the process $(C_n, \overline{C}_n)$. For an alternative derivation of the asymptotic distribution of similar rank order statistics for Spearman's rho, see also Stepanova [107]. If the copula $C$ is radially symmetric, it follows that $\sigma_1^2 = \sigma_2^2$. The asymptotic variances can only be explicitly computed for a few copulas of simple form. For example in case of stochastic independence (i.e. $C = \Pi$), we obtain (cf. [94])

$$\sigma_1^2 = \sigma_2^2 = \frac{(d+1)^2(3(4/3)^d - d - 3)}{3(1+d-2^d)^2}.$$

As Schmid and Schmidt [92] show, the asymptotic variances can consistently be estimated by a nonparametric bootstrap method otherwise. Tests for stochastic independence based on various multivariate versions of Spearman's rho with regard to their asymptotic relative efficiency are considered by Stepanova [107] and Quessy [85].

### 10.3.2 Kendall's Tau

Let $(X_1, X_2)$ and $(Y_1, Y_2)$ be independent and identically distributed random vectors with distribution function $F$. In the bivariate case, the population version of Kendall's tau is defined as the probability of concordance minus the probability of discordance (see [60]):

$$\tau(X_1, X_2) = P\{(X_1 - Y_1)(X_2 - Y_2) > 0\} - P\{(X_1 - Y_1)(X_2 - Y_2) < 0\}. \quad (10.8)$$

If $F$ has the bivariate copula $C$, this is equal to

$$\tau(C) = 4 \int_{[0,1]^2} C(u,v) dC(u,v) - 1, \quad (10.9)$$

see e.g. Nelsen [79]. For (bivariate) Archimedean copulas, Kendall's tau can directly be calculated from the generator $\phi_C$ of the copula through [35, 36]

$$\tau(C) = 1 + 4 \int_0^1 \frac{\phi_C(t)}{\phi_C'(t)} dt.$$

For the relationship between Kendall's tau and Spearman's rho in the bivariate case, see Genest and Nešlehová [38] and references therein. Multivariate versions of Kendall's tau are considered in Nelsen [76, 78], Joe [57], and Taylor [109]. Let $\mathbf{X}$ and $\mathbf{Y}$ be two independent $d$-dimensional random vectors with distribution function $F$ and let $D_j = X_j - Y_j$, $j = 1,\dots,d$. Joe [57] suggests the following family of generalizations of Kendall's tau

$$\tau_1(\mathbf{X}) = \sum_{k=d'}^{d} w_k P\{(D_1,\dots,D_d) \in B_{k,d-k}\}, \tag{10.10}$$

with $d' = \lfloor (d+1)/2 \rfloor$ and $B_{k,d-k}$ being the subset of $\mathbf{x} = (x_1,\dots,x_d)$ in $\mathbb{R}^d$ with $k$ positive components and $d-k$ negative or $k$ negative components and $d-k$ positive. Some technical conditions on the coefficients $w_k$ such that $\tau_1$ satisfies **N1**, **N3**, **M3**, **T2**, **R**, and **TP** are given in Joe [57] and Taylor [109]. Hence, for certain choices of $w_k$, the above generalization of Kendall's tau is a multivariate measure of concordance according to Taylor [109], who also gives an alternative representation of $\tau_1$ in terms of the copula $C$ of $F$. Note that the family studied by Joe [57] includes both the average pairwise Kendall's tau and the following generalization, given by

$$\tau_2(C) = \frac{1}{2^{d-1}-1}\left\{2^d \int_{[0,1]^d} C(\mathbf{u})dC(\mathbf{u}) - 1\right\},$$

which is also considered in Nelsen [76, 78]. For dimension $d = 2$, the latter reduces to Kendall's tau as given in (10.9). According to Nelsen [76], a lower bound for $\tau_2$ is $-1/(2^{d-1}-1)$, which is also best possible and attained if at least one of the bivariate margins of the copula $C$ equals $W$ as shown by Úbeda-Flores [111]. The measure $\tau_2$ equals the average of the pairwise Kendall's tau for dimension $d = 3$ (cf. [76]).

Based on a random sample $(\mathbf{X}_j)_{j=1,\dots,n}$ from $\mathbf{X}$ with distribution function $F$, the sample version of (10.10) is

$$\hat{\tau}_1 := \tau_1(\mathbf{X}_1,\dots,\mathbf{X}_n) = \sum_{k=d'}^{d} \frac{2w_k}{n(n-1)} \sum_{i<j} \mathbf{1}_{B_{k,d-k}}(\mathbf{X}_i - \mathbf{X}_j).$$

In case $C = \Pi$, $\hat{\tau}_1$ is asymptotically normally distributed. Joe [57] calculates the asymptotic variance of $\hat{\tau}_1$ in this case and calculates corresponding asymptotic relative efficiencies for different families of copulas when the $\hat{\tau}_1$'s are used as test statistics for multivariate independence, see also Stepanova [107]. Note that a natural estimator for $\tau_2$ is given by

$$\tau_2(\hat{C}_n) = \frac{1}{2^{d-1}-1}\left\{2^d \int_{[0,1]^d} \hat{C}_n(\mathbf{u})d\hat{C}_n(\mathbf{u}) - 1\right\},$$

with empirical copula $\hat{C}_n$. According to Gänßler and Stute [33], $\tau_2(\hat{C}_n)$ is asymptotically normally distributed for dimension $d = 2$; for a discussion regarding $d \geq 2$ see Barbe et al. [3]. Other multivariate (sample) versions of Kendall's tau are discussed in Simon [104, 105], Chop and Marden [11], El Maache and Lepage [25], and Taskinen et al. [108], mainly in the context of tests for stochastic independence. For further nonparametric statistical analysis of Kendall's tau and related tests for (serial) independence, see Genest et al. [41] and references therein.

### 10.3.3 Blomqvist's Beta

Blomqvist [5] suggested a simple measure of association which is commonly referred to as Blomqvist's beta or the medial correlation coefficient. If $X_1$ and $X_2$ are two continuous random variables with medians $\tilde{x}_1$ and $\tilde{x}_2$, the population version of Blomqvist's beta is given by

$$\beta = P\{(X_1 - \tilde{x}_1)(X_2 - \tilde{x}_2) > 0\} - P\{(X_1 - \tilde{x}_1)(X_2 - \tilde{x}_2) < 0\}.$$

It can be expressed in terms of the copula $C$ of $(X_1, X_2)$ via

$$\begin{aligned}\beta(C) &= 2P\{(X_1 - \tilde{x}_1)(X_2 - \tilde{x}_2) > 0\} - 1 = 4C(1/2,1/2) - 1 \\ &= \frac{C(1/2,1/2) - \Pi(1/2,1/2) + \overline{C}(1/2,1/2) - \overline{\Pi}(1/2,1/2)}{M(1/2,1/2) - \Pi(1/2,1/2) + \overline{M}(1/2,1/2) - \overline{\Pi}(1/2,1/2)}. \end{aligned} \tag{10.11}$$

As Eq. (10.11) implies, Blomqvist's beta can be interpreted as a normalized difference between the copula $C$ and the independence copula at $(1/2, 1/2)$. Various extensions of Blomqvist's beta to the multivariate case have been considered in Joe [57], Nelsen [78], Taskinen et al. [108], Úbeda-Flores [111], and Schmid and Schmidt [93]. The following multivariate version is motivated by Eq. (10.11):

$$\begin{aligned}\beta(C) &= \frac{C(\mathbf{1/2}) - \Pi(\mathbf{1/2}) + \overline{C}(\mathbf{1/2}) - \overline{\Pi}(\mathbf{1/2})}{M(\mathbf{1/2}) - \Pi(\mathbf{1/2}) + \overline{M}(\mathbf{1/2}) - \overline{\Pi}(\mathbf{1/2})} \\ &= h_\beta(d)\left\{C(\mathbf{1/2}) + \overline{C}(\mathbf{1/2}) - 2^{1-d}\right\}, \end{aligned} \tag{10.12}$$

with $h_\beta(d) := 2^{d-1}/(2^{d-1} - 1)$ and $\mathbf{1/2} := (1/2, \ldots, 1/2)$. It satisfies the properties **N1**, **N3**, and **M3**. Úbeda-Flores [111] shows that the lower bound $-1/(2^{d-1} - 1)$, which is attained if at least one of the bivariate margins of $C$ equals $W$, is best-possible. Further, $\beta$ equals the average of pairwise Blomqvist's beta in dimension $d = 3$. Note that if the copula $C$ is radially symmetric (i.e. $C = \widehat{C}$), the expression in (10.12) reduces to

$$\frac{2^d C(\mathbf{1/2}) - 1}{2^{d-1} - 1},$$

which coincides with the multivariate version originally introduced in Nelsen [78]. According to Taylor [109], this version also satisfies the properties **R** and **TP**. Schmid and Schmidt [93] studied more general extensions of Blomqvist's beta, which measure the association in the tail region of the copula (cf. Sect. 10.6) and which include $\beta$ as defined in (10.12).

A natural estimator for $\beta$ is obtained by replacing the copula $C$ and the survival function $\overline{C}$ in the defining Eq. (10.12) with their empirical counterparts, i.e.

$$\hat{\beta}_n := \beta(\hat{C}_n) = h_\beta(d)\left\{\hat{C}_n(\mathbf{1/2}) + \hat{\overline{C}}_n(\mathbf{1/2}) - 2^{1-d}\right\},$$

where $\hat{\overline{C}}_n$ denotes the empirical survival function as defined in Eq. (10.3). Under weak assumptions on the copula $C$ and the survival function $\overline{C}$, Schmid and Schmidt [93] establish asymptotic normality and consistency of $\hat{\beta}_n$. Namely, if the $i$-th partial derivatives $D_iC$ and $D_i\overline{C}$ exist and are continuous at the point $\mathbf{1/2}$, we have

$$\sqrt{n}\,(\beta(\hat{C}_n) - \beta(C)) \xrightarrow{w} Z \quad \text{with} \quad Z \sim N(0,\sigma^2).$$

The variance $\sigma^2$ is given by $\sigma^2 = h_\beta(d)^2 E[\{\mathbb{G}_C(\mathbf{1/2}) + \mathbb{G}_{\overline{C}}(\mathbf{1/2})\}^2]$ with the tight Gaussian processes $\mathbb{G}_C$ and $\mathbb{G}_{\overline{C}}$ as stated in Eqs. (10.2) and (10.4). One main advantage of Blomqvist's beta over other copula-based measures such as Spearman's rho or Kendall's tau is that the asymptotic variance of its estimator can explicitly be calculated whenever the copula and its partial derivatives are of explicit form (see Schmid and Schmidt [93] for related examples). For example if $C = \Pi$, we have

$$\sigma^2 = \frac{1}{2^{d-1}-1}.$$

In case the copula is of more complicated form, it can be shown that a nonparametric bootstrap method can be applied to estimate the asymptotic variance. This makes it possible to use (standardized) Blomqvist's beta as test statistic for testing stochastic independence or more general dependence structures.

### *10.3.4 Gini's Gamma*

Another measure of association is Gini's gamma (or Gini's rank association coefficient), which was proposed by Gini [43]. Its population version is quite similar to Spearman's rho, which can be rewritten in the bivariate case as (cf. [78])

$$\rho(C) = 3\int_{[0,1]^2} \{(u+v-1)^2 - (u-v)^2\}\,dC(u,v).$$

Gini's gamma now focuses on absolute values rather than on squares:

$$\begin{aligned} \gamma(C) &= 2\int_{[0,1]^2} (|u+v-1| - |u-v|)\,dC(u,v) \\ &= 4\int_{[0,1]^2} \{M(u,v) + W(u,v)\}\,dC(u,v) - 2, \end{aligned} \tag{10.13}$$

see Nelsen [77, 78]. A multivariate extension of Gini's gamma has recently been considered by Behboodian et al. [4]. By defining the function $A(\mathbf{u}) = \{M(\mathbf{u}) + W(\mathbf{u})\}/2$, $\mathbf{u} \in [0,1]^d$, with corresponding survival function $\overline{A}$, the expression in Eq. (10.13) is equal to

$$\gamma(C) = 4[\int_{[0,1]^2} \{A(u,v) + \overline{A}(u,v)\} dC(u,v) - \int_{[0,1]^2} \{A(u,v) + \overline{A}(u,v)\} d\Pi(u,v)],$$

as $A(u,v) + \overline{A}(u,v) = 1 - u - v + 2A(u,v)$ for every $(u,v) \in [0,1]^2$. A multivariate version of Gini's gamma is then defined as

$$\gamma(C) = \frac{1}{b(d) - a(d)} \left[ \int_{[0,1]^d} \{A(\mathbf{u}) + \overline{A}(\mathbf{u})\} dC(\mathbf{u}) - a(d) \right], \tag{10.14}$$

with normalization constants $a(d)$ and $b(d)$ of the form

$$a(d) = \int_{[0,1]^d} \{A(\mathbf{u}) + \overline{A}(\mathbf{u})\} d\Pi(\mathbf{u}) = \frac{1}{d+1} + \frac{1}{2(d+1)!} + \sum_{i=0}^{d} (-1)^i \binom{d}{i} \frac{1}{2(i+1)!},$$

and

$$b(d) = \int_{[0,1]^d} \{A(\mathbf{u}) + \overline{A}(\mathbf{u})\} dM(\mathbf{u}) = 1 - \sum_{i=1}^{d-1} \frac{1}{4i}.$$

It immediately follows from the above definition that $\gamma = 0$ if $C = \Pi$ and $\gamma = 1$ if $C = M$; thus, **N1** and **N3** hold. For dimension $d = 3$, $\gamma$ equals the average of pairwise Gini's gamma. Another multivariate generalization is discussed by Taylor [109] in the context of multivariate measures of concordance. Behboodian et al. [4] also provide a sample version for $\gamma$ as defined in (10.14). In the bivariate case, a sample version based on the empirical copula is considered in Nelsen [77] which coincides with the traditional sample version of Gini's gamma. The latter plays an important role in the context of tests for stochastic independence and has been discussed by many authors. We refer to Genest et al. [39], Cifarelli and Regazzini [13] and references therein (see also [12], who establish asymptotic normality of a generalized class of bivariate statistics including Gini's gamma under suitable conditions). An asymptotic theory for $d \geq 3$ is not yet available to our knowledge.

## 10.4 Information-Based Measures of Multivariate Association

Relative entropy (also known as Kullback-Leibler divergence, see [66, 67]) is a measure of multivariate association that originated from information theory. This section focuses on a solely copula-based representation that is therefore independent of the marginal distributions. We will review theoretical aspects and consider nonparametric estimation techniques.

Joe [54, 56] introduced relative entropy as a measure of multivariate association in a random vector $\mathbf{X} = (X_1, \ldots, X_d)$. It is defined as

$$\delta(\mathbf{X}) = \int_{\mathbb{R}^d} \log \left[ \frac{f(\mathbf{x})}{\prod_{i=1}^{d} f_i(x_i)} \right] f(\mathbf{x}) d\mathbf{x}, \tag{10.15}$$

where $f$ is the density of the distribution of $\mathbf{X}$ (which is assumed to exist) and $f_i$ are the densities of the respective marginal distributions.

It is easy to prove that

$$\delta(\mathbf{X}) = \delta(C) = \int_{[0,1]^d} \log\left[c(\mathbf{u})\right] c(\mathbf{u}) d\mathbf{u},$$

where $c$ is the density of the copula $C$ of $\mathbf{X}$. $\delta$ therefore does not depend on the marginal distributions of $\mathbf{X}$ but only on its copula $C$ via its density $c$. If a density of $\mathbf{X}$ does not exist $\delta$ is usually set to infinity and thus satisfies **W** and **P**. It is well known that $\delta = 0$ if and only if $c(\mathbf{u}) \equiv 1$, i.e. if $C = \Pi$. Therefore properties **N1** and **N2** are satisfied.
The invariance of copulas under increasing and continuous transformations implies **T1**, because the respective densities are invariant under these transformations as well. It is also easy to prove that properties **T2** and **T3** as well as **A2** and **I** are satisfied. For a sequence of copula densities $(c_n)_{n\in\mathbb{N}}$ converging uniformly to a copula density $c$ one can see that **C** holds as well.

Relative entropy can be calculated explicitly for selected distributions. For the Gaussian distribution it is given by

$$\delta(\mathbf{X}) = -\frac{1}{2} \log\left[|\Sigma|\right],$$

with $|\Sigma|$ being the determinant of the correlation matrix $\Sigma$. In case of an equicorrelated Gaussian distribution (where $-\frac{1}{d-1} < \rho < 1$ and $\Sigma = \rho(11') + (1-\rho)I_d$) we have

$$\delta(\mathbf{X}) = -\frac{1}{2} \log\left[(1-\rho)^{d-1}(1+(d-1)\rho)\right], \tag{10.16}$$

which reduces to $\delta = -(\log[1-\rho^2])/2$ in the bivariate case. One can see that **N5** is satisfied for $0 \le \rho < 1$ for a general $d$. As $\delta$ is $[0,\infty]$-valued a normalization $\delta^*$ is introduced by solving Eq. (10.16) for $\rho$; therefore $\delta^* = |\rho|$ (in case of an equicorrelated Gaussian copula). For $d > 2$ this has to be done numerically; in the bivariate case, the normalization function is given explicitly as $\delta^* = [1-\exp(-2\delta)]^{1/2}$. **N3** is satisfied asymptotically for the normalized relative entropy.
$\delta$ can be calculated not only for the Gaussian, but also for the Student's $t$ distribution with $\nu$ degrees of freedom (see [44, 45]).

If we expand the function $g(x) = x \log x$ into a Taylor series at the point $x^* = 1$, we get under suitable regularity conditions

$$\delta(C) = \int_{[0,1]^d} g(c(\mathbf{u})) d\mathbf{u} = \sum_{t=2}^{\infty} \frac{(-1)^t}{t(t-1)} \int_{[0,1]^d} (c(\mathbf{u}) - 1)^t \, d\mathbf{u}.$$

The integral in the first summand is $\int_{[0,1]^d}(c(\mathbf{u})-1)^2\,d\mathbf{u}$ and can be regarded as a measure of the deviation of the copula density from the density of the independence copula $\Pi$. It is easy to see that

$$\int_{[0,1]^d}(c(\mathbf{u})-1)^2\,d\mathbf{u}=\int_{[0,1]^d}c^2(\mathbf{u})d\mathbf{u}-1=\int_{[0,1]^d}\frac{f^2(\mathbf{x})}{\prod_{i=1}^{d}f_i(x_i)}d\mathbf{x}-1$$

by substituting the densities on $\mathbb{R}^d$ for the copula density. This is the multivariate version of Pearson's Phi-Square as given in Joe [56] (see also [82]).

Estimation of $\delta$ can be based on $n^{-1}\sum_{j=1}^{n}\log[c(\mathbf{U}_j)]$ or $\int_{[0,1]^d}\log[c(\mathbf{u})]c(\mathbf{u})d\mathbf{u}$. In the latter case we have $\delta=\int_{[0,1]^d}\log[c(\mathbf{u})]c(\mathbf{u})d\mathbf{u}=E_C(\log[c(\mathbf{U})])$ where $E_C$ denotes the expectation with respect to the copula $C$ with corresponding density $c$. An estimator for $\delta$ is therefore given in both cases by $\hat{\delta}_n=n^{-1}\sum_{j=1}^{n}\log[\hat{c}(\hat{\mathbf{U}}_{j,n})]$, where $\hat{c}$ is an estimate of the copula density $c$ based on pseudo-observations $\hat{\mathbf{U}}_{j,n}=(\hat{U}_{1j,n},\ldots,\hat{U}_{dj,n})$ for $j=1,\ldots,n$. As copula densities have compact support, conventional kernel density estimators are subject to boundary bias and thus have to be complemented by boundary correction schemes. It would be preferable using estimators that have compact support themselves.

Probably the best known estimator with compact support is the histogram (cf. [100])

$$\hat{c}_{\mathrm{h}}(\mathbf{u})=\frac{N_k}{nh^d}$$

for $\mathbf{u}\in B_k$ with hyper-rectangular bins $B_k$ $(k=1,\ldots,m;\ m\in\mathbb{N})$. For the histogram we have the equality $n^{-1}\sum_{j=1}^{n}\log[\hat{c}(\mathbf{U}_j)]=\int_{[0,1]^d}\log[\hat{c}(\mathbf{u})]\hat{c}(\mathbf{u})d\mathbf{u}$ and thus the equivalence of both estimation approaches.
Another possible estimator is the $k$-nearest neighbour estimator

$$\hat{c}_{\mathrm{knn}}(\mathbf{u})=\frac{k/n}{\varepsilon},$$

where $\varepsilon=(2d_k)^d$ and $d_k$ denotes the distance in the maximum norm from $\mathbf{u}$ to its $k$-nearest neighbour (cf. [103]). However, this estimator is not restricted to the unit cube especially if $k$ is large and if $\mathbf{u}$ is near the boundary. We therefore suggest truncating the neighbourhood of $\mathbf{u}$ at the boundary. The modified estimator denoted by $\hat{c}_{trunc}$ differs from $\hat{c}_{knn}$ by the definition of the denominator, which is given for the truncated estimator by

$$\varepsilon=\prod_{i=1}^{d}\{d_k+d_k\mathbf{1}_{\{u_i-d_k\geq 0\}}(u_i)\mathbf{1}_{\{u_i+d_k\leq 1\}}(u_i)\\ +u_i\mathbf{1}_{\{u_i-d_k<0\}}(u_i)+(1-u_i)\mathbf{1}_{\{u_i+d_k>1\}}(u_i)\}.$$

The histogram and the nearest neighbour estimator suffer from the disadvantage of being discontinuous. Additionally, the latter integrates to infinity (cf. [103]).

There are, however, other estimators which combine the properties of continuity and compact support with finite integral such as the Beta estimator, developed by Chen [10]. It is given as

$$\hat{c}_{\text{beta}}(\mathbf{u}) = \frac{1}{n}\sum_{j=1}^{n}\prod_{i=1}^{d} K\left(\hat{U}_{ij,n}, \frac{u_i}{h}+1, \frac{1-u_i}{h}+1\right),$$

where

$$K(x,\alpha,\beta) = \frac{x^{\alpha-1}(1-x)^{\beta-1}}{B(\alpha,\beta)}$$

for some $x \in [0,1]$ denotes the univariate p.d.f. of the Beta distribution.
Sancetta and Satchell [90] proposed using the density of the empirical Bernstein copula as this estimator is itself a copula density; it is given in Bouezmarni et. al. [8] as

$$\hat{c}_{\text{bst}}(\mathbf{u}) = \frac{1}{n}\sum_{j=1}^{n} K_h(\mathbf{u}, \hat{\mathbf{U}}_{j,n}),$$

where

$$K_h(\mathbf{u}, \hat{\mathbf{U}}_{j,n}) = h^d \sum_{v_1=0}^{h-1} \cdots \sum_{v_d=0}^{h-1} \mathbf{1}_{\{\hat{\mathbf{U}}_{j,n}\in B_v\}} \prod_{i=1}^{d} \binom{h-1}{v_i} u_i^{v_i}(1-u_i)^{h-v_i},$$

with $B_v = \left[\frac{v_1}{h}, \frac{v_1+1}{h}\right] \times \cdots \times \left[\frac{v_d}{h}, \frac{v_d+1}{h}\right]$.

The performance of the different estimators with regard to the unnormalized relative entropy is compared in Blumentritt and Schmid [7]. The results indicate a good performance of the truncated nearest neighbour estimator with respect to bias and standard deviation.
Other simulation studies based on Kullback's and Leibler's original definition (10.15) of $\delta$ are due to Kraskov et al. [65] and Darbellay and Vajda [15]. Joe [55] as well as Hall and Morton [47] give results for the estimation of the Shannon entropy.

## 10.5 Measures of Multivariate Association Based on $L_p$-Distances

Hoeffding [49] was the first to consider measures of association based on a $L_p$-type distance between a copula $C$ and the independence copula $\Pi$. His work focuses on $p = 2$ and was extended by Schweizer and Wolff [99] who introduce $L_1$- and $L_\infty$-based measures of bivariate association. We first outline the multivariate generalizations of these measures and describe their properties. Secondly, we discuss their estimation and asymptotic behaviour.

### 10.5.1 $\Phi^2$ as a $L_2$-Distance-Based Measure

Gaißer et al. [34] define a generalized multivariate version of Hoeffding's $\Phi^2$ by

$$L_2^2(C) = \Phi^2(C) := h_2(d) \int_{[0,1]^d} (C(\mathbf{u}) - \Pi(\mathbf{u}))^2 \, d\mathbf{u}.$$

The normalization factor $h_2(d)$ is given by

$$\begin{aligned} h_2(d) : &= \left( \int_{[0,1]^d} (M(\mathbf{u}) - \Pi(\mathbf{u}))^2 \, d\mathbf{u} \right)^{-1} \\ &= \left( \frac{2}{(d+1)(d+2)} - \frac{1}{2^d} \frac{d!}{\prod_{i=0}^{d} \left( i + \frac{1}{2} \right)} + \left( \frac{1}{3} \right)^d \right)^{-1}. \end{aligned}$$

The latter explicit expression for $h_2(d)$ is derived in Gaißer et al. [34]. Note that for dimension $d = 2$, $h_2(2) = 90$ and $\Phi^2(C)$ reduces to the (bivariate) measure originally considered by Hoeffding [49]. Extracting the square root, we obtain $L_2(C) = \Phi(C) := +\sqrt{\Phi^2(C)}$. This measure allows for an interpretation as the normalized distance between the copula $C$ and the independence copula $\Pi$ with respect to the $L_2$-norm.

Due to their structure, all $L_p$-distance-based measures share a set of common properties. Irrespective of the particular choice of $p$, the measures satisfy **W** and **P**. They further possess the strong property that they are zero if and only if $\Pi$ is the copula of $X$, thus **N1** and **N2** hold. Normalizing by means of the upper Fréchet-Hoeffding bound, **N3** is assured. Consider a multivariate normal random vector $\mathbf{X}$ for which all pairwise correlations $\rho_{ij}$ of $X_i$ and $X_j$ are either nonnegative or nonpositive. Analogously to Wolff [114], it can be shown that all $L_p$-distance-based measures are a strictly increasing function of the absolute value of each of the pairwise correlations. Thus **N5** is valid. In general, the $L_p$-distance-based measures further satisfy **M1**, **C**, **I** and **T1**. For dimension $d \geq 3$, **T2** usually does not hold except in case the copula $C$ is radially symmetric, i.e. $C = \widehat{C}$.

We discuss some important analytical properties of $\Phi^2$ next; analogous results hold for $\Phi$ (the respective proofs are given in Gaißer et al. [34]). The measure satisfies **N4**. With regard to **N5**, it is an open problem to determine the explicit form of the function, cf. Schweizer and Wolff [99]. However, in the bivariate case a power series expansion for $\Phi^2$ given $\rho$ is provided by Hoeffding [49].

Regarding property **T**, $\Phi^2$ satisfies **T2** and **T3** in dimension $d = 2$. In higher dimensions, $\Phi^2$ is invariant under strictly decreasing transformations of one component $X_i$, if one of the following two conditions holds: the remaining $(d-1)$ components are either independent, i.e. their copula is $\Pi$, or they are independent of the transformed component.

In the particular case that an independent component $X_{d+1}$ is added to a $d$-dimensional random vector $\mathbf{X} = (X_1,...,X_d)$ with copula $C$, $\Phi^2(X_1,...,X_{d+1})$ can be expressed as a function of the $d$-dimensional measure:

$$\Phi^2(X_1,...,X_{d+1}) = \frac{1}{3}\frac{h_2(d+1)}{h_2(d)}\Phi^2(X_1,...,X_d) < \Phi^2(X_1,\dots,X_d).$$

Thus, criterion **A1** is satisfied, meaning that an independent variable $X_{d+1}$ reduces overall association in the enlarged vector.

Based on a random sample $(\mathbf{X}_j)_{j=1,...,n}$ from $\mathbf{X}$, the estimation of $\Phi^2(C)$ can be performed by replacing the copula $C$ with the empirical copula $\hat{C}_n$:

$$\begin{aligned}\Phi^2(\hat{C}_n) &= h_2(d)\int_{[0,1]^d}\left(\hat{C}_n(\mathbf{u}) - \Pi(\mathbf{u})\right)^2 du \\ &= h_2(d)\left\{\left(\frac{1}{n}\right)^2\sum_{j=1}^{n}\sum_{k=1}^{n}\prod_{i=1}^{d}\left(1-\max\left\{\hat{U}_{ij},\hat{U}_{ik}\right\}\right)\right. \\ &\left.-\frac{2}{n}\left(\frac{1}{2}\right)^d\sum_{j=1}^{n}\prod_{i=1}^{d}(1-\hat{U}_{ij}^2)+\left(\frac{1}{3}\right)^d\right\}.\end{aligned}$$

The estimate is therefore easy to calculate even for large $d$. A bias reduction for $\Phi^2(\hat{C}_n)$ has been suggested in Gaißer et al. [34]. Simulations have shown that the estimator works well for various copula families. Obviously, we obtain an estimator for the alternative measure $\Phi$ by $\Phi(\hat{C}_n) = +\sqrt{\Phi^2(\hat{C}_n)}$.

The asymptotic theory for $\Phi^2(\hat{C}_n)$ is derived from the asymptotic behaviour of the empirical copula process $\sqrt{n}(\hat{C}_n(\mathbf{u}) - C(\mathbf{u}))$ as provided by Proposition 10.1.1. Then, asymptotic normality of the estimator $\Phi^2(\hat{C}_n)$ can be derived by means of the functional delta method (see e.g. [113], p. 389). Under the assumptions of Proposition 10.1.1 and the additional presumption that $C \neq \Pi$ it follows that

$$\sqrt{n}\{\Phi^2(\hat{C}_n) - \Phi^2(C)\} \xrightarrow{w} Z_{\Phi^2},$$

where $Z_{\Phi^2} \sim N(0,\sigma^2_{\Phi^2})$ and

$$\sigma^2_{\Phi^2} = \{2h_2(d)\}^2\int_{[0,1]^d}\int_{[0,1]^d} E\Big\{\{C(\mathbf{u})-\Pi(\mathbf{u})\}\mathbb{G}_C(\mathbf{u})\mathbb{G}_C(\mathbf{v})\{C(\mathbf{v})-\Pi(\mathbf{v})\}\Big\}d\mathbf{u}d\mathbf{v}.$$

Regarding the alternative measure $\Phi$ we have

$$\sqrt{n}(\Phi(\hat{C}_n) - \Phi(C)) \xrightarrow{w} Z_\Phi$$

with $Z_\Phi \sim N(0,\sigma^2_\Phi)$ and

$$\sigma_{\Phi}^{2}=\frac{\sigma_{\Phi^2}^{2}}{4\Phi^2}=h_2(d)\frac{\int_{[0,1]^d}\int_{[0,1]^d}E\Big[\{C(\mathbf{u})-\Pi(\mathbf{u})\}\mathbb{G}_C(\mathbf{u})\mathbb{G}_C(\mathbf{v})\{C(\mathbf{v})-\Pi(\mathbf{v})\}\Big]d\mathbf{u}d\mathbf{v}}{\int_{[0,1]^d}\{C(\mathbf{u})-\Pi(\mathbf{u})\}^2d\mathbf{u}}.$$

The proof is given in Gaißer et al. [34]. The above assumption $C \neq \Pi$ guarantees that the limiting random variable is nondegenerate as implied by the form of the variance $\sigma_{\Phi^2}^2$; the limiting behaviour of $\Phi^2(\hat{C}_n)$ in case $C = \Pi$ is considered in Gaißer et al. [34].

### 10.5.2 $\sigma$ as a $L_1$-Distance-Based Measure

Wolff [114] generalizes the $L_1$-distance-based measure of Schweizer and Wolff [99] to the multivariate case. It is defined by

$$L_1(C)=\sigma(C):=h_1(d)\int_{[0,1]^d}|C(\mathbf{u})-\Pi(\mathbf{u})|\,d\mathbf{u},$$

where the normalizing factor $h_1(d)$ is given by

$$h_1(d):=\left(\tfrac{1}{d+1}-\tfrac{1}{2^d}\right)^{-1}.$$

The measure satisfies **N4**. With regard to **N5**, an explicit form of the function is derived in Schweizer and Wolff [99] for the bivariate case: $\sigma(C_\rho)=\frac{6}{\pi}\arcsin\left(\left|\frac{\rho}{2}\right|\right)$. Except for taking the absolute value, this functional form matches the one that can be derived for Spearman's $\rho$, illustrating that the two measures are closely related. A similar calculation as before shows that $\sigma$ satisfies **A1**, too:

$$\sigma(X_1,...,X_{d+1})=\frac{1}{2}\frac{h_1(d+1)}{h_1(d)}\sigma(X_1,...,X_d)<\sigma(X_1,\ldots,X_d).$$

The estimation of $L_1(C)$ has not yet been considered in detail. Various estimators for this measure can be obtained by replacing $C$ in the defining formulas with the empirical copula $\hat{C}_n$. However, no explicit expressions (as e.g. for $\Phi^2(\hat{C}_n)$) are available and the estimate must be determined numerically, which can be demanding for large dimension $d$.

### 10.5.3 $\kappa$ as a $L_\infty$-Distance-Based Measure

A $L_\infty$-distance-based multivariate measure is derived in Wolff [114] and investigated in detail by Fernández-Fernández and González-Barrios [30]. The measure is defined by

$$L_\infty(C)=\kappa(C):=h_\infty(d)\sup_{\mathbf{u}\in[0,1]^d}|C(\mathbf{u})-\Pi(\mathbf{u})|.$$

Fernández-Fernández and González-Barrios [30] do not normalize the population version of the measure. We add a normalization factor $h_\infty(d)$ in order to assure comparability with alternative measures, which is given by

$$h_\infty(d) := \left( \left(\tfrac{1}{d}\right)^{\frac{1}{d-1}} \left(1 - \frac{1}{d}\right) \right)^{-1}.$$

Wolff [114] proves that the measure satisfies all normalization criteria except for **N4**. This is due to the fact that there exist other copulas than the upper Fréchet-Hoeffding bound for which the measure attains its maximal value. With regard to **N5**, an explicit form of the function is derived in Schweizer and Wolff [99] for the bivariate case: $\kappa(C_\rho) = \frac{2}{\pi}\arcsin(|\rho|)$. With respect to the addition of further components, the measure behaves differently than the measures discussed before. It generally holds that

$$0 \leq \kappa(X_1, X_2) \leq \kappa(X_1, X_2, X_3) \leq \ldots \leq \kappa(X_1, \ldots, X_d).$$

In particular, the measure satisfies **A2** if an independent component is added to a $d$-dimensional random vector $\mathbf{X}$, i.e.

$$\kappa(X_1, ..., X_{d+1}) = \kappa(X_1, ..., X_d).$$

Estimation of $\kappa(C)$ from a sample $(\mathbf{X}_j)_{j=1,\ldots,n}$ from $\mathbf{X}$ can analogously be performed by replacing all distribution functions with their empirical counterparts:

$$\kappa(\hat{C}_n) = \frac{\sup_{\mathbf{u}\in[0,1]^d} \left|\hat{C}_n(\mathbf{u}) - \prod_{j=1}^d U_n(u_j)\right|}{\max_{0\leq i\leq n} \left(\frac{i}{n} - \left(\frac{i}{n}\right)^d\right)},$$

where $U_n$ denotes the (univariate) distribution function of a uniformly distributed random variable on the set $\{\frac{1}{n}, \ldots, \frac{n}{n}\}$. In order to reduce bias, the independence copula is replaced by its discretized version $\prod_{j=1}^d U_n(u_j)$. Fernández-Fernández and González-Barrios [30] prove a strong law of large numbers for the unnormalized statistic. An explicit asymptotic theory for this estimator is not available.

The measures introduced in this section offer a range of applications, whereas a substantial strand of literature considers tests of stochastic independence: Hoeffding [51] defines a test of independence based on $\Phi^2$ in the bivariate case. Blum et al. [6], Genest and Rémillard [40] as well as Genest et al. [42] define related statistics for testing multivariate independence.

## 10.6 Multivariate Tail Dependence

This section gives an overview of various measures of multivariate tail dependence. Here, tail dependence quantifies the degree of dependence in the joint tail of a multivariate distribution function, i.e. the dependence between extreme events. For a bivariate distribution, tail dependence is commonly defined as the limiting proportion of exceedance of one margin over a certain threshold given that the other margin has already exceeded that threshold. More precisely, the coefficient of lower tail dependence $\lambda_L$ ([102]) is defined by

$$\begin{aligned}\lambda_L(C) &:= \lim_{u\downarrow 0}\frac{C(u,u)}{u} = \lim_{u\downarrow 0} P(X_1 \le F_1^{-1}(u) \mid X_2 \le F_2^{-1}(u)) \qquad (10.17)\\ &= \lim_{u\downarrow 0} P(U_1 \leqslant u \mid U_2 \leqslant u) = \lim_{u\downarrow 0} P(U_2 \leqslant u \mid U_1 \leqslant u).\end{aligned}$$

where $\mathbf{X} = (X_1, X_2)$ is a bivariate random vector with distribution function $F$ and inverse marginal distribution functions $F_1^{-1}$, $F_2^{-1}$. Further, $U_i = F_i(X_i)$, $i = 1,2$. Equivalently, the coefficient of upper tail dependence $\lambda_U$ is

$$\lambda_U(C) := \lim_{u\uparrow 1}\frac{1-2u+C(u,u)}{1-u} = \lim_{u\uparrow 1} P(U_1 > u \mid U_2 > u)$$

if the above limits exist. Observe that $0 \leqslant \lambda_L, \lambda_U \leqslant 1$. We say $C$ is lower (orthant) tail dependent if $\lambda_L > 0$ or is upper (orthant) tail dependent if $\lambda_U > 0$. Similarly $C$ is called lower and upper tail independent if $\lambda_L = 0$ and $\lambda_U = 0$, respectively.

Joe [58] derives the coefficient of tail dependence for various families of bivariate distributions. Tail dependence of elliptically contoured distributions and copulas is discussed in Hult and Lindskog [53], Schmidt [96], Abdous et al. [1], Klüppelberg et al. [63], and Chan and Li [9]. Other copulas are for example considered in Schmidt [97], Li [71, 72], Joe et al. [59], see also reference therein. The natural nonparametric estimator for $\lambda_L$ from a random sample $(\mathbf{X}_j)_{j=1,\dots,n}$ of $\mathbf{X}$ is $\hat{\lambda}_{L,n,k} = \hat{C}_n\left(\frac{k}{n},\frac{k}{n}\right) / \left(\frac{k}{n}\right)$ with suitably chosen parameter $k = k(n)$. The statistical properties of $\hat{\lambda}_{L,n,k}$ have been investigated by several authors using techniques from extreme value theory; we mention Huang [52], Ledford and Tawn [70], Dobrić and Schmid [20], Frahm et al. [31], and Schmidt and Stadtmüller [98]. Coles et al. [14] and Draisma et al. [23] investigate the case of tail independence. For an overview and background reading see also Falk et al. [27], and de Haan and Ferreira [16], Chap. 7.

A natural way to model and analyze tail dependence is by considering extreme value distributions which arise as the limiting distribution of linearly normalized (sample) componentwise maxima, as the sample size tends to infinity; we refer to the monograph by de Haan and Ferreira [16] for a detailed treatment. In particular, a $d$-dimensional random vector $\mathbf{X}$ with distribution function $F$ is in the domain of attraction of a $d$-dimensional extreme value distribution $G$, if there exist constants $a_{mi} > 0$ and $b_{mi} \in \mathbb{R}$, $i = 1,\dots,d$, such that for all $(x_1,\dots,x_d)' \in \mathbb{R}^d_+$

$$\lim_{m\to\infty} F^m(a_{m1}x_1 + b_{m1},\dots,a_{md}x_d + b_{md}) = G(x_1,\dots,x_d).$$

The copula function of a $d$-dimensional extreme value distribution $G$ is given by ([84])

$$C_G(u_1,\dots,u_d) = \exp\Big[-V\Big\{-1/\log(u_1),\dots,-1/\log(u_d)\Big\}\Big], \tag{10.18}$$

where the function $V$ is homogeneous of order $-1$ and called the exponent measure function. For a comprehensive discussion regarding extreme value copulas see Gudendorf and Segers [46]. It can be shown that the following relationship holds between the coefficient of upper tail dependence $\lambda_U$ and a bivariate extreme value distribution $G$ with marginal distribution functions $G_1$ and $G_2$ :

$$\lambda_U = 2+\log G\Big\{\Big(\frac{1}{-\log G_1}\Big)^{-1}(1),\Big(\frac{1}{-\log G_2}\Big)^{-1}(1)\Big\}. \tag{10.19}$$

Equation (10.19) can be rewritten as follows

$$\lambda_U = 2-\lim_{t\to\infty} t\Big\{1-C_F\Big(1-\frac{1}{t},1-\frac{1}{t}\Big)\Big\} = 2+\log\Big\{C_G\Big(\frac{1}{e},\frac{1}{e}\Big)\Big\} = 2-V(1,1), \tag{10.20}$$

where $C_F$ and $C_G$ denote the copula of $F$ and $G$. Note that $1 \le V(1,1) \le 2$.

Equations (10.19) and (10.20) yield various possibilities to generalize the coefficient of bivariate tail dependence to a multidimensional tail-dependence measure. For example, the findings of Eq. (10.20) suggest to consider the copula $C_G$ of a multivariate extreme value distribution $G$, which is defined in (10.18), and evaluate it at a particular point such as $(1/e,\dots,1/e)$. Alternatively we may consider the multivariate version of the homogeneous function $V$ in (10.18) and evaluate it at $(1,\dots,1)$. Appropriate normalization then yields a multivariate measure of tail dependence (or extremal dependence) with values between 0 and 1. Similarly to considering extreme value distributions, alternatively one may consider so-called tail-dependence functions which are e.g. discussed in Huang [52], Schmidt and Stadtmüller [98], de Haan et al. [17], Einmahl et al. [24], Klüppelberg et al. [64], and Joe et al. [59], see also reference therein.

In the following, we focus on the lower tail-dependence coefficient $\lambda_L$, noting however that analogue definitions and results can be established for $\lambda_U$. In particular, the copula $C$ is upper (or lower) tail dependent if and only if the survival copula $\widehat{C}$ is lower (or upper) tail dependent.

Suppose again that $\mathbf{X} = (X_1,\dots,X_d)$ is a $d$-dimensional random vector with distribution function $F$ and copula $C$. Set $U_i = F_i(X_i)$. An evident generalization of $\lambda_L$, as defined in the bivariate case (10.17), is given by (cf. [72, 96])

$$\lambda_{L,I}(C) = \lim_{u\downarrow 0} P(U_j \leqslant u, j\notin I \mid U_i \leqslant u, i\in I) = \lim_{u\downarrow 0}\frac{C(u\mathbf{1})}{C(\mathbf{u}_{(I)})}$$

for every $I \subset \{1,...,d\}, I \notin \emptyset$ and $C$ is said to be lower tail dependent if $\lambda_{L,I} > 0$ for some $I$. The vector $\mathbf{u}_{(I)}$ denotes the vector where all coordinates, except the ith coordinate $(i\in I)$ of $u\mathbf{1}$, are replaced by 1. In the case of lower tail independence, i.e. $\lambda_{L,I} = 0$, the following multivariate measure $\eta_{L,I}$ is useful

$$C(u\mathbf{1}) = P(U_1 \leqslant u,\dots,U_d \leqslant u)$$

$$\sim \mathscr{L}(u)\{P(U_i \leqslant u, i \in I)\}^{1/\eta_{L,I}(C)} = \mathscr{L}(u)\{C(\mathbf{u}_{(I)})\}^{1/\eta_{L,I}(C)}$$

for $u \downarrow 0$. The function $\mathscr{L}(u)$ is slowly varying as $u \downarrow 0$. This type of tail-dependence measure has been considered in Ledford and Tawn [69], Coles et al. [14], and Heffernan [48] in the bivariate case. Corresponding statistical estimation is addressed in Peng [83]. For an alternative multivariate measure of tail dependence of similar type, we refer to Martins and Ferreira [73].

The following multivariate generalization of $\lambda_L$ is considered in Frahm [32]:

$$\lambda_L(C) = \lim_{u \downarrow 0} P\Big(\max\{U_1,...,U_d\} \leqslant u \mid \min\{U_1,...,U_d\} \leqslant u\Big) = \lim_{u \downarrow 0} \frac{C(u\mathbf{1})}{1 - \overline{C}(u\mathbf{1})},$$

where $\overline{C}(u,...,u) = P(U_1 > u,...,U_d > u)$ denotes the survival function of $C$. Note that the relationship between the survival copula $\widehat{C}$ and the survival function is as follows: $\widehat{C}(u,...,u) = \overline{C}(1-u,...,1-u)$, cf. Durante and Sempi [22] for related discussions.

Schmid and Schmidt [93, 95] define multivariate generalizations of $\lambda_L$ which are based on conditional versions of Spearman's rho and Blomqvist's beta. Given the following $d$-dimensional conditional version of Spearman's rho

$$\rho_p(C) = \frac{\int_{[0,p]^d} C(\mathbf{u})\, d\mathbf{u} - (p^2/2)^d}{p^{d+1}/(d+1) - (p^2/2)^d} \qquad \text{with} \quad 0 < p \leqslant 1,$$

a coefficient of multivariate lower tail dependence $\rho_L$ can be defined by

$$\rho_L(C) := \lim_{p \downarrow 0} \rho_p(C) = \lim_{p \downarrow 0} \frac{d+1}{p^{d+1}} \int_{[0,p]^d} C(\mathbf{u}) d\mathbf{u}$$

in case the limit exists. Obviously $0 \leqslant \rho_L \leqslant 1$. A possible estimator for $\rho_L$ is

$$\rho_L(\hat{C}_n) = \rho_{\frac{k}{n}}(\hat{C}_n)$$

with appropriate value $k = k(n)$, chosen by the statistician, and

$$\rho_p(\hat{C}_n) := \left\{\frac{1}{n}\sum_{j=1}^{n}\prod_{i=1}^{d}\left(p - \hat{U}_{ij,n}\right)^+ - \left(\frac{p^2}{2}\right)^d\right\} \Big/ \left\{\frac{p^{d+1}}{d+1} - \left(\frac{p^2}{2}\right)^d\right\}.$$

Asymptotic normality of $\sqrt{n}\left(\rho_L(\hat{C}_n) - \rho_L(C)\right)$ can be established if $k = k(n) \to \infty$ and $k/n \to 0$ as $n \to \infty$. The asymptotic variance can be estimated using bootstrap techniques.

In a similar spirit, a $d$-dimensional conditional version of Blomqvist's beta is defined by

$$\beta_{\mathbf{u},\mathbf{v}}(C) := h_{\mathbf{u},\mathbf{v}}(d)\left[\{C(\mathbf{u}) + \overline{C}(\mathbf{v})\} - g_{\mathbf{u},\mathbf{v}}(d)\right]$$

for $\mathbf{u}, \mathbf{v} \in [0,1]^d$ where $\mathbf{u} \leqslant \mathbf{1/2} \leqslant \mathbf{v}$ and normalization is asssured by $h_{\mathbf{u},\mathbf{v}}(d)$ and $g_{\mathbf{u},\mathbf{v}}(d)$. A coefficient of lower tail dependence can now be defined by

$$\beta_L(C) := \lim_{p\downarrow 0} \beta_{p\mathbf{1},\mathbf{1}}(C) = \lim_{p\downarrow 0} \frac{C(p\mathbf{1}) - p^d}{p + p^d}$$

if the limit exists.

Since tail dependence is limit-based, comparisons to the measures introduced in previous sections are only possible with constraints. The tail dependence measures presented generally satisfy **W**, **N1**, **N3**, **T1**, and **I**.

## References


1. Abdous, B., Fougères, A.L., Ghoudi, K.: Extreme behaviour for bivariate elliptical distributions. Can. J. Stat. **33**(3), 317–334 (2005)
2. Bakirov, N.K., Rizzo, M.L., Székely, G.J.: A multivariate nonparametric test of independence. J. Multivar. Anal. **97**(8), 1742–1756 (2006)
3. Barbe, P., Genest, C., Ghoudi, K., Rémillard, B.: On Kendall's process. J. Multivar. Anal. **58**(2), 197–229 (1996)
4. Behboodian, J., Dolati, A., Úbeda-Flores, M.: A multivariate version of Gini's rank association coefficient. Stat. Pap. **48**(2), 295–304 (2007)
5. Blomqvist, N.: On a measure of dependence between two random variables. Ann. Math. Stat. **21**(4), 593–600 (1950)
6. Blum, J.R., Kiefer, J., Rosenblatt, M.: Distribution free tests of independence based on the sample distribution function. Ann. Math. Stat. **32**(2), 485–498 (1961)
7. Blumentritt, T., Schmid, F.: Mutual information as a measure of multivariate association: Analytical properties and statistical estimation. Working paper, University of Cologne, Cologne (2010)
8. Bouezmarni, A., Rombouts, J.V.K., Taamouti, A.: Asymptotic properties of the Bernstein density copula estimator for $\alpha$-mixing data. J. Multivar. Anal. **101**(1), 1–10, (2010)
9. Chan, Y., Li, H.: Tail dependence for multivariate t-copulas and its monotonicity. Insur. Math. Econ. **42**(2), 763–770 (2008)
10. Chen, S.X.: Beta kernel estimators for density functions. Comp. Stat. Data Anal. **31**(2), 131–145 (1999)
11. Chop, K., Marden, J.: A multivariate version of Kendall's $\tau$ . J. Nonparametr. Stat. **9**(3), 261–293 (1998)
12. Cifarelli, D.M., Conti, P.L., Regazzini, E.: On the asymptotic distribution of a general measure of monotone dependence. Ann. Stat. **24**(3), 1386–1399 (1996)
13. Cifarelli, D.M., Regazzini, E.: On a distribution-free test of independence based on Gini's rank correlation coefficient. In: Barra, J.R. et al. (eds.) Recent Developments in Statistics, pp. 375–385. North-Holland, Amsterdam (1977)
14. Coles, S., Heffernan, J., Tawn, J.: Dependence measures for extreme value analyses. Extremes **2**(4), 339–365 (1999)
15. Darbellay, G.A., Vajda, I.: Estimation of the information by an adaptive partitioning of the observation space. IEEE Trans. Inf. Theory **45**(4), 1315–1321 (1999)
16. de Haan, L., Ferreira, A.: Extreme Value Theory: An Introduction. Springer, Boston, MA (2006)
17. de Haan, L., Neves, C., Peng, L.: Parametric tail copula estimation and model testing. J. Multivar. Anal. **99**(6), 1260–1275 (2008)
18. Deheuvels, P.: La fonction de dépendance empirique et ses propriétés: Un test non paramétrique d'indépendance. Acad. Roy. Belg. Bull. Cl. Sci. **65**(5), 274–292 (1979)
19. Denuit, M., Lambert, P.: Constraints on concordance measures in bivariate discrete data. J. Multivar. Anal. **93**(1), 40–57 (2005)

20. Dobrić, J., Schmid, F.: Nonparametric estimation of the lower tail dependence $\lambda L$ in bivariate copulas. J. Appl. Stat. **32**(4), 387–407 (2005)
21. Dolati, A., Úbeda-Flores, M.: On measures of multivariate concordance. J. Prob. Stat. Sci. **4**(2), 147–164 (2006)
22. Durante, F., Sempi, C.: Copula theory: an introduction. In: Durante, F., Härdle, W., Jaworski, P., Rychlik, T.: (eds.) Copula Theory and Its Applications, Proceedings of the Workshop, Warsaw, 25–26 Sept 2009. Springer, Dordrecht (2010)
23. Draisma, G., Drees, H., Ferreira, A., de Haan, L.: Bivariate tail estimation: dependence in asymptotic independence. Bernoulli **10**(2), 251–280 (2004)
24. Einmahl, J.H.J., Krajina, A., Segers, J.: A method of moments estimator of tail dependence. Bernoulli **14**(4), 1003–1026 (2008)
25. El Maache, H., Lepage, Y.: Spearman's rho and Kendall's tau for Multivariate Data Sets. Mathematical Statistics and Applications: Festschrift for Constance van Eeden, IMS Lecture Notes-Monograph Series, vol. 42, pp. 113–130 (2003)
26. Embrechts, P., McNeil, A., Straumann, D.: Correlation and dependency in risk management: properties and pitfalls. In: Dempster, M.A.H. (ed.) Risk Management: Value at Risk and Beyond, pp. 176–223. Cambridge University Press, Cambridge (2002)
27. Falk, M., Hüsler, J., Reiß, R.D.: Laws of Small Numbers: Extremes and Rare Events, 2nd revised. Birkhäuser, Basel (2004)
28. Feidt, A., Genest, C., Nešlehová, J.: Asymptotics of joint maxima for discontinuous random variables. Extremes **13**(1), 35–53 (2010)
29. Fermanian, J.-D., Radulović, D., Wegkamp, M.: Weak convergence of empirical copula processes. Bernoulli **10**(5), 847–860 (2004)
30. Fernández-Fernández, B., González-Barrios, J.M.: Multidimensional dependency measures. J. Multivar. Anal. **89**(2), 351–370 (2004)
31. Frahm, G., Junker, M., Schmidt, R.: Estimating the tail-dependence coefficient: properties and pitfalls. Insur. Math. Econ. **37**(1), 80–100 (2005)
32. Frahm, G.: On the extremal dependence coefficient of multivariate distributions. Stat. Probab. Lett. **76**(14), 1470–1481 (2006)
33. Gänßler, P., Stute, W.: Seminar on Empirical Processes. DMV-Seminar, vol. 9. Birkhäuser, Basel (1987)
34. Gaißer, S., Ruppert, M., Schmid, F.: A multivariate version of Hoeffding's phi-square. Working paper, University of Cologne, Cologne (2009)
35. Genest, C., MacKay, R.J.: Copules archimédiennes et familles de lois bidimensionnelles dont les marges sont données. Can. J. Stat. **14**(2), 145–159 (1986)
36. Genest, C., MacKay, R.J.: The joy of copulas: Bivariate distributions with uniform marginals. Am. Stat. **40**(4), 280–285 (1986)
37. Genest, C., Nešlehová, J.: A primer on copulas for count data. Astin Bull. **37**(2), 475–515 (2007)
38. Genest, C., Nešlehová, J.: Analytical proofs of classical inequalities between Spearman's $\rho$ and Kendall's $\tau$. J. Stat. Plann. Inf. **139**(11), 3795–3798 (2009)
39. Genest, C., Nešlehová, J., Ben Ghorbal, N.: Spearman's footrule and Gini's gamma: A review with complements. J. Nonparametric Stat. 22, in press (2010)
40. Genest, C., Rémillard, B.: Tests of independence and randomness based on the empirical copula process. Test **13**(2), 335–369 (2004)
41. Genest, C., Quessy, J.-F., Rémillard, B.: Tests of serial independence based on Kendall's process. Can. J. Stat. **30**, 441–461 (2002)
42. Genest, C., Quessy, J.-F., Rémillard, B.: Asymptotic local efficiency of Cramér-von Mises tests for multivariate independence. Ann. Stat. **35**(1), 166–191 (2007)
43. Gini, C.: L'ammontare e la composizione della ricchezza delle nazioni. F. Bocca, Torino (1914)
44. Guerrero-Cusumano, J.-L.: An asymptotic test of independence for multivariate $t$ and Cauchy random variables with applications. Inform. Sci. **92**(1–4), 33–45 (1996)

45. Guerrero-Cusumano, J.-L.: A measure of total variability for the multivariate $t$ distribution with applications to finance. Inform. Sci. **92**(1–4), 47–63 (1996)
46. Gudendorf, G., Segers, J.: Extreme-value copulas. In: Jaworski, P., Durante, F., Härdle, W., Rychlik, T.: (eds.) Copula Theory and Its Applications, Proceedings of the Workshop, Warsaw, 25–26 Sept 2009, Springer, Dordrecht (2010)
47. Hall, P., Morton, S.C.: On the estimation of entropy. Ann. Inst. Stat. Math. **45**(1), 69–88 (1993)
48. Heffernan, J.E.: A Directory of coefficients of tail dependence. Extremes **3**(3), 279–290 (2001)
49. Hoeffding, W.: Massstabinvariante Korrelationstheorie, Schriften des Mathematischen Seminars und des Instituts für Angewandte Mathematik der Universität Berlin **5**(3), 181–233 (1940)
50. Hoeffding, W.: Stochastische Abhängigkeit und funktionaler Zusammenhang, Skand. Aktuar. Tidskr. **25**, 200–227 (1942)
51. Hoeffding, W.: A non-parametric test of independence. Ann. Math. Stat. **19**(4), 546–557 (1948)
52. Huang, X.: Statistics of bivariate extreme values. Ph.D. Thesis, Erasmus University Rotterdam Publishers, Tinbergen Institute, Research Series 22 (1992)
53. Hult, H., Lindskog, F.: Multivariate extremes, aggregation and dependence in elliptical distributions. Adv. Appl. Probab. **34**(3), 587–608 (2002)
54. Joe, H.: Majorization, randomness and dependence for multivariate distributions. Ann. Probab. **15**(3), 1217–1225 (1987)
55. Joe, H.: Estimation of entropy and other functionals of a multivariate density. Ann. Inst. Stat. Math. **41**(4), 683–697 (1989)
56. Joe, H.: Relative entropy measures of multivariate dependence. J. Am. Stat. Assoc. **84**(405), 157–164 (1989)
57. Joe, H.: Multivariate concordance. J. Multivar. Anal. **35**(1), 12–30 (1990)
58. Joe, H.: Multivariate Models and Dependence Concepts. Chapman & Hall, London (1997)
59. Joe, H., Li, H., Nikoloulopoulos, A.K.: Tail dependence functions and vine copulas. J. Multivar. Anal. **101**(1), 252–270 (2010)
60. Kendall, M.G.: A new measure of rank correlation. Biometrika **30**(1/2), 81–93 (1938)
61. Kendall, M.G.: Rank Correlation Methods. Griffin, London (1970)
62. Kimeldorf, G., Sampson, A.R.: Monotone Dependence. Ann. Stat. **6**(4), 895–903 (1978)
63. Klüppelberg, C., Kuhn, G., Peng, L.: Estimating the tail dependence of an elliptical distribution. Bernoulli **13**(1), 229–251 (2007)
64. Klüppelberg, C., Kuhn, G., Peng, L.: Semi-parametric models for the multivariate tail dependence function – the asymptotically dependent case. Scand. J. Stat. **35**, 701–718 (2008)
65. Kraskov, A., Stögbauer, H., Grassberger, P.: Estimating mutual information. Phys. Rev. E **69**(6) Pt. 2, 066138–1-066138–16 (2004)
66. Kullback, S., Leibler, R.A.: On information and sufficiency. Ann. Math. Stat. **22**(1), 79–86 (1951)
67. Kullback, S.: Information Theory and Statistics. Wiley, New York, NY (1959)
68. Lancaster, H.: Correlation and complete dependence of random variables. Ann. Math. Stat. **34**(4), 1315–1321 (1963)
69. Ledford, A.W., Tawn, J.A.: Statistics for near independence in multivariate extreme values. Biometrika **83**(1), 169–187 (1996)
70. Ledford, A.W., Tawn, J.A.: Modelling dependence within joint tail regions. J. R. Stat. Soc. Ser. B Methods **59**(2), 475–499 (1997)
71. Li, H.: Tail dependence comparison of survival Marshall-Olkin copulas. Methodol. Comput. Appl. Probab. **10**(1), 39–54 (2008)
72. Li, H.: Orthant tail dependence of multivariate extreme value distributions. J. Multivar. Anal. **100**(1), 243–256 (2009)
73. Martins, A.P., Ferreira, H.: Measuring the extremal dependence. Stat. Probab. Lett. **73**(2), 99–103 (2005)

74. McNeil, A., Frey, R., Embrechts, P.: Quantitative Risk Management: Concepts, Techniques and Tools. Princeton University Press, Princeton (2005)
75. Mesfioui, M., Tajar, A.: On the properties of some nonparametric concordance measures in the discrete case. J. Nonparametr. Stat. **17**(5), 541–554 (2005)
76. Nelsen, R.B.: Nonparametric measures of multivariate association. In: Distribution with Fixed Marginals and Related Topics. IMS Lecture Notes – Monograph Series, vol. 28, pp. 223–232. Institute of Mathematical Statistics, Hayward, CA (1996)
77. Nelsen, R.B.: Concordance and Gini's measure of association. J. Nonparametr. Stat. **9**(3), 227–238 (1998)
78. Nelsen, R.B.: Concordance and copulas: a survey. In: Cuadras, C.M., Fortiana, J., Rodriguez-Lallena, J.A. (eds.) Distributions with Given Marginals and Statistical Modelling, pp. 169–178. Kluwer, Dordrecht (2002)
79. Nelsen, R.B.: An Introduction to Copulas, 2nd edn. Springer, New York, NY (2006)
80. Nešlehová, J.: Dependence of non-continuous random variables. Ph.D. Thesis, Carl von Ossietzky Universität Oldenburg, Oldenburg (2004)
81. Nešlehová, J.: On rank correlation measures for non-continuous random variables. J. Multivar. Anal. **98**(3), 544–567 (2007)
82. Pearson, K.: Mathematical contributions to the theory of evolution, XIII: On the theory of contingency and its relation to association and normal correlation. Draper's Company Research Memoirs (Biometric Series I), University College, London (1904) [reprinted in: Early Statistical papers, Cambridge University Press, Cambridge (1948)]
83. Peng, L.: Estimation of the coefficient of tail dependence in bivariate extremes. Stat. Probab. Lett. **43**(4), 399–409 (1999)
84. Pickands, J.: Multivariate extreme value distributions. Bull. Int. Stat. Inst. **49**(2), 859–878 (1981)
85. Quessy, J.F.: Theoretical efficiency comparisons of independence tests based on multivariate versions of Spearman's rho. Metrika **70**(3), 315–338 (2009)
86. Rényi, A.: On measures of dependence. Acta. Math. Acad. Sci. Hung. **10**(3/4), 441–451 (1959)
87. Rodriguez-Lallena, J.A., Úbeda-Flores, M.: A new class of bivariate copulas. Stat. Probab. Lett. **66**(3), 315–325 (2004)
88. Rüschendorf, L.: Asymptotic distributions of multivariate rank order statistics. Ann. Stat. **4**(5), 912–923 (1976)
89. Ruymgaart, F.H., van Zuijlen, M.C.A.: Asymptotic normality of multivariate linear rank statistics in the non-i.i.d. case. Ann. Stat. **6**(3), 588–602 (1978)
90. Sancetta, A., Satchell, S.: The Bernstein copula and its applications to modelling and approximations of multivariate distributions. Econom. Theory **20**(3), 535–562 (2004)
91. Scarsini, M.: On measures of concordance. Stochastica **8**(3), 201–218 (1984)
92. Schmid, F., Schmidt, R.: Bootstrapping Spearman's multivariate rho. In: Rizzi, A., Vichi, M. (eds.) Proceedings of COMPSTAT 2006, pp. 759–766 (2006)
93. Schmid, F., Schmidt, R.: Multivariate conditional versions of Spearman's rho and related measures of tail dependence. J. Multivar. Anal. **98**(6), 1123–1140 (2007)
94. Schmid, F., Schmidt, R.: Multivariate extensions of Spearman's rho and related statistics. Stat. Probab. Lett. **77**(4), 407–416 (2007)
95. Schmid, F., Schmidt, R.: Nonparametric inference on multivariate versions of Blomqvist's beta and related measures of tail dependence. Metrika **66**(3), 323–354 (2007)
96. Schmidt, R.: Tail dependence for elliptically contoured distributions. Math. Methods Oper. Res. **55**(2), 301–327 (2002)
97. Schmidt, R.: Tail dependence. In: P. Cizek, W. Härdle, R. Weron (eds.) Statistical Tools for Finance and Insurance. Springer, New York, NY (2005)
98. Schmidt, R., Stadtmüller, U.: Non-parametric estimation of tail dependence. Scand. J. Stat. **33**(2), 307–335 (2006)
99. Schweizer, B., Wolff, E.F.: On nonparametric measures of dependence for random variables. Ann. Stat. **9**(4), 879–885 (1981)

100. Scott, D.W.: Multivariate Density Estimation. Wiley, New York, NY (1992)
101. Siburg, K.F., Stoimenov, P.A.: A measure of mutual complete dependence. Metrika **71**(2), 239–251 (2010)
102. Sibuya, M.: Bivariate extreme statistics. Ann. Inst. Stat. Math. **11**(3), 195–210 (1960)
103. Silverman, B.W.: Density Estimation for Statistics and Data Analysis. Chapman & Hall, London (1986)
104. Simon, G.: Multivariate Generalization of Kendall's Tau with application to data reduction. J. Am. Stat. Assoc. **72**(358), 367–376 (1977)
105. Simon, G.: A nonparametric test of total independence based on Kendall's tau. Biometrika **64**(2), 277–282 (1977)
106. Spearman, C.: The proof and measurement of association between two things. Am. J. Psychol. **15**(1), 72–101 (1904)
107. Stepanova, N.A.: Multivariate rank tests for independence and their asymptotic efficiency. Math. Methods Stat. **12**(2), 197–217 (2003)
108. Taskinen, S., Oja, H., Randles, R.H.: Multivariate nonparametric tests of independence. J. Am. Stat. Assoc. **100**(471), 916–925 (2005)
109. Taylor, M.D.: Multivariate measures of concordance. Ann. Inst. Stat. Math. **59**(4), 789–806 (2007)
110. Tsukahara, H.: Semiparametric estimation in copula models. Can. J. Stat. **33**(3), 357–375 (2005)
111. Úbeda-Flores, M.: Multivariate versions of Blomqvist's beta and Spearman's footrule. Ann. Inst. Stat. Math. **57**(4), 781–788 (2005)
112. Vandenhende, F., Lambert, P.: Improved rank-based dependence measures for categorical data. Stat. Probab. Lett. **63**(2), 157–163 (2003)
113. van der Vaart, A.W., Wellner, J.A.: Weak Convergence and Empirical Processes. Springer, New York, NY (1996)
114. Wolff, E.F.: N-dimensional measures of dependence. Stochastica **4**(3), 175–188 (1980)

# Chapter 11
# Semi-Copulas and Interpretations of Coincidences Between Stochastic Dependence and Ageing

Fabio Spizzichino

**Abstract** We aim at providing probabilistic explanations of equivalences, between conditions of positive dependence and of univariate ageing, that have been pointed out in the literature. To this purpose we consider bivariate survival functions $\overline{F}(x,y)$ and properties of them that are respectively invariant under transformations of the type $\overline{F}(\varphi(x),\varphi(y))$ and $\psi\left(\overline{F}(x,y)\right)$, for $\varphi,\psi:[0,1]\to[0,1]$ increasing bijections. Bivariate Schur-constant survival models will have a central role in our discussion.

## 11.1 Introduction

Probabilistic theory of Reliability is essentially based on the study of non-negative random variables that have the meaning of life-times, time-to-failures ... and so on. An important part of literature in this area is devoted to introducing and analyzing notions of ageing and stochastic dependence; these notions are then employed to obtain useful inequalities in reliability computation and estimation. An introduction to these topics can in particular be found in [3, 7, 22, 25, 28, 37] and references contained therein. We address the readers to such a bibliography for what concerns the heuristic meaning, main mathematical properties, and applications of dependence and ageing. For the ease of the non-specialist reader, we here limit ourselves just to recall the definitions of the basic notions that are needed in the discussion.

Roughly speaking, the term *dependence* means that we deal with different units (components of a reliability system, living beings, industrial products, ...), whose life-times cannot be, realistically, modelled as stochastically independent random variables; by the term *ageing* we mean that not all life-times of the units involved in a problem can have, marginally, exponential distributions.

Fabio Spizzichino
Department of Mathematics, University La Sapienza, Rome, Italy
e-mail: Fabio.Spizzichino@uniroma1.it

P. Jaworski et al. (eds.), *Copula Theory and Its Applications*, Lecture Notes in Statistics 198,
DOI 10.1007/978-3-642-12465-5_11, 

The theory of Reliability is indeed permeated by the phenomena of stochastic dependence and ageing. The same also happens for our everyday life. We can say however that, within the realm of Reliability, not only such phenomena are studied in terms of exact probabilistic concepts but even, as a more stringent feature of this theory, various connections between dependence and ageing emerge in a natural way, at different conceptual levels and in the frame of different approaches (see e.g. [2, 28, 37, 41, 42] and references therein).

Up to a certain extent, we can even claim that it there exists a sort of formal identification between some notions of dependence and some notions of ageing. Our main purpose here is just to present a discussion on this specific point. Namely, we will see in which sense ageing and dependence are different avatars of same properties. We also propose probabilistic interpretations of facts that might appear of a pure analytic character.

It is well-known to the reader that the notion of copula has a natural and useful role in the description of stochastic dependence. In these proceedings the reader can find a detailed review about the use of copulas for the topic of coherent systems, in Reliability Theory (see Chap. 9 of this volume [35]). A fact of interest here is that copulas can also have a role in the description of stochastic ageing. Actually, as we shall briefly recall in the following sections, it is the more general notion of *semi-copula* that emerges in connection with the study of ageing. But we have even more to say, concerning the use of copulas in the description of dependence: it has been just such an approach that inspired the use of semi-copulas for problems of applied probability and suggested the study of the concepts, related to ageing, that will be reviewed here.

Let us then come to explain more specifically the topic and the purposes of our discussion.

We denote by $C_\phi$ the bivariate Archimedean copula with *additive generator* $\phi$, i.e. we put, for $0 \le u, v \le 1$,

$$C_\phi(u,v) := \phi^{-1}\left[\phi(u) + \phi(u)\right].$$

$\phi : (0,1] \to [0,+\infty)$ must be a continuous, convex, decreasing function such that $\phi(1) = 0$. We assume, even if not always strictly required, that $\phi$ is strictly decreasing and $\lim_{u\to 0^+} \phi(u) = +\infty$. $\phi^{-1} : [0,+\infty) \to (0,1]$ is a continuous, convex, strictly decreasing function such that $\phi^{-1}(0) = 1$ and can be seen as a *univariate survival function,* i.e. we can find non-negative random variables $X$ such that, for $x \ge 0$,

$$\phi^{-1}(x) = \overline{F}_X(x) := P\{X > x\}.$$

$C_\phi$, being a copula (even if one of a special type, i.e. an Archimedean one), is the natural object for which properties of stochastic dependence can be defined. $\phi^{-1}$, being a univariate survival function, is the natural object for which one can define properties of univariate stochastic ageing.

Several results presented in the recent literature show how dependence properties of Archimedean copulas are related to analytic properties of their generators $\phi$;

see in particular the papers by Muller and Scarsini [33], and Averous and Dortet-Bernadet [4].

It is remarkable for our purposes that an equivalence can be established between some notions of positive dependence (such as, say, *Left Tail Decreasing*) and corresponding notions of negative univariate ageing (such as *Decreasing Failure Rate*), in the following sense:

As pointed out in [4], $C_\phi$ satisfies one of such properties of positive dependence if and only if $\phi^{-1}$ satisfies a corresponding property of negative univariate ageing. We shall report these results in Sect. 11.2, together with the necessary definitions of properties of ageing and dependence. We shall also see that, by introducing a simple and natural notion of *dual* of a (univariate) survival function, the results in [4] can be formulated under equivalent forms that are more adapt for our purposes: positive dependence of $C_\phi$ is equivalent to corresponding positive ageing of the dual of $\phi^{-1}$. Furthermore these results will be used to extend the definitions of dependence properties to Archimedean semi-copulas, what is of interest for the subsequent discussion.

Section 11.3 will be essentially devoted to the so-called *Schur-constant* models. These are special exchangeable survival models where the survival copula is Archimedean and the joint distribution is simply determined by the univariate marginal. More exactly the marginal survival function has also the role of inverse of the generator.

After recalling essential terminology (survival model, survival function, survival copula, Schur-constant property, ...) and showing some aspects of the Archimedean copulas in our specific context, we will analyze properties of ageing and of dependence for Schur-constant models. Since, for these models, the joint distribution is completely determined by the univariate marginal, it does not come as a surprise that properties of the survival copula are determined by the behavior of the marginal. We shall see in fact that some of the results about dependence properties that can be given for these models coincide with the characterizations presented in [4].

In Sect. 11.4 we sketch how ageing can be connected to the level sets of a joint survival function and reasons way semi-copulas enter in the play. We will then shortly review a concept of duality among multivariate survival models. These arguments provide the basis for our interpretation of relations among ageing and dependence. The discussion will be based on a brief review of topics from some past and some recent papers and on a few related comments and remarks. It will, in particular, emerge that Archimedean (semi-)copulas can be used as natural objects to describe univariate ageing and to extend the characterizations in [4] to the cases when $\phi^{-1}$ is not convex. By focusing attention on the i.i.d. and on the Schur-constant models, we shall also explain the substantial probabilistic motivations for such a role of Archimedean (semi-)copulas.

In Sect. 11.5 we present a short summary of our discussion and some concluding remarks. We also give an hint for future research that is suggested by the fact that all the models with Archimedean survival copulas can be obtained by the i.i.d. and the Schur-constant models by means of simple transformations.

## 11.2 Univariate Ageing and Dependence Properties of Archimedean Semi-Copulas

For a scalar, non-negative, random variable $X$ (life-time), the *survival function* (or reliability function) is defined as

$$\overline{G}_X(x) := P(X > x),\ x \geq 0.$$

In our treatment we shall consider, as survival functions, the functions $\overline{G}$ : $[0,+\infty) \to (0,1]$ that are continuous, strictly decreasing, strictly positive all over $[0,+\infty)$ and such that $\overline{G}(0) = 1, \lim_{x\to\infty} \overline{G}(x) = 0$. For fixed $\overline{G}$, we also set

$$\rho(x) := -\log \overline{G}(x).$$

Then $\rho : [0,+\infty) \to [0,+\infty)$ is a continuous, increasing function such that $\rho(0) = 0$.

When $\overline{G}$ is absolutely continuous ($\overline{G}(x) = \int_x^{+\infty} g(\xi)d\xi$), the derivative

$$r(x) := \rho'(x) = -\frac{d}{dx}\log \overline{G}(x) = \frac{g(x)}{\overline{G}(x)}$$

has the meaning of *failure rate function.* The exponential distributions, characterized by the condition $r(x) =$ constant, are those with the *memory-less* or *no-ageing* property.

For a given survival function $\overline{G}$, it is of interest to consider the ratio

$$\overline{G}_t(x) := \frac{\overline{G}(t+x)}{\overline{G}(t)} = \exp\{-[\rho(x+t) - \rho(t)]\}. \tag{11.1}$$

In the absolutely continuous case, we can also write

$$\overline{G}_t(x) = \exp\{-\int_t^{t+x} r(\xi)d\xi\}.$$

Obviously, if $\overline{G}$ is the survival function of $X$, where $X$ is the lifetime of a unit $U$, then we can write

$$\overline{G}_t(x) = P\{X > x+t \mid X > t\},$$

i.e. $\overline{G}_t(x)$ is interpreted as the *survival probability*, for an extra time $x$, for $U$ when $U$ reached the *age* $t$.

The different notions of univariate ageing concern, more or less directly, the qualitative behavior of the family of functions $\overline{G}_t$, for increasing values of $t \geq 0$. The following ones are among the most well-known notions.

$\overline{G}$ is *New Better than Used* (NBU) if $\overline{G}(x+y) \leq \overline{G}(x)\cdot\overline{G}(y)$, i.e. $\rho(x+y) \geq \rho(x)\cdot\rho(y)$ or

$$\overline{G}(x) \geq \overline{G}_t(x), \forall x,t \geq 0.$$

$\overline{G}$ is *Increasing Failure Rate in Average* (IFRA) if $\frac{\rho(t)}{t}$ is increasing in $t$.

$\overline{G}$ is *Increasing Failure Rate* (IFR) if, for $0 \le t' < t''$,

$$\overline{G}_{t'}(x) \ge \overline{G}_{t''}(x), \forall x \ge 0.$$

Notice that an absolutely continuous $\overline{G}$ is IFR if and only if the failure rate $r$ is non-decreasing (whence, actually, the origin of the term IFR).

The following chain of implications holds

$$\text{IFR} \Rightarrow \text{IFRA} \Rightarrow \text{NBU}.$$

Each of the afore-mentioned notions is one of *positive ageing*. The corresponding notions of negative ageing, respectively called *New Worse that Used (NWU), Decreasing Failure Rate in Average (DFRA), Decreasing Failure Rate (DFR)*, are obtained by reversing the above inequalities.

An absolutely continuous $\overline{G}$ is *Strongly Unimodal* (SU) if $g$ is log-concave. This can be seen as a strong property of positive univariate ageing and, in particular, it implies that $\overline{G}$ is IFR. The opposite condition, $g$ is log-convex, is a property of negative univariate ageing.

We now recall some notions of positive dependence for a bivariate copula $C$. For simplicity sake, we limit ourselves to the case of exchangeable copulas, where there is a situation of symmetry between the two variables. The extension to the non-symmetric case is straightforward.

$C$ is *Positive Quadrant Dependent* (PQD) if

$$C(u,v) \ge u \cdot v.$$

$C$ is *Left Tail Decreasing* (LTD) if, for any $v \in (0,1)$ and $0 \le u' < u'' \le 1$

$$\frac{C(u',v)}{u'} \ge \frac{C(u'',v)}{u''}.$$

A differentiable $C$ is *Stochastic Increasing* (SI) if, for $u, v \in [0,1]$,

$$\frac{\partial C}{\partial u} \text{is non-decreasing in } u.$$

An Archimedean bivariate copula $C_\phi$ is *Positive K-Dependent* (PKD) if, for all $0 < v \le 1$,

$$\frac{\phi(v)}{\phi'(v^+)} \ge v \log v.$$

As mentioned in the Introduction, for an Archimedean copula $C_\phi$ the above positive dependence properties correspond to negative ageing properties of $\phi^{-1}$. More precisely, for convex $\phi$, the following equivalences hold (see [4] and references cited therein):

(a) $C_\phi$ is PQD if and only if $\phi^{-1}$ is NWU;
(b) $C_\phi$ is PKD if and only if $\phi^{-1}$ is DFRA;

(c) $C_\phi$ is LTD if and only if $\phi^{-1}$ is DFR;
(d) $C_\phi$ is SI if and only if $\phi^{-1}$ has a log-convex density.

Of course, even if $\phi : (0,1] \to R_+ \cup \{0\}$ is not convex, $\phi^{-1}$ may well be a properly-defined survival function, for which it makes sense to consider properties of ageing such as NWU, IFRA, DFR, and so on. For this reason we can use the above equivalences (a) - (d) in order to formally *define* properties of positive dependence for functions of the form $C_\phi$, even when $\phi$ is not convex.

If $\phi$ is not convex, then $C_\phi$ is not a copula; actually it is an Archimedean *semi-copula* or a *t-norm* (see [12, 27]). Qualitative properties such as PQD, PKD, LTD, SI cannot anymore be properly interpreted as properties of stochastic dependence between two random variables. However it is still interesting for our discussion to consider, along with Archimedean copulas, Archimedean semi-copulas and to extend to them those properties that, when restricted to copulas, become *bona-fide* properties of dependence.

For the sake of our discussion it is also useful to introduce the following

**Definition 11.2.1.** Let $\overline{G}(x) = \exp\{-\rho(x)\}$ be a univariate survival function. Then

$$\overline{G}^*(x) := \exp\{-\rho^{-1}(x)\} \tag{11.2}$$

is a univariate survival function as well, that we call the *dual* of $\overline{G}(x)$.

The proof of the following Proposition amounts just to a simple verification.

**Proposition 11.2.1.** *A univariate survival function $\overline{G}(x)$ is NBU (IFRA, IFR, SU) if and only if $\overline{G}^*(x)$ is NWU (DFRA, DFR, with log-convex density).*

For a univariate survival function $\overline{G}(x)$ we now let, for $0 \le u, v \le 1$,

$$A_{\overline{G}}(u,v) := C_{\overline{G}^{*-1}}(u,v) =$$

$$= \exp\{-\rho^{-1}(\rho(-\log u) + \rho(-\log v))\}. \tag{11.3}$$

By collecting the arguments above we can now state:

**Proposition 11.2.2.**

$(\alpha)$ $A_{\overline{G}}$ *is PQD if and only if $\overline{G}$ is NBU;*
$(\beta)$ $A_{\overline{G}}$ *is PKD if and only if $\overline{G}$ is IFRA;*
$(\gamma)$ $A_{\overline{G}}$ *is LTD if and only if $\overline{G}$ is IFR;*
$(\delta)$ $A_{\overline{G}}$ *is SI if and only if $\overline{G}$ is SU.*

In other words, positive "dependence properties" of $A_{\overline{G}}$ correspond to positive ageing properties of $\overline{G}$ (remember however that $A_{\overline{G}}$ is a copula if and only if $\overline{G}^{*-1}$ is convex).

We shall present a probabilistic interpretation of (*a*)–(*d*) in Sect. 11.3 and, as a related consequence, we shall examine the meaning of ($\alpha$)–($\delta$) in Sect. 11.4.

## 11.3 Dependence and Univariate Ageing in Schur-Constant Models

We start this section by introducing some notation and a few well-known notions about multivariate survival models.

By the term *multivariate survival model* we simply mean the joint probability distribution of a $n$-tuple of non-negative random variables (or life-times) $X_1,...,X_n$. A multivariate survival model is commonly described by its *joint survival function*

$$\overline{F}(x_1,...,x_n) := P\{X_1 > x_1,...,X_n > x_n\}.$$

For $i = 1,2,...,n$, the *marginal survival function* $\overline{G}_i$ is defined by

$$\overline{G}_i(x) := P\{X_i > x\} = \overline{F}(0,...,0,x,0,...0)$$

($x$ being here the $i$-th coordinate of the vector $(0,...,0,x,0,...,0)$) and the *survival copula* is given by

$$K(u_1,...,u_n) = \overline{F}\left(\overline{G}_1^{-1}(u_1),...,\overline{G}_n^{-1}(u_n)\right).$$

Many positive or negative dependence notions for $\overline{F}$ are actually properties of $K$. We say that a survival model is a *Schur-constant model* if it has the form

$$\overline{F}(x_1,...,x_n) = \overline{G}\left(\sum_{i=1}^{n} x_i\right), \tag{11.4}$$

where $\overline{G}$ is a positive, continuous, and strictly decreasing one-dimensional survival function (such that actually $\overline{F}$ turns out to be a $n$-dimensional survival function). Notice that $\overline{F}$ is obviously exchangeable and that $\overline{G}$ has the meaning of the one-dimensional marginal survival function of $\overline{F}$. It is easy to see that, in the case when $\overline{F}$ is absolutely continuous with a joint density function $f(x_1,...,x_n)$, i.e. when

$$\overline{F}(x_1,...,x_n) = \int_{x_1}^{+\infty}\int_{x_2}^{+\infty}...\int_{x_n}^{+\infty} f(\xi_1,...,\xi_n)d\xi_1...d\xi_n,$$

$\overline{F}$ is Schur-constant if and only if $f$ has the form $f(x_1,...,x_n) = \varkappa(\sum_{i=1}^{n} x_i)$, with $\varkappa(x) = (-1)^n \frac{d^n}{dx^n}\overline{G}(x)$.

Schur-constant models arise as a generalization of the joint distributions of i.i.d. exponential or conditionally i.i.d. exponential variables. A slightly more general notion is the one of multivariate $l_1$-norm symmetric distribution; a detailed treatment of this notion is provided in [23] and some of the references indicated therein. A survival function of the form (11.4) is simultaneously Schur-concave and Schur-convex (see [29]). The term *Schur-constant models* was originally used in the reliability field, where such models are considered (see, in particular [5, 6, 8, 38–40] ) in connection with a multivariate version of the *memory-less* or *no-ageing* property. For more details about this interpretation, related properties and characterizations see [41]; see also [16] for further characterizations.

Schur-constant models are special cases of survival models of the form

$$\overline{F}(x_1,...,x_n) = W\left[R(x_1) + ... + R(x_n)\right], \tag{11.5}$$

where $W$ is a continuous and strictly decreasing one-dimensional survival function and $R : [0,+\infty) \to [0,+\infty)$ is a strictly increasing function such that $R(0) = 0$ and $\lim_{x\to\infty} R(x) = \infty$. A $n$-variate survival model, with a continuous one-dimensional marginal survival function strictly decreasing where positive, has the form (11.5) if and only if its survival copula is Archimedean; more precisely we have

$$K(u_1,...,u_n) = W\left[W^{-1}(u_1) + ... + W^{-1}(u_n)\right]. \tag{11.6}$$

Notice that $K$ is not influenced by $R$.

Models defined by (11.5) and by a continuous and strictly decreasing, positive, one-dimensional marginal survival function, have also been called *Time-Transformed Exponential (TTE)* models. They are exchangeable. A strictly related notion is the one of Frailty Models introduced in [30]. In the special case (11.4) (i.e. $R(x) = x, W = \overline{G}$) (11.6) becomes

$$K(u_1,...,u_n) = \overline{G}\left[\overline{G}^{-1}(u_1) + ... + \overline{G}^{-1}(u_n)\right].$$

It is to be noticed that we can find different models still satisfying (11.6) by dropping the above assumption on the one-dimensional marginal (see [33]).

For our purposes we can now limit our attention to the Schur-constant case with $n = 2$:

$$\overline{F}(x,y) = \overline{G}(x+y). \tag{11.7}$$

We look at $\overline{F}$ as at the joint survival function of a pair of exchangeable life-times $X$ and $Y$ and $\overline{G}$ is their (common) marginal survival function. In order to let $\overline{F}$ to be properly defined as a bivariate survival function, $\overline{G}$ must be convex so that, also,

$$K(u,v) = C_{\overline{G}^{-1}}(u,v) = \overline{G}\left[\overline{G}^{-1}(u) + \overline{G}^{-1}(v)\right], 0 \le u,v \le 1, \tag{11.8}$$

is an (Archimedean) copula with the meaning of survival copula of $\overline{F}$. $K(u,v)$ can be seen as (the restriction to $[0,1]\times[0,1]$ of) the joint distribution function of the pair $\left(\overline{G}(X), \overline{G}(Y)\right)$.

Notions of dependence for $K$, such as those recalled in Sect. 11.1 (i.e. PQD, PKD, LTD, SI), can be given a more transparent probabilistic meaning in terms of conditions of correlation between $X$ and $Y$, as recalled in the statements presented below. For a wider specific analysis about dependence properties of Schur-constant models, see [13, 14], [41, Chap. 3], [15, 34]. For general discussions about the role of copulas to describe stochastic dependence see e.g. [17, 26, 34] .

It is easy to see that

$K$ is PQD if and only if $X$ and $Y$ are PQD in the sense

$$P\{X > x \mid Y > y\} \geq P\{X > x\}$$

$K$ is LTD if and only if $X$ and $Y$ are Right-Tail Increasing (RTI) in the sense

$$P\{X > x \mid Y > y\} \text{ increasing in } y$$

$K$ is SI if and only if $X$ and $Y$ are Stochastically Increasing (SI) in the sense

$$P\{X > x \mid Y = y\} \text{ increasing in } y.$$

Recall that $X,Y$, with joint survival function given in (11.7) are exchangeable and then all the dependence conditions considered above are symmetric in $x,y$.

The notion of PKD is interpreted in terms of the behavior of the *Kendall distribution* $V(t) := P\{\overline{F}(X,Y) \leq t\}$.

The following remarks show why bivariate Schur-constant models are relevant in our discussion.

*Remark 11.3.1.* Any Archimedean copula can be seen as the survival copula of a suitable Schur-constant model.

This fact was pointed out in [9] as a property of basic interest for the purposes of that paper and it is also of interest in the present discussion. A detailed analysis on this property is also presented on [31].

As shown by Eq. (11.8), for Schur-constant models the survival copula is determined by the marginal survival function $\overline{G}$. Then we can expect that properties of dependence are characterized in terms of properties of $\overline{G}$. In this respect we can obviously write

$$P\{X > x, Y > y\} = \overline{G}(x+y) = P\{X > x+y\},$$

$$P\{X > x \mid Y > t\} = \frac{\overline{G}(x+t)}{\overline{G}(t)} = P\{X > x+t \mid X > t\}. \tag{11.9}$$

In the absolutely continuous case, something analog to (11.9) can also be said for the function $P\{X > x \mid Y = t\}$. In fact we point out the following property.

**Proposition 11.3.1.** *Let $(X,Y)$ be distributed according to a (jointly) absolutely continuous Schur-constant model with a positive marginal density $g$. Then*

$$P\{X > x \mid Y = t\} = \frac{g(x+t)}{g(t)}. \tag{11.10}$$

*Proof.* In the Schur-constant case, the joint distribution of $(X,Y)$ is absolutely continuous if and only if $\overline{G}$ admits a second derivative $\varkappa$ and the joint density is given by

$$f_{X,Y}(x,y) = \varkappa(x+y).$$

In such a case, on the other hand, the one-dimensional marginal density is given by

$$g(t) = -\frac{d}{dy}\overline{G}(y)_{|y=t} = \int_t^{+\infty} \varkappa(w)dw. \tag{11.11}$$

Then we can write

$$\int_x^{+\infty} f_{X,Y}(\xi,t)d\xi = \int_x^{+\infty} \varkappa(\xi+t)d\xi =$$

$$= \int_{x+t}^{+\infty} \varkappa(w)dw = -\frac{d}{dy}\overline{G}(y)_{|y=x+t} = g(x+t). \tag{11.12}$$

In view of (11.12), Eq. (11.10) can be readily obtained by noticing that, for a pair of jointly absolutely continuous life-times $A,B$, we can generally write

$$P\{A > a \mid B = t\} = \frac{\int_a^{+\infty} f_{A,B}(\xi,t)d\xi}{f_B(t)}.$$

*Remark 11.3.2.* Different properties of dependence for a pair of lifetimes $X,Y$ (and then for their survival copula) can be described in terms of behavior of the functions $P\{X > x \mid Y > t\}$ and $P\{X > x \mid Y = t\}$. As shown by Eq. (11.9) and Proposition 11.3.1, on the other hand properties of $P\{X > x \mid Y > t\}$ and $P\{X > x \mid Y = t\}$ respectively coincide, for the Schur-constant case, with properties of $\frac{\overline{G}(x+t)}{\overline{G}(t)}$ and $\frac{g(x+t)}{g(t)}$.

By recalling the form (11.8) of the survival copula of a bivariate Schur-constant model, we see that the two remarks above heuristically explain the reason why those properties of $\overline{G}$, that are related to dependence properties of the Archimedean copula $C_{\overline{G}^{-1}}$, can just be seen as univariate ageing properties.

More precisely, by using the arguments above, the following equivalences can easily be checked:

(a') $X$ and $Y$ are PQD if and only if they are marginally NWU;
(c') $X$ and $Y$ are RTI if and only if they are marginally DFR;
(d') $X$ and $Y$ are SI if and only if they have marginal log-convex density.

Notice that (a'), (c'), (d') just imply, respectively, the equivalences (a), (c), (d) stated in the previous Section.

Also for the equivalence (b) a probabilistic justification could be given, in terms of bivariate Schur-constant models and Kendall distributions. This topic, that we omit here for brevity sake, will be developed in a next paper.

## 11.4 Level Curves, B functions, Duality, and Interpretation of Coincidence Between Ageing and Dependence

Let $\overline{G}(x)$ be a univariate survival function (not necessarily convex). In order to give an interpretation of the claims in Proposition 11.2.2 we discuss the meaning of the t-norm $A_{\overline{G}}$ and employ a specific concept of duality between pairs of survival models. Actually we want to briefly review the relations among arguments contained in some previous and some recent papers (limiting attention to bivariate distributions), with the addition of a few pertinent comments and remarks.

All the story starts from the search of multivariate definitions of Increasing Failure Rate, that could, in a certain sense, be appropriate for a Bayesian analysis. In particular, interest was concentrated on properties that can be shared by both the joint distributions of i.i.d. IFR life-times and of conditionally i.i.d. IFR life-times.

In this search one is, of course, supposed to start from an analysis of the univariate notion of IFR and then to look for appropriate multivariate extensions. Actually IFR admits several different characterizations and, as it was pointed out in [5], the following characterization [29] is of remarkable interest: $\overline{G}$ is IFR if and only if the bivariate function $\overline{F}(x,y) := \overline{G}(x)\overline{G}(y)$ is Schur-concave, i.e., for $0 \le x \le x+\tau \le y-\tau \le y$, we have

$$\overline{F}(x+\tau, y-\tau) \ge \overline{F}(x,y).$$

Consider now the conditionally i.i.d. IFR case where $\overline{H}(x,y)$ has the form

$$\overline{H}(x,y) = \int_{\Theta} \overline{G}(x \mid \theta)\overline{G}(y \mid \theta)\Pi(d\theta), \tag{11.13}$$

$\Pi$ is a probability measure on $\Theta$, and $\overline{G}(x \mid \theta)$ is IFR, $\forall\theta \in \Theta$, $\Theta$ being a parameter space. Even if $\overline{G}(x \mid \theta)$ is IFR, $\forall\theta \in \Theta$, the marginal survival function of $\overline{H}$,

$$\overline{G_{\overline{H}}}(x) = \int_{\Theta} \overline{G}(x \mid \theta)\Pi(d\theta),$$

is not generally IFR. On the contrary, $\overline{G}(x \mid \theta)\overline{G}(y \mid \theta)$ being Schur-concave $\forall\theta \in \Theta$, guarantees that $\overline{H}(x,y)$ in (11.13) is still Schur-concave. This was a reason to consider Schur-concavity of an exchangeable joint survival function $\overline{F}$ as a useful notion of bivariate IFR. In addition, a characterization of this property in terms of comparisons between residual life-times, given in [39], shows the reliability meaning of this position. For more details on this topic see e.g. [41], Chap. 4.

The idea of describing IFR of $\overline{G}$, i.e. a univariate ageing property, by means of a property of the bivariate function $\overline{G}(x)\overline{G}(y)$ already contains, *in nuce*, the developments that lead to the description of ageing in terms of semi-copulas.

In [8] it was noticed that Schur-concavity of an exchangeable bivariate survival function $\overline{F}$ is actually a property of the family $\mathscr{L}_{\overline{F}}$ of the level curves (or of the family of level sets) of $\overline{F}$. In [9] it was more generally argued that other possible properties of bivariate ageing can be characterized in terms of the behavior of the family $\mathscr{L}_{\overline{F}}$.

Let us denote by $\mathscr{I}$ the set of increasing bijections $\psi : [0,1] \to [0,1]$, such that $\psi(0) = 0, \psi(1) = 1$, and notice that two different bivariate survival functions, say $\overline{F}(x,y)$ and $\overline{H}(x,y)$, are such that $\mathscr{L}_{\overline{F}} = \mathscr{L}_{\overline{H}}$ if and only if one has

$$\overline{H}(x,y) = \psi\left(\overline{F}(x,y)\right) \tag{11.14}$$

for a given $\psi \in \mathscr{I}$.

In order to describe $\mathscr{L}_{\overline{F}}$ Bassan and Spizzichino [9] proposed to employ the function $B_{\overline{F}} : [0,1] \times [0,1] \to [0,1]$ defined by

$$B_{\overline{F}}(u,v) = \exp\{-\overline{G}^{-1}\left(\overline{F}(-\log u, -\log v)\right)\}. \tag{11.15}$$

On the one hand in fact we have that two survival functions $\overline{F}$ and $\overline{H}$ satisfy $\mathscr{L}_{\overline{F}} = \mathscr{L}_{\overline{H}}$ if and only if $B_{\overline{H}} = B_{\overline{F}}$. On the other hand $B_{\overline{F}}$, being an increasing function from $[0,1] \times [0,1]$ to $[0,1]$, can be compared with the survival copula (that is apt to describe the dependence properties of $\overline{F}$).

Saying that a property # of $\overline{F}$ is actually a property of $\mathscr{L}_{\overline{F}}$ means that $\overline{F}$ has the property # in common with all the different joint survival functions $\overline{H}$ of the form (11.14) and then it is a property of $B_{\overline{F}}$. In conclusion: we concentrate our attention on bivariate ageing properties of $\overline{F}$ that can be characterized in terms of $B_{\overline{F}}$. $B_{\overline{F}}$ has been also called *bivariate ageing function*; we also refer to it (generally, for the multivariate case) with the term *B-function*. For our purposes it is useful to notice here that it is Archimedean if and only if $K_{\overline{F}}$ is such; for other details about $B_{\overline{F}}$ see [9–12, 21]. Actually, however, it turns out that $B_{\overline{F}}$ is not generally a copula. Then it was introduced the term *semi-copula* and this circumstance stimulated interest in the formalization of the appropriate generalization of the notion of copula [19, 20]. In a different framework, some aspects of bivariate ageing for Archimedean copulas have been analyzed in [32].

Let us come to consider now the case of i.i.d. variables.

Common sense suggests that, in the i.i.d. case, any plausibly defined notion of multivariate ageing should coincide with the corresponding univariate notion. In particular we can expect that, in the case $\overline{F}(x,y) = \overline{G}(x)\overline{G}(y)$, $\overline{F}$ has a bivariate ageing property # (such as NBU, IFRA, IFR,...) if and only if $\overline{G}$ admits the univariate ageing property #.

Then, as far as properties of bivariate ageing are properties of $B_{\overline{F}}$, we have that univariate ageing properties of $\overline{G}$ correspond to properties of $B_{\overline{F}}$ for $\overline{F}(x,y) = \overline{G}(x)\overline{G}(y)$.

But in this case, as it is immediately seen, $B_{\overline{F}}$ just turns out to coincide with the Archimedean semi-copula

$$B_{\overline{F}} = A_{\overline{G}} = C_{\overline{G}^{*-1}}!$$

We obtained thus substantial motivations to see way ageing properties of $\overline{G}$ reflect on properties of $A_{\overline{G}}$, according to the equivalences $(\alpha)$–$(\delta)$ listed in Sect. 11.2.

In this respect, we also point out that the properties of $A_{\overline{G}}$, that are involved in the equivalences, have just the form of *dependence* properties. This can be explained in view of Proposition 11.2.1 and of the formal extension of dependence notions to Archimedean semi-copulas, suggested in Sect. 11.2.

Some more insight on the above considerations can be obtained by extending the concept of multivariate survival model and introducing a suitable concept of duality. We start from a bivariate, exchangeable, survival model characterized by a survival copula $K$ and a marginal survival function $\overline{G}$, i.e. $\overline{F}(x,y) = K\left[\overline{G}(x), \overline{G}(y)\right]$. We then set

$$\overline{F}^{*}(x,y) := B_{\overline{F}}\left[\overline{G}^{*}(x), \overline{G}^{*}(y)\right], \tag{11.16}$$

where $B_{\overline{F}}$ is the ageing function of $\overline{F}$, and $\overline{G}^{*}$ is, as a univariate survival function, the dual of $\overline{G}$.

Notice that, if $B_{\overline{F}}$ is not a bivariate copula, $\overline{F}^{*}$ is not a bivariate survival function, in the sense that it does not satisfy the *rectangular inequality* or, in other words, the *2-increasingness property*.

In this context we shall however maintain for $\overline{F}^{*}$ the use of the term "survival function", even if in such an extended sense. Actually a bivariate function of the form $S\left[\overline{H}(x), \overline{H}(y)\right]$ (with $S$ a semi-copula and $\overline{H}$ a univariate survival function) can formally be seen as the bivariate survival function associated to a *capacity* (in place of a probability measure), with formal survival copula $S$ and marginal $\overline{H}$; see in particular [21, 36].

In any case, by formally applying to $\overline{F}^{*}$ definitions (11.15) and (11.16), we can consider the function $B_{\overline{F}^{*}}$ and obtain $B_{\overline{F}^{*}} = K$, $\left(\overline{F}^{*}\right)^{*} = \overline{F}$, by slightly extending arguments in [43].

We will refer to $\overline{F}^{*}$ as to the bivariate survival model *dual* of $\overline{F}$.

If $\overline{F}$ has marginally a positive (negative) ageing property, then $\overline{F}^{*}$ has, marginally, the corresponding negative (positive) ageing property, as an immediate consequence of Proposition 11.2.1.

If the marginal of $\overline{F}$ is standard exponential then $\overline{F}^{*} = \overline{F}$. $\overline{F}^{*}$ is a TTE model if and only if $\overline{F}$ is a TTE model. More in particular

*Remark 11.4.1.* Schur-constant models are characterized by the condition $B_{\overline{F}}(u,v) = u \cdot v$. The bivariate model of a pair of i.i.d life-times with marginal $\overline{G}$ is dual to the bivariate Schur-constant model with marginal $\overline{G}^{*}$.

We argued above that a property # of $\overline{F}$ only depends on $B_{\overline{F}}$ (or on $\mathscr{L}_{\overline{F}}$) if and only if it is in common among all the (possibly extended) survival functions $\overline{H}$ of the form (11.14), for arbitrary functions $\psi \in \mathscr{I}$.

On the other hand, properties of stochastic dependence for $\overline{F}$ that can be described as properties of its survival copula $K_{\overline{F}}$, can be regarded as properties that $\overline{F}$ has in common with all the survival functions of the form

$$\overline{L}(x,y) = \overline{F}\left[\varphi(x), \varphi(y)\right], \tag{11.17}$$

with $\varphi \in \mathscr{I}$.

We can recognize a sort of duality between transformations of the type (11.14) and those of the type (11.17); and then a sort of duality between properties of dependence (that are invariant under transformation of the type (11.17)) and properties of ageing (that are invariant under transformation of the type (11.14)).

In this respect, the following remark is relevant.

**Proposition 11.4.1.** *Let $\overline{F}$ be a bivariate survival model and consider $\overline{M}(x,y) = \psi\left(\overline{F}(x,y)\right)$, for $\psi \in \mathscr{I}$. Then*

$$\overline{M}^*(x,y) = \overline{F}^*\left[\varphi(x), \varphi(y)\right] \tag{11.18}$$

*with $\varphi(x) := -\log \psi^{-1}(e^{-x})$.*

*Proof.* Let $B$ be the ageing function of $\overline{F}$ and $\overline{G}(x) = \exp\{-\rho(x)\}$ its marginal survival function. From $\overline{M} = \psi\left(\overline{F}\right)$, we can immediately obtain

$$B_{\overline{M}} = B, \overline{G}_{\overline{M}}(x) = \psi\left(\exp\{-\rho(x)\}\right).$$

Now we observe that the equation $B_{\overline{M}} = B$ implies $K_{\overline{M}^*} = B = K_{\overline{F}^*}$. More precisely, by Definition (11.16), $\overline{M}^*$ is characterized by the conditions

$$K_{\overline{M}^*} = B_{\overline{M}} = B,$$

$$\overline{G}_{\overline{M}^*}(x) = \exp\{-\left[\log(\psi\left(\exp\{-\rho\}\right]^{-1}(x))\right]\} = \exp\{-\rho^{-1}\left[-\log\psi^{-1}(e^{-x})\right]\},$$

whence

$$\overline{M}^*(x,y) = B\left(\exp\{-\rho^{-1}(-\log\left[\psi^{-1}\left(e^{-x}\right)\right]\right), \exp\{-\rho^{-1}(-\log\left[\psi^{-1}\left(e^{-y}\right)\right]\right)\right). \tag{11.19}$$

On the other hand, again by applying the Definition (11.16), we have

$$\overline{F}^*(x,y) = B\left(\exp\{-\rho^{-1}(x)\}, \exp\{-\rho^{-1}(y)\}\right). \tag{11.20}$$

A direct comparison between (11.19) and (11.20) immediately yields the identity (11.18).

Some more aspects of duality between ageing and dependence also emerge in the papers [12, 24]. Bassan and Spizzichino [12] concerns with properties of bivariate ageing and their relations with univariate ageing and dependence. The results therein can be used to obtain sufficient or necessary for conditions of stochastic dependence in terms of conditions of univariate ageing for the marginal distribution for exchangeable models. In [24] some analogies and differences between $B$ and $K$ (between ageing and dependence, in other words) have been analyzed for exchangeable survival models in the dynamic context of the family of conditional models for *residual life-times* $X-t, Y-t$ given the survival data $(X>t, Y>t)$ at increase of the *age* $t \in [0,\infty)$.

## 11.5 Summary and Concluding Remarks

In this paper we first recalled and then discussed some relations between positive dependence and univariate ageing properties of a convex univariate survival function $\overline{G}$. We interpreted these relations in the frame of Schur-constant models.

Then we described the ageing of $\overline{G}$ (not necessarily convex) in terms of properties of the bivariate survival function $\overline{F}(x,y) := \overline{G}(x)\overline{G}(y)$. More precisely, in such an approach, ageing of $\overline{G}$ is seen as a property of the family $\mathscr{L}_{\overline{F}}$ of the level curves of $\overline{F}$. $\mathscr{L}_{\overline{F}}$ can generally be described in terms of the ageing function $B_{\overline{F}}$ defined in (11.15); a focal point is that, in the particular i.i.d. case, $B_{\overline{F}}$ turns out to coincide with the Archimedean semi-copula $A_{\overline{G}}$, defined in (11.3).

Further, we observed that ageing of $\overline{G}$ can actually be seen just as a property of dependence for $A_{\overline{G}}$. To this purpose, we extended properties of positive dependence from Archimedean copulas to Archimedean semi-copulas. This extension becomes natural when considering results, about the ageing properties of the inverse of the generator, that are valid in the case of Archimedean copulas.

Generally, for an exchangeable survival model $\overline{F}$, the relation between $B_{\overline{F}}$ and $K_{\overline{F}}$ is one of the form

$$B_{\overline{F}}(u_1,...,u_n) = h^{-1}\left(K_{\overline{F}}(h(u_1),...,h(u_n))\right). \tag{11.21}$$

with $h \in \mathscr{I}$.

The interest toward transformations of the type $h^{-1}\left(S\left(h(u_1),...,h(u_n)\right)\right)$ for a function $S : [0,1] \times [0,1] \to [0,1]$ emerges in several different fields. For a study of such transformations related with the topics illustrated here see in particular [1, 18] and cited references. The relation (11.21) in particular shows that $B_{\overline{F}}$ is Archimedean if and only if $K_{\overline{F}}$ is such.

The i.i.d. and the Schur-constant models, on which our discussion has been based, are special cases of TTE models and, in fact, $B_{\overline{F}}$ and $K_{\overline{F}}$ are both Archimedean for such two types of models.

Thus we have been allowed to limit attention to the case of bivariate Archimedean (semi-)copulas.

As it is now clear (see also e.g. [9, 26, 31]), the special role of i.i.d. and Schur-constant models is also explained by the following characterizations. Let $\overline{F}$ be an exchangeable bivariate survival function with the afore-mentioned assumptions on the marginal; then the conditions

(i) $\overline{F}$ has an Archimedean survival copula $K_{\overline{F}}$;
(ii) $\overline{F}$ shares its survival copula $K_{\overline{F}}$ with a Schur-constant model;
(iii) $\overline{F}$ shares the system of its level curves $\mathscr{L}_{\overline{F}}$ (i.e. it shares $B_{\overline{F}}$) with an i.i.d. model;

are equivalent.

These characterizations could be also useful in more general discussions about multivariate ageing. In particular they show the possibility, for TTE models, of directly defining multivariate ageing in terms of univariate ageing (see [9]). Furthermore, since (see in particular [14] and [41]) different properties of positive dependence become equivalent each other in the case of Schur-constant models, different notions of dependence collapse into only one notion in the case of Archimedean copula and of general TTE models.

Any TTE model can be obtained from a transformation of the type,

$$\psi\left(\overline{F}(\tau(x_1),...,\tau(x_n))\right), \tag{11.22}$$

starting from a Schur-constant or a i.i.d. model $\overline{F}$, with a suitable choice of $\tau, \psi$.

For future research, here we point out the interest for a more general study of the class of the survival models obtained by means of transformations of the form (11.22), starting from a fixed model $\overline{F}$.

**Acknowledgements** I like to thank Fabrizio Durante and Rachele Foschi for useful discussions and comments. Partially supported by Progetto Ateneo La Sapienza 2008, No. 8411288, Interazione e Dipendenza nei Modelli Stocastici.

## References

1. Alvoni, E., Papini, P.L., Spizzichino, F.: On a class of transformations of copulas and quasi-copulas. Fuzzy Sets Syst. **160**(3), 334–343 (2009)
2. Arjas, E.: Survival Models and Martingale Dynamics. Scand. J. Stat. **16**, 177–225 (1989)
3. Aven, T., Jensen, U.: Stochastic Models in Reliability. Applications of Mathematics, vol. 41. Springer, New York, NY (1999)
4. Averous, J., Dortet-Bernadet, J.-L.: Dependence for Archimedean copulas and aging properties of their generating functions. Sankhya **66**, 607–620 (2004)
5. Barlow, R.E., Mendel, M.B.: de Finetti-type representations for life distributions. J. Am. Stat. Assoc. **87**, 1116–1122 (1992)
6. Barlow, R.E., Mendel, M.B.: Similarity as a probabilistic characteristic of aging. In: Barlow, R.E., Clarotti, C.A., Spizzichino, F. (eds.) Reliability and Decision Making. Chapman & Hall, London (1993)
7. Barlow, R.E., Proschan, F.: Statistical Theory of Reliability and Life Testing. Holt, Rinehart and Winston, New York, NY (1975)
8. Barlow, R.E., Spizzichino, F.: Schur-concave survival functions and survival analysis. J. Comp. Appl. Math. **46**, 437–447 (1993)
9. Bassan, B., Spizzichino, F.: Dependence and multivariate aging: the role of level sets of the survival function. In: Hayakawa, Y., Irony, T., Xie, M. (eds.) System and Bayesian Reliability, pp. 229–242. World Scientific Publishing, River Edge, NJ (2001)
10. Bassan, B., Spizzichino, F.: On some properties of dependence and aging for residual lifetimes in the exchangeable case. Mathematical and Statistical Methods in Reliability, World Scientific Publishing, Singapore (2003)
11. Bassan, B., Spizzichino, F.: Bivariate survival models with Clayton aging functions. Insur. Math. Econ. **37**(1), 6–12 (2005)
12. Bassan, B., Spizzichino, F.: Relations among univariate aging, bivariate aging and dependence for exchangeable lifetimes. J. Multivar. Anal. **93**(2), 313–339 (2005)

13. Caramellino, L., Spizzichino, F.: Dependence and aging properties of life-times with Schur-constant survival functions. Prob. Eng. Inform. Sci. **4**, 103–111 (1994)
14. Caramellino, L., Spizzichino, F.: WBF property and stochastic monotonicity of the Markov process associated to Schur-constant survival functions. J. Multivar. Anal. **56**, 153–163 (1996)
15. Chi, Y., Yang, J., Qi, Y.: Decomposition of a Schur-constant model and its applications. Insur. Math. Econ. **44**, 398–408 (2009)
16. Chick, S.E., Mendel, M.B.: New characterizations of the no-aging property and the $l1$-isotropic model. J. Appl. Probab. **35**, 903–910 (1998)
17. Drouet Mari, D., Kotz, S.: Correlation and Dependence. Imperial College Press, London (2001)
18. Durante, F., Foschi, R., Sarkoci, P.: Distorted copulas: constructions and tail dependence. Commun. Stat. Theory Methods (2010). In press
19. Durante, F., Quesada-Molina, J.J., Sempi, C.: Semicopulas: characterizations and applicability. Kybernetika **42**, 287–302 (2006)
20. Durante, F., Sempi, C.: Semicopulæ. Kybernetika **41**, 315–328 (2005)
21. Durante, F., Spizzichino, F.: Semi-copulas, capacities and families of level curves. Fuzzy Sets Syst. **161**(2), 269–276 (2009)
22. Faltin, F.W., Kennett, R., Ruggeri, F. (eds.): Encyclopedia of Statistics for Quality and Reliability. Wiley, Chichester (2007)
23. Fang, K.-T., Kotz, S., Ng, K-W.: Symmetric Multivariate and Related Distributions. Chapman and Hall, London (1990)
24. Foschi, R., Spizzichino, F.: Semigroups of semicopulas and evolution of dependence at increase of age. Mathware Soft Comput. **XV**(1), 95–111 (2008)
25. Gertsbakh, I.: Reliability Theory. With Applications to Preventive Maintenance. Springer, Berlin (2000)
26. Joe, H.: Multivariate Models and Dependence Concepts. Monographs on Statistics and Applied Probability, vol. 73. Chapman & Hall, London (1997)
27. Klement, E.P., Mesiar, R., Pap, E.: Triangular Norms. Trends in Logic—Studia Logica Library, vol. 8. Kluwer Academic Publishers, Dordrecht (2000)
28. Lai, C.-D., Xie, M.: Stochastic Ageing and Dependence for Reliability. Springer, New York, NY (2006)
29. Marshall, A., Olkin, I.: Inequalities: Theory of Majorization and Its Applications. Academic Press, New York, NY (1979)
30. Marshall, A.W, Olkin, I.: Families of multivariate distributions. J. Am. Stat. Soc. **83**, 834–841 (1988)
31. McNeil, A., Nešhelová, A.: Multivariate Archimedean copulas, d-monotone functions and $\lambda_1$-norm symmetric distributions. Ann. Stat. **37**, 3059–3097 (2009)
32. Mulero, J., Pellerey, F.: Bivariate aging properties under Archimedean dependence structures. Commun. Stat. Theory Methods (2010). In press
33. Müller, A., Scarsini, M.: Archimedean copulae and positive dependence. J. Multivar. Anal. **93**, 443–445, (2005)
34. Nelsen, R.: Some properties of Schur-constant survival models and their copulas. Braz. J. Probab. Stat. **19**, 179–190 (2005)
35. Rychlik, T.: Copulae in reliability theory (order statistics, coherent systems). In: Jaworski, P., Durante, F., Härdle, W., Rychlik, T. (eds.) Copula Theory and Its Applications, Proceedings of the Workshop, Warsaw, 25–26 Sept 2009. Springer, Dordrecht (2010)
36. Scarsini, M.: Copulae of capacities on product spaces. In: Distributions with Fixed Marginals and Related Topics. IMS Lecture Notes Monograph Series, vol. 28, pp. 307–318. Institute of Mathematical Statistics, Hayward, CA (1996)
37. Shaked, M., Shanthikumar, J.G.: Stochastic Orders and Their Applications. Springer, New York, NY (2007)
38. Spizzichino, F.: Symmetry conditions on opinion assessment leading to time-transformed exponential models. In: Clarotti, C.A., Lindley, D.V. (eds.) Accelerated Life Testing and Experts' Opinions in Reliability. North Holland, Amsterdam (1988)

39. Spizzichino, F.: Reliability decision problems under conditions of ageing. In: Bernardo, J., Berger, J., Dawid, A.P., Smith, A.F.M. (eds.) Bayesian Statistic, vol. 4, pp. 803–811. Clarendon Press, Oxford (1992)
40. Spizzichino, F.: A unifying approach to optimal design of life-testing and burn-in. In: Barlow, R.E., Clarotti, C.A., Spizzichino, F. (eds.) Reliability and Decision Making. Chapman & Hall, London (1993)
41. Spizzichino, F.: Subjective Probability Models for Life-Times. Chapman and Hall/CRC, Boca Raton, FL (2001)
42. Spizzichino, F.: Ageing and positive dependence. In: Ruggeri, F., Kennett, R., Faltin, F.W. (eds.) Encyclopedia of Statistics for Quality and Reliability, pp. 82–95. Wiley, Chichester (2007)
43. Spizzichino, F.: A concept of duality for multivariate exchangeable survival models. Fuzzy Sets Syst. **160**, 325–333 (2009)

# Part II
# Contributed Papers

# Chapter 12
# A Copula-Based Model for Spatial and Temporal Dependence of Equity Markets

Umberto Cherubini, Fabio Gobbi, Sabrina Mulinacci and Silvia Romagnoli

**Abstract** In this contribution we provide a consistent pricing setting for multivariate equity derivatives. Consistently with the prescriptions of the Efficient Market Hypothesis and of the martingale pricing approach, we provide a model in which prices are martingales both with respect to their own filtration and to the enlarged multivariate filtration. We show that if the log-prices follow processes with independent increments and each one of them is not Granger caused by the others, the pricing procedure can be performed by simply: (i) generating time series of each asset; (ii) linking assets at each time with a prescribed copula function. We provide applications to multivariate digital options and spread options.

## 12.1 Introduction

In standard applications of copula functions to option pricing, basket derivative prices are obtained by applying copulas to marginal distributions at a given maturity. An open question is the consistency relationship between multivariate claim prices with different maturities. Actually, this relationship is a mandatory require-

Umberto Cherubini
Department of Mathematical Economics, University of Bologna, Bologna, Italy
e-mail: umberto.cherubini@unibo.it

Fabio Gobbi
Department of Mathematical Economics, University of Bologna, Bologna, Italy
e-mail: fabio.gobbi@unibo.it

Sabrina Mulinacci
Department of Mathematical Economics, University of Bologna, Bologna, Italy
e-mail: sabrina.mulinacci@unibo.it

Silvia Romagnoli
Department of Mathematical Economics, University of Bologna, Bologna, Italy
e-mail: silvia.romagnoli@unibo.it

P. Jaworski et al. (eds.), *Copula Theory and Its Applications*, Lecture Notes in Statistics 198,
DOI 10.1007/978-3-642-12465-5_12, 

ment if one wants to impose the well known martingale restriction that ensures to rule out arbitrage opportunities. In this chapter we address this topic proposing an algorithm for the evaluation of multivariate equity derivatives in a very general discrete time setting. The model uses copula functions to represent dependence both in the spatial and temporal perspective. The market model is based on three assumptions: (i) prices follow Markov processes with independent increments, according to the *Efficient Market Hypothesis*; (ii) marginal distributions are assigned a mean equal to the corresponding forward price, so that the analysis is consistently carried out under the so called Equivalent Martingale Measure, as required by the standard no-arbitrage assumption; (iii) the forecast of each stock price depends on its current value only, and cannot be improved upon by using the history of the other prices, a property that is known as no-Granger causality in the econometric literature: this requirement ensures that each price is a martingale both with respect to its natural filtration and the enlarged filtration generated by all the prices. Thus, this setting provides a general no-arbitrage pricing environment in which the evaluation of multivariate derivative contracts can be suitably accomplished in a *bottom-up* approach, in which the martingale requirement is firstly imposed on each univariate claim and then copula functions are applied to make the process multivariate. Beyond this approach, a more general pricing methodology, that we could call *top-down*, would require to impose the martingale requirement on the general filtration, and then to price the univariate claims under that filtration. This way, we provide a justification of the pricing techniques for multivariate digital, basket and spread options typically traded in the equity markets.

## 12.2 A market Model in Discrete Time

Formally, we assume a filtered probability space $\{\Omega, \mathscr{F}_t, \mathbb{P}\}$ satisfying the usual conditions. As usual, we work with logarithm of prices. So, if $S$ is the price of the asset, we deal with $X$, defined as

$$S = \exp(X)$$

The model is in discrete time, so we consider a set of periods limited by dates $\{0, t_1, t_2, ..., t_n\}$. The model is multivariate, so that we denote with $X_i^j$, $j = 1, 2, ..., m$, the log-prices of assets in the economy. As for the modelling strategy, we focus our attention on the price increments of asset $X^j$, namely: $\left\{Y_1^j, Y_2^j, ..., Y_n^j\right\}$. Obviously, we have $X_i^j = X_{i-1}^j + Y_i^j$.

Here we describe the main assumptions of the market model:

- Assumption 1 – The logarithm of prices follows a Markov process with independent increments.
- Assumption 2 – Each asset is not Granger-caused by any of the others.

It is now clear that the requirements above corresponds to imposing the price dynamics prescribed by the standard Efficient Market Hypothesis. The idea is that all information is embedded in current prices and no news can be used to predict future innovations. Processes with independent increments have been selected as the natural choice to accomplish these requirements, even though the selection has been further restricted to models in continuous time. Here we propose a model that encompasses the processes with independent increments proposed in the literature (namely Lévy processes and additive processes) within a more general class of models specified in discrete time.

In order to recover prices under the model described above, and to prevent arbitrage opportunities, it is well known that the price of each asset is a martingale process with respect to the overall information set. In a multivariate setting, this requires that the price has to be a martingale not only with respect to its natural filtration, but also the filtration generated by the other prices. In general each and every price should be transformed into a martingale with respect to the enlarged filtration, including all available information, an approach that we could define *top-down*. This is not what is typically done in copula pricing applications, in which the price of each asset is ensured to be a martingale with respect to its own filtration only, with no regard to information generated in other markets, and then multivariate derivatives are computed linking this information together. We call this technique *bottom-up*. We show that in the market model proposed in this paper, the latter technique is actually grounded and corresponds to the *top-down* approach.

In the pricing approach described below we first change each price process to a martingale with respect to its own filtration, and then prove that under no-Granger causality, the process remains a martingale with respect to the enlarged filtration.

## 12.3 The Martingale Property

For each asset (we drop superscript $j$ for convenience) we assume that the log-price increment $Y_i$ be endowed with a probability distribution $F_{Y_i}$. Likewise, we denote $F_{X_i}$ the set of distributions of log-prices. Of course, we have $F_{Y_1} = F_{X_1}$. We also assume a set of copula functions $C_{X_{i-1},Y_i}$ representing the dependence structure between the value of the asset at the beginning of the period and its increment in that period.

Given the independent increments structure, we may use the Cherubini, Mulinacci and Romagnoli [3] approach to recover the distribution of $X_i$ from that of $X_{i-i}$ and $Y_i$. The relationship is given by

$$F_{X_i}(t) = \int_0^1 F_{Y_i}\left(t - F_{X_{i-1}}^{-1}(w)\right) dw. \tag{12.1}$$

By the same token, the temporal dependence structure of the process, that is the dependence between $X_{i-1}$ and $X_i$ is given by the copula function (see [3])

$$G_{X_{i-1},X_i}(u,v) = \int_0^u F_{Y_i}\left(F_{X_i}^{-1}(v) - F_{X_{i-1}}^{-1}(w)\right) dw. \tag{12.2}$$

To impose positivity and martingale restriction of the prices, we now want to ensure that under the probability measure used for pricing the prices of $S = \exp(X)$ are martingales. So we normalize the process as

$$S_t = \frac{e^{X_t}}{\mathbb{E}(e^{X_t})}.$$

In fact it is straightforward to verify that $\forall s > t$

$$\begin{aligned}
\mathbb{E}(S_s|\mathscr{F}_t) &= \mathbb{E}\left[\frac{e^{X_s}}{\mathbb{E}(e^{X_s})}|\mathscr{F}_t\right] \\
&= \frac{e^{X_t}}{\mathbb{E}(e^{X_t+\Delta X})}\mathbb{E}(e^{\Delta X}|\mathscr{F}_t) \\
&= \frac{e^{X_t}}{\mathbb{E}(e^{X_t})\mathbb{E}(e^{\Delta X})}\mathbb{E}(e^{\Delta X}) \\
&= S_t
\end{aligned}$$

where $\Delta X = X_s - X_t$.
Since $S$ is, by construction, a Markov process, its temporal dependence structure satisfies the representation in [8]

$$G_{t_1 \dots t_n} = G_{t_1,t_2} \star G_{t_2,t_3} \star .. \star G_{t_{n-1},t_n}$$

where $G_{t_1 \dots t_n}$ is the copula of $(S_{t_1}, \dots, S_{t_n})$ and $G_{t_{k-1},t_k}$ is the copula of $(S_{t_{k-1}}, S_{t_k})$. The $\star$ product is defined as

$$A \star B(u_1,u_2,u_3) \equiv \int_0^{u_2} \frac{\partial A(u_1,t)}{\partial t}\frac{\partial B(t,u_3)}{\partial t} dt.$$

In order to ensure that each price process be a martingale with respect to the filtration generated by all the prices, it is sufficient to introduce the assumption of no-Granger causaliy.

**Definition 12.3.1.** $X^1, \dots, X^{i-1}, X^{i+1}, \dots, X^m$ do not Granger cause $X^i$ if

$$\mathbb{P}[X^i_{t_{k+1}} \leq x|\mathscr{F}_{t_k}^{X^1,\dots,X^m}] = \mathbb{P}[X^i_{t_{k+1}} \leq x|\mathscr{F}_{t_k}^{X^i}]$$

for any $t_k$ and $x$.

Remember that a process, Markov with respect to a given filtration, is not in general Markov with respect to a larger filtration. We show that this is in fact guaranteed by no-Granger causality.

**Theorem 12.3.1.** *The following are equivalent:*

*1. $X^i$ is not Granger caused by $X^1,\dots,X^{i-1},X^{i+1},\dots,X^m$;*
*2. if $X^i$ is an $\mathscr{F}_{t_k}^{X^i}$–Markov process, then it is an $\mathscr{F}_{t_k}^{X^1,\dots,X^m}$–Markov process, as well.*

*Proof.* 1. ⇒ 2.:1. implies

$$\mathbb{P}[X^i_{t_{k+1}} \le x|\mathscr{F}_{t_k}^{X^1,\dots,X^m}] = \mathbb{P}[X^i_{t_{k+1}} \le x|\mathscr{F}_{t_k}^{X^i}].$$

for every $x \in \mathbb{R}$. By hypothesis

$$\mathbb{P}[X^i_{t_{k+1}} \le x|\mathscr{F}_{t_k}^{X^i}] = \mathbb{P}[X^i_{t_{k+1}} \le x|X^i_{t_k}]$$

and the thesis follows. The other implication is trivial.

We recall that the concept of no-causality allows to analyze the stability of a martingale property with respect to increasing filtration. More specifically it is proved (see [1, 10]) that if $S^1,\dots,S^m$ are stochastic processes and $S^i$ is an $\mathscr{F}_{t_k}^{S^i}$– martingale, it is an $\mathscr{F}_{t_k}^{S^1,\dots,S^m}$–martingale as well iff $S^1,\dots,S^{i-1},S^{i+1},\dots,S^m$ do not Granger cause $S^i$ for every $t_k$ (that is, in our setting, iff $X^1,\dots,X^{i-1},X^{i+1},\dots,X^m$ do not Granger cause $X^i$ for every $t_k$).

Given the structure above, a general consistent algorithm to price multivariate equity derivatives can be summarized as follows

- construct the joint distribution of the dynamics of each asset,
- assume a copula $C_t$ linking the price of assets at the same date.

## 12.4 Applications

We now introduce applications to equity multivariate derivatives which account for the general market model described in the sections above. The setting allows consistently to price multivariate options on the same basket of underlying assets and with different strikes and maturities. In this way we are able to recover "price surfaces" spotting no arbitrage relationships across different dates and states.

For illustrative purposes, below we present applications to the most standard products, namely digital and basket options.

### *12.4.1 Multivariate Digital Options*

Multivariate digital options, also called *Altiplanos*, pay a digital payoff conditional on the event that the price of a set of underlying assets be above (or below) a given strike defined at the beginning of the contract. The reference event may be specified in different ways, depending whether it refers to prices of the assets at a given future date or at several *reset* dates. Cherubini and Romagnoli [4, 5] proposed a taxonomy distinguishing three kinds of products:

- European Altiplanos: the digital payoff is paid if at the exercise date all assets are above the corresponding strike;
- Barrier Altiplanos: the digital payoff is paid if all assets remain above the corresponding barrier at a set of given *reset* dates;
- Altiplanos with Memory: a set of digital payoffs, accrued on a corresponding set of payment dates, is paid the first time the event takes place.

Pricing Altiplanos was the first, straightforward application of copulas to equity derivatives. In fact, due to the digital payoff, their price amounts to the computation of a copula function. Furthermore, price consistency between call and put prices are enforced by the duality between copula functions and survival copulas [2, 5]. As for the behavior of prices across different maturities, van der Goorberg, Genest and Werker [14] propose the use of a dynamic copula linking the price of the underlying assets at the same future dates. All these applications are examples of the *bottom-up* approach described above, and are grounded only under the assumptions made in the model presented in this paper. Figure 12.1 report the pricing surface of an European Digital Altiplano, that is the change of the option price across different strikes and maturities. Each surface refers to a different level of dependence measured in terms

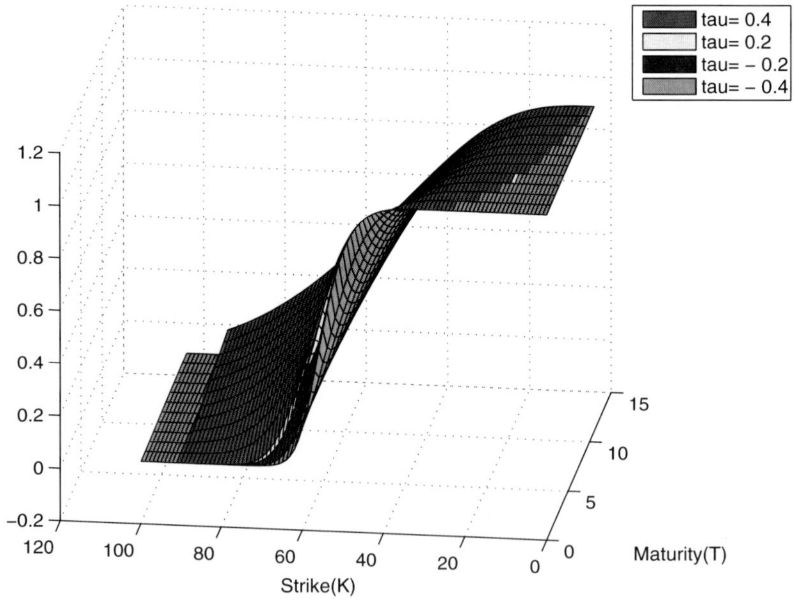


**Fig. 12.1** Prices surface of an European Digital Altiplano with log-normal marginals (30% volatility) and Frank copula with several dependence parameters.

of Kendall's $\tau$. As expected the price increases for higher levels of dependence, meaning that the product is "long correlation".

### 12.4.2 Basket and Spread Options

Basket options are derivative contracts written on the sum of a set of underlying assets. In the simplest bivariate example, we consider an option on $S_t^1 + S_t^2$. The distribution of the sum $S_t^1 + S_t^2$ is given by the $C$-convolution

$$F_{S_t^1+S_t^2}(z) = F_{S_t^1} \overset{C_t}{*} F_{S_t^2}(z) = \int_0^1 D_1 C_t[w, F_{S_t^2}(z - F_{S_t^1}^{-1}(w))]dw$$

as proved in [3].

The same technique can be applied to spread options which are written on the difference $S_t^1 - S_t^2$. The extension is immediately obtained considering that the dependence structure between $S_t^1$ and $-S_t^2$ is $\overline{C}_t(u, 1-v) = u - C_t(u,v)$.

Figures 12.2 and 12.3 report the pricing surface of spread options, that is the change of the option price across different strikes and maturities. Each surface refers to a different level of dependence measured in terms of Kendall's $\tau$. As expected the

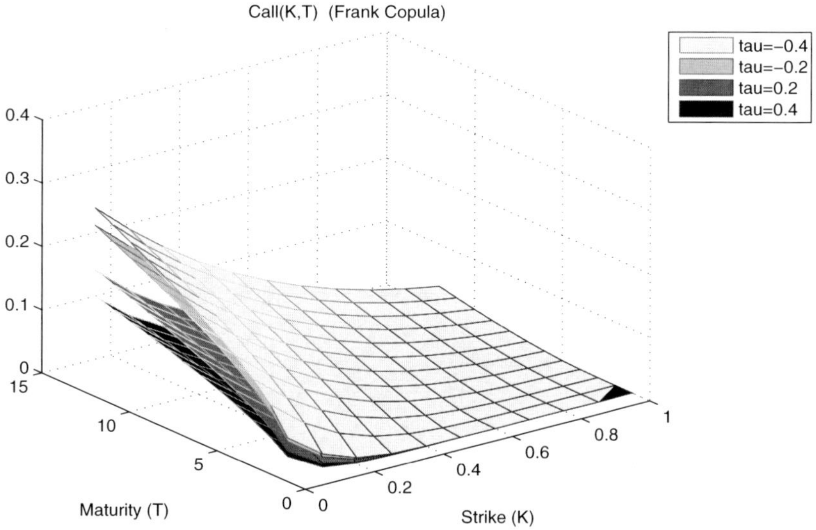


**Fig. 12.2** Prices surface of a spread call option with log-normal marginals (30% volatility) and Frank copula with several dependence parameters.

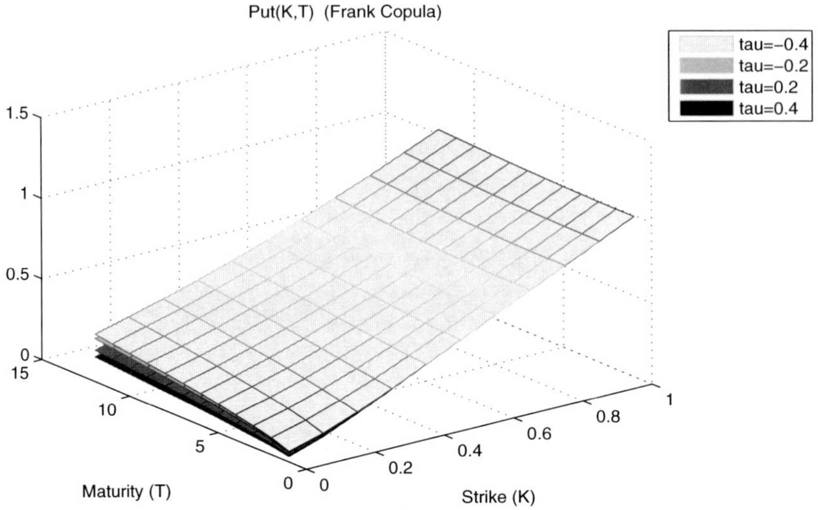


**Fig. 12.3** Prices surface of a spread put option with log-normal marginals (30% volatility) and Frank copula with several dependence parameters.

price increases for lower levels of dependence, meaning that the product is "short correlation": an increase in correlation causes a loss of value.

## References

1. Brémaud, P., Yor, M.: Changes of filtrations and of probability measures. Z. Wahrscheinl. **45**, 269–295 (1978)
2. Cherubini, U., Luciano, E.: Bivariate option pricing with copulas. Appl. Math. Finance **9**, 69–82 (2002)
3. Cherubini, U., Mulinacci, S., Romagnoli, S.: A copula-based model of speculative price dynamics in discrete time. Submitted (2009)
4. Cherubini, U., Romagnoli, S.: Multivariate digital options with memory. European Journal of Finance, In press (2010)
5. Cherubini, U., Romagnoli, S.: Computing copula volume in n-dimensions. Appl. Math. Finance, **16**(4), 307–314 (2009)
6. Cherubini, U., Romagnoli, S.: The dependence structure of running maxima and minima: theoretical results and option pricing application. Math. Finance **20**(1) 35–58 (2010)
7. Dambis, K.E.: On the decomposition of continuous submartingales. Theor. Prob. Appl. **10**, 401–410 (1965)
8. Darsow, W.F., Nguyen, B., Olsen, E.T.: Copulas and Markov processes. Ill. J. Math. **36**(4), 600–642 (1992)
9. Embrechts, P.: Copulas: a personal view. J. Risk Ins. **76**(3), 639–650 (2009)
10. Florens, J.P., Fougère, D.: Non-causality in continuous time: applications to counting processes. Cahier du GREMAQ 91.b, Université des Sciences Sociales de Toulouse (1991)

11. Ibragimov, R.: Copula based characterization and modeling for time series. Harvard Institute of Economic Research, Discussion paper no. 2094 (2005)
12. Li, D.: On default correlation: a copula function approach. J. Fixed Inc. **9**, 43–54 (2001)
13. Nelsen, R.B.: An introduction to copulas. Springer Series in Statistics, 2nd edn. Springer, New York, NY (2006)
14. van der Goorberg, R., Genest, C., Werker, B.: Bivariate option pricing using dynamic copula models. Insur. Math. Econ. **37**, 101–114 (2005)

# Chapter 13
# Nonparametric and Semiparametric Bivariate Modeling of Petrophysical Porosity-Permeability Dependence from Well Log Data

Arturo Erdely and Martin Diaz-Viera

**Abstract** Assessment of rock formation permeability is a complicated and challenging problem that plays a key role in oil reservoir modeling, production forecast, and the optimal exploitation management. Generally, permeability evaluation is performed using porosity-permeability relationships obtained by integrated analysis of various petrophysical measurements taken from cores and wireline well logs. Dependence relationships between pairs of petrophysical variables, such as permeability and porosity, are usually nonlinear and complex, and therefore those statistical tools that rely on assumptions of linearity and/or normality and/or existence of moments are commonly not suitable in this case. But even expecting a single copula family to be able to model a complex bivariate dependency seems to be still too restrictive, at least for the petrophysical variables under consideration in this work. Therefore, we explore the use of the Bernstein copula, and we also look for an appropriate partition of the data into subsets for which the dependence strucure was simpler to model, and then a conditional gluing copula technique is applied to build the bivariate joint distribution for the whole data set.

## 13.1 Introduction

Assessment of rock formation permeability is a complex and challenging problem that plays a key role in oil reservoir modeling, production forecast, and the opti-

Arturo Erdely
Programa de Actuaría, División de Matemáticas e Ingeniería
Facultad de Estudios Superiores Acatlán, Universidad Nacional Autónoma de México
e-mail: arturoerdely@yahoo.com.mx

Martín Díaz-Viera
Programa de Investigación en Recuperación de Hidrocarburos, Instituto Mexicano del Petróleo, México D.F., México
e-mail: mdiazv@imp.mx

P. Jaworski et al. (eds.), *Copula Theory and Its Applications*, Lecture Notes in Statistics 198,
DOI 10.1007/978-3-642-12465-5_13, 

mal exploitation management. Generally, permeability evaluation is performed using porosity-permeability relationships obtained by an integrated analysis of various petrophysical measurements taken from cores and wireline well logs. In particular, in carbonate double-porosity formations with an heterogeneous structure of pore space this problem becomes more difficult because the permeability usually does not depend on the total porosity, but on classes of porosity, such as vuggular and fracture porosity (secondary porosity). Even more, in such cases permeability is directly related to the connectivity degree of the pore system structure. This fact makes permeability prediction a challenging task.

Dependence relationships between pairs of petrophysical random variables, such as permeability and porosity, are usually nonlinear and complex, and therefore those statistical tools that rely on assumptions of linearity and/or normality and/or existence of moments are commonly not suitable in this case. The use of copulas for modeling petrophysical dependencies is not new [3] and $t$-copulas have been used for this purpose. But expecting a single copula family to be able to model any kind of bivariate dependency seems to be still too restrictive, at least for the petrophysical variables under consideration in this work. Therefore, we first adopted a non-parametric approach by the use of the Bernstein copula [15, 16] and estimating the quantile function by Bernstein polynomials [12]. Later, we adopted a semiparametric approach by looking for a partition of the data into subsets for which the dependence structure was simpler to model by means of parametric families of copulas, and then a conditional gluing copula technique is applied to build the bivariate joint distribution for the whole data set.

Among many others, two main tasks are usually of interest in petrophysical modeling: to reproduce the underlying dependence structure by data simulation, and to explain a variable of interest in terms of an allegedly predictive variable (regression).

## 13.2 Methodology

Let $\mathscr{S} := \{(x_1,y_1),\ldots,(x_n,y_n)\}$ be independent and identically distributed bivariate observations of a random vector $(X,Y)$. In this work, the $(x_k,y_k)$ paired values represent porosity-permeability measurements. We may obtain empirical estimates for the marginal distributions of $X$ and $Y$ by means of

$$F_n(x) = \frac{1}{n}\sum_{k=1}^{n} \mathbb{I}\{x_k \leq x\}\,, \qquad G_n(y) = \frac{1}{n}\sum_{k=1}^{n} \mathbb{I}\{y_k \leq y\}\,, \tag{13.1}$$

where $\mathbb{I}$ stands for an indicator function which takes a value equal to 1 whenever its argument is true, and 0 otherwise. It is well-known [1] that the empirical distribution $F_n$ is a consistent estimator of $F$, that is, $F_n(t)$ converges almost surely to $F(t)$ as $n \to \infty$, for all $t$.

We model vuggy porosities as an absolutely continuous random variable $X$ with unknown marginal distribution function $F$, and permeability as an absolutely continuous random variable $Y$ with unknown marginal distribution function $G$. From [11] we have bivariate observations from the random vector $(X,Y)$, see Figs. 13.1 (left) and 13.2, and Fig. 13.3 and Table 13.1 for marginal descriptive statistics. For simulation of continuous random variables, the use of the empirical distribution function estimates (13.1) is not appropriate since $F_n$ is a step function, and therefore discontinuous, so a smoothing technique is needed. Since one of our main goals is simulation of porosity-permeability paired variates, it will be better to have a smooth estimation of the marginal quantile function $Q(u) = F^{-1}(u) = \inf\{x : F(x) \geq u\}$, $0 \leq u \leq 1$, which is possible by means of Bernstein-Kantorovic polynomials as in [12]:

$$\widetilde{Q}_n(u) = \sum_{k=0}^{n} \frac{x_{(k)} + x_{(k+1)}}{2} \binom{n}{k} u^k (1-u)^{n-k}, \tag{13.2}$$

where the $x_{(k)}$ are the order statistics, and the analogous case for marginal $G$ in terms of values $y_{(k)}$.

Similarly, we have the *empirical copula* [2], a function $C_n$ with domain $\{\frac{i}{n} : i = 0,1,\ldots,n\}^2$ defined as

$$C_n\left(\frac{i}{n}, \frac{j}{n}\right) = \frac{1}{n} \sum_{k=1}^{n} \mathbb{I}\{rank(x_k) \leq i \,,\, rank(y_k) \leq j\} \tag{13.3}$$

and its convergence to the true copula $C$ has also been proved [7, 14]. The empirical copula is not a copula, since it is only defined on a finite grid, not in the whole unit square $[0,1]^2$, but by Sklar's Theorem $C_n$ may be extended to a copula. Moreover, a *smooth* extension is possible by means of the Bernstein copula [15, 16]:

$$\widetilde{C}(u,v) = \sum_{i=0}^{n} \sum_{j=0}^{n} C_n\left(\frac{i}{n}, \frac{j}{n}\right) \binom{n}{i} u^i (1-u)^{n-i} \binom{n}{j} v^j (1-v)^{n-j} \tag{13.4}$$

for every $(u,v)$ in the unit square $[0,1]^2$, and where $C_n$ is as defined in (13.3). In order to simulate replications from the random vector $(X,Y)$ with the dependence structure inferred from the observed data $\mathscr{S} := \{(x_1,y_1),\ldots,(x_n,y_n)\}$, accordingly to [13] we have the following:

**Algorithm 1**

1. *Generate two independent and continuous Uniform* $(0,1)$ *random variates* $u,t$.
2. *Set* $v = c_u^{-1}(t)$ *where*

$$c_u(v) = \frac{\partial \widetilde{C}(u,v)}{\partial u}, \tag{13.5}$$

   *and* $\widetilde{C}$ *is obtained by (13.4).*
3. *The desired pair is* $(x,y) = (\widetilde{Q}_n(u), \widetilde{R}_n(v))$, *where* $\widetilde{Q}_n$ *and* $\widetilde{R}_n$ *are the smoothed estimated quantile functions of* $X$ *and* $Y$, *respectively, accordingly to (13.2).*

For a value $x$ in the range of the random variable $X$ and $0 < \alpha < 1$ let $y = \varphi_\alpha(x)$ denote a solution to the equation $\mathbb{P}(Y \leq y \,|\, X = x) = \alpha$. Then the graph of $y = \varphi_\alpha(x)$

is the *α-quantile regression curve* of $Y$ conditional on $X = x$. Recalling (13.5), we have that

$$\mathbb{P}(Y \leq y \,|\, X = x) \;=\; c_u(v)\Big|_{u=F(x),\, v=G(y)}, \tag{13.6}$$

and this result leads to the following algorithm [13] to obtain the $\alpha$-quantile regression curve of $Y$ conditional on $X = x$ :

**Algorithm 2**

1. *Set* $c_u(v) = \alpha$.
2. *Solve for the regression curve* $v = g_\alpha(u)$.
3. *Replace* $u$ *by* $\widetilde{Q}_n^{-1}(x)$ *and* $v$ *by* $\widetilde{R}_n^{-1}(y)$.
4. *Solve for the regression curve* $y = \varphi_\alpha(x)$.

So far (13.2) and (13.4) constitute a completely nonparametric approach for modeling a jointly continuous random vector $(X,Y)$, without imposing restrictive conditions on the dependence structure and/or marginal behavior. Of course, a "no-free-lunch" principle applies, and a price is paid in terms of a "noisy" regression, as it will be seen later, and the fact that the Bernstein copula is unable to model *tail dependence.*

The nature of the data that will be analyzed in the following section, makes it plausible that certain value ranges for the predictor variable (porosity) are related to different dependence structures (copulas). Under this assumption, we recall the gluing copula technique [17] for the particular case of vertical section gluing and bivariate copulas. For example, given two bivariate copulas $C_1$ and $C_2$, and a fixed value $0 < \theta < 1$, we may scale $C_1$ to $[0,\theta] \times [0,1]$ and $C_2$ to $[\theta,1] \times [0,1]$ and *glue* them into a single copula:

$$C_{1,2,\theta}(u,v) \;=\; \begin{cases} \theta C_1(\frac{u}{\theta}, v), & 0 \leq u \leq \theta, \\ (1-\theta) C_2(\frac{u-\theta}{1-\theta}, v) + \theta v, & \theta \leq u \leq 1. \end{cases} \tag{13.7}$$

For a more specific example, in the particular case $C_1(u,v) = \Pi(u,v) = uv$, $C_2(u,v) = W(u,v) = \max\{u+v-1, 0\}$, and $\theta = \frac{1}{2}$, the graph of the diagonal section $\delta_{\Pi,W,1/2}(t) = C_{W,\Pi,1/2}(t,t)$ is shown in Fig. 13.4. This particular kind of copula construction may easily lead to discontinuities in the derivative of the diagonal section at the value $t = \theta$. Therefore, if the empirical diagonal $\delta_n(i/n) = C_n(i/n, i/n)$ exhibits a behavior that suggests a discontinuity of the diagonal derivative at certain points, one may ask if this is possibly due to an abrupt change of the dependence structure at those points, and that being the case, a gluing copula procedure may help to explain the whole dependence structure in terms of two or more simpler dependence models (copulas), in a piecewise manner. It is beyond the scope of this paper to make an exhaustive discussion of different nonparametric techniques to detect possible derivative discontinuity points, but for the data analyzed in this work, using as an heuristic method the Dierckx cubic spline knot detection algorithm [4] has led to a dependence (copula) decomposition that was easier to model by means of known parametric families of copulas, and that is compatible with a petrophysical interpretation. Under this scheme, Algorithms 1 and 2 are used piecewise, with (estimated) parametric copulas instead of $\widetilde{C}$ (the nonparametric Bernstein copula).

## 13.3 Data Analysis

In order to find knot candidates (or *gluing points*) we applied the Dierckx cubic spline knot detection algorithm [4], which is available as an R contributed package [5], to the empirical diagonal $\delta_n$, obtaining the partition shown in Fig. 13.1 (Right), suggesting to decompose the total sample into three subsamples. Since we are willing to explain permeability $(Y)$ in terms of porosity $(X)$, without loss of information about the joint distribution of the random vector $(X,Y)$ we will consider the total sample $\mathscr{S} := \{(x_1,y_1),\ldots,(x_n,y_n)\}$ to be ordered in the $x_k$ values, that is $x_1 < x_2 < \cdots < x_n$, and so the partition will be induced in the observed $x_k$ values total range $[\min x_k, \max x_k]$.

**Fig. 13.1** *Left:* Scatterplot of porosity-permeability data. *Right:* Empirical diagonal (*thick line style*) with suggested gluing points $\theta_1 = 0.46$ and $\theta_2 = 0.68$ (*dashed vertical lines*), Fréchet-Hoeffding diagonal bounds and independence diagonal (*thin line style*).

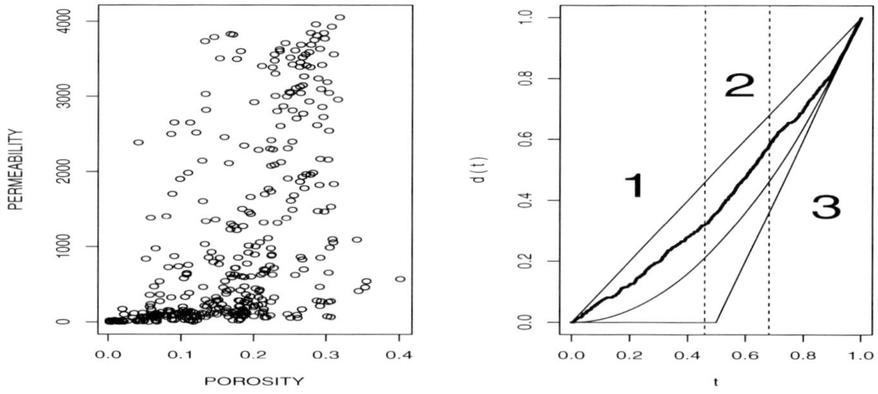


In Table 13.2 we present a summary of empirical values for the *concordance measures* known as Spearman's rho $(\rho_S)$ and Kendall's tau $(\tau_K)$, and an empirical version of the *dependence measure* $\Phi_H$ which is just the square root of Hoeffding's *dependence index* [10]. We notice that the concordance and dependence values of the total sample are basically due to subsample 1, since subsamples 2 and 3 clearly exhibit significantly lower values. Also, it should be noticed that the values of $\rho_S$ and $\Phi_H$ are quite similar under subsample 1, in contrast with subsamples 2 and 3.

**Fig. 13.2** Scatterplot of porosity-permeability data ranks, rescaled to $[0,1]^2$.

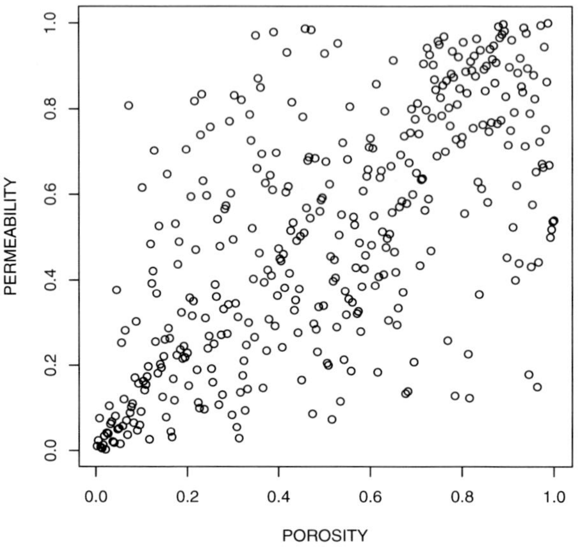


**Fig. 13.3** Frequency histograms of porosity and permeability data.

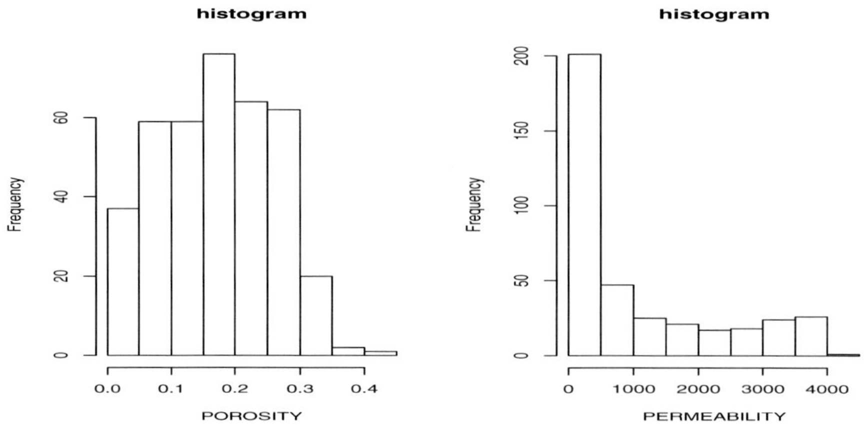

**Fig. 13.4** Diagonal section (*thick line style*) of gluing copulas $\Pi(u,v) = uv$ and $W(u,v) = \max\{u+v-1,0\}$ with $\theta = 1/2$, along with Fréchet-Hoeffding diagonal bounds (*thin lines*).

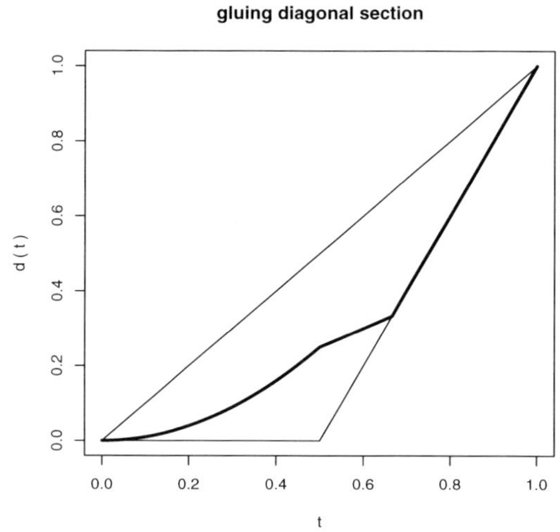


**Table 13.1** Summary statistics of porosity and permeability data.

| Variable | Min. | 1st quartile | Median | Mean | 3rd quartile | Max. |
|---|---|---|---|---|---|---|
| Porosity | 0.001855 | 0.098670 | 0.176700 | 0.169700 | 0.236100 | 0.401300 |
| Permeability | 0.05584 | 0.01006 | 412.3 | 1,050 | 1,826 | 4,054 |

This last observation may have as an explanation either the effect of randomness due to smaller subsample sizes or that there could be some kind of dependence that neither Spearman's rho nor Kendall's tau are being able to detect (recall that if a concordance measure is equal to zero this does not imply independence). Therefore, a powerful nonparametric test of independence *à la Deheuvels* based on a Cramér–von Mises statistic [8] has been applied, see Table 13.2 for $p$-values, clearly rejecting independence for both the total sample and subsample 1, not rejecting independence for subsample 2, but with some doubts about rejecting independence in case of subsample 3, since there is no rule to decide if 0.1274 is a sufficiently low $p$-value to reject the null hypothesis. Fortunately, a nonparametric symmetry test [6] has been helpful for taking a final decision on subsample 3: by rejecting symmetry we have to reject independence.

In terms of choosing a parametric copula, strong evidence against symmetry is challenging since there is not such a huge catalog of parametric asymmetric copulas as there is indeed for the symmetric case. For the particular data under consideration, it has been possible to transform the data in order to "remove" the asymmetry, by means of Theorem 2.4.4 (1) in [13] and the particular case

$$C_{X,Y}(u,v) = u - C_{X,-Y}(u,1-v)\,, \tag{13.8}$$

**Table 13.2** Empirical concordance and dependence values, and $p$-values for nonparametric tests of independence and symmetry.

| Subsample | Size | $\rho_S$ | $\tau_K$ | $\Phi_H$ | Independence test $p$-value | Symmetry test $p$-value |
|---|---|---|---|---|---|---|
| Total | 380 | +0.6579 | +0.4843 | 0.6294 | 0.0000 | 0.9249 |
| 1 | 174 | +0.6072 | +0.4379 | 0.5819 | 0.0000 | 0.8893 |
| 2 | 85 | +0.0538 | +0.0487 | 0.1617 | 0.4661 | 0.3194 |
| 3 | 121 | +0.0004 | +0.0143 | 0.1795 | 0.1274 | 0.0014 |

with which, fortunately in this case, it was not possible to reject symmetry for the observations of the random vector $(X,-Y)$ from the transformed subsample 3 (denoted as 3T), see Table 13.3, and Fig. 13.5 for level curves of the empirical copulas for subsamples 3 and 3T. But still we mantain our rejection of independence for the transformed subsample 3T since the independence copula $\Pi(u,v)=uv$ is invariant under transformation (13.8).

**Table 13.3** Empirical concordance and dependence values, and $p$-values for nonparametric tests of independence and symmetry for transformed subsample 3T.

| Subsample | Size | $\rho_S$ | $\tau_K$ | $\Phi_H$ | Independence test $p$-value | Symmetry test $p$-value |
|---|---|---|---|---|---|---|
| 3T | 121 | −0.0004 | −0.0143 | 0.1795 | 0.1274 | 0.4040 |
| 3Ta | 75 | −0.2668 | −0.1870 | 0.2795 | 0.0358 | 0.8323 |
| 3Tb | 46 | +0.2771 | +0.1903 | 0.3149 | 0.0803 | 0.7869 |

The close to zero values for the concordance measures $\rho_S$ and $\tau_K$ under subsample 3 or 3T, but the not so close to zero value for $\Phi_H$ along with the independence rejection for subsample 3 and 3T, might have as an explanation that under non-monotone dependence, positive and negative quadrant dependencies "cancel out" under concordance measures (for a theoretical example see Example 5.18 in [13]), but not under a dependence measure such as $\Phi_H$. Therefore, we searched for a partition of subsample 3T to see if it was possible to detect a *gluing point* for positive and negative dependence, again with the aid of the Dierckx cubic spline knot detection algorithm [4], which suggested $\theta_{3T}=0.615$, see Fig. 13.6, and with such partition into subsamples 3Ta and 3Tb it was possible to decompose subsample 3T into negative and positive dependence, with better $p$-values to reject independence, and with $p$-values far away from willing to reject symmetry, see Table 13.3.

So far, it has been possible to decompose the total sample into subsamples 1,2,3Ta, and 3Tb, which do not exhibit strong evidence against symmetry, but with strong evidence against the hypothesis of independence for all cases except subsample 2, and therefore for subsample 2 the final decision is to be modeled by independence. For the remaining subsamples we looked for an appropriate fit among the

**Fig. 13.5** Level curves of empirical copulas for subsamples 3 and 3T.

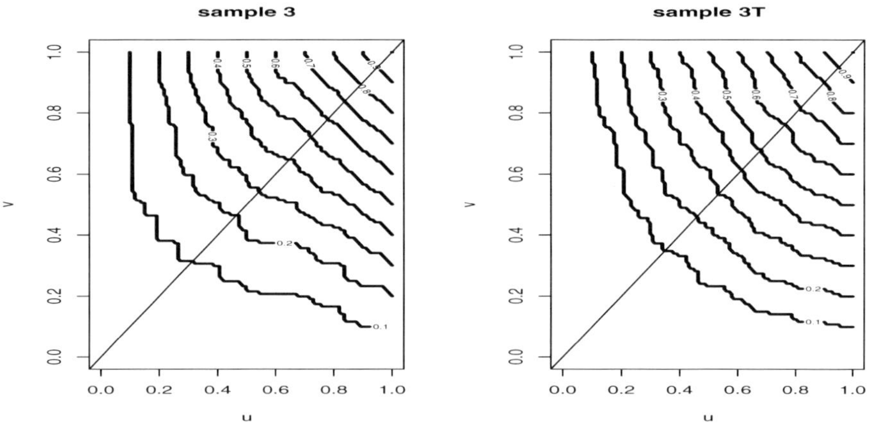


**Fig. 13.6** In *thin line style*, Fréchet-Hoeffding diagonal bounds and independence diagonal section. In *thick line style*: empirical diagonals for subsample 3 (*Left*) and subsample 3T (*Right*).

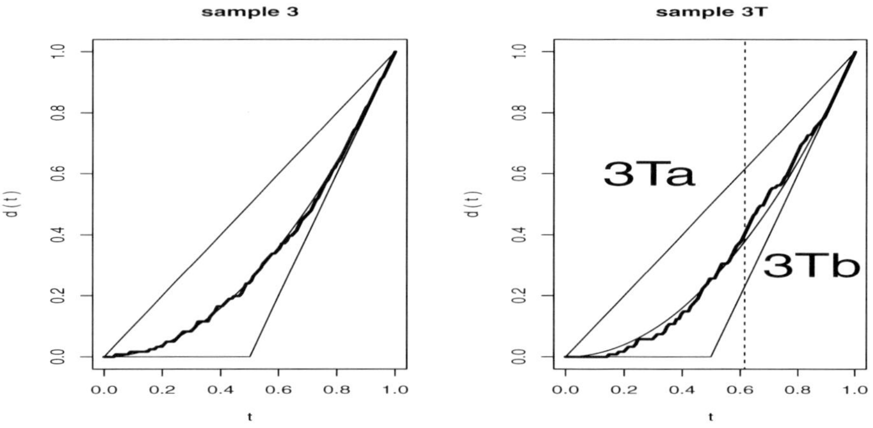

known catalog of symmetric parametric copulas (see for example [13]) by means of a goodness-of-fit test as in [9], with the results summarized in Table 13.4.

**Table 13.4** Parameter estimation and goodness-of-fit $p$-values for the selected copulas.

| Subsample | Copula | Parameter | GoF $p$-value |
|---|---|---|---|
| 1 | Clayton | +1.5581 | 0.7880 |
| 3Ta | Clayton | −0.3151 | 0.9102 |
| 3Tb | A-M-H | +0.6869 | 0.7725 |

With the above results, it is now possible to perform simulations and regression in a semiparametric fashion (parametric copula and nonparametric Bernstein marginals as described in Sect. 13.2). We also show analog results under a totally nonparametric fashion, as described in Sect. 13.2, see Figs. 13.7 and 13.8.

**Fig. 13.7** Simulated porosity-permeability paired values, sample size $n = 380$. *Left:* Under Bernstein copula and Bernstein marginals. *Right:* Under a Gluing copula (Clayton + Independence + asymmetric Clayton + asymmetric Ali-Mikhail-Haq) and Bernstein marginals.

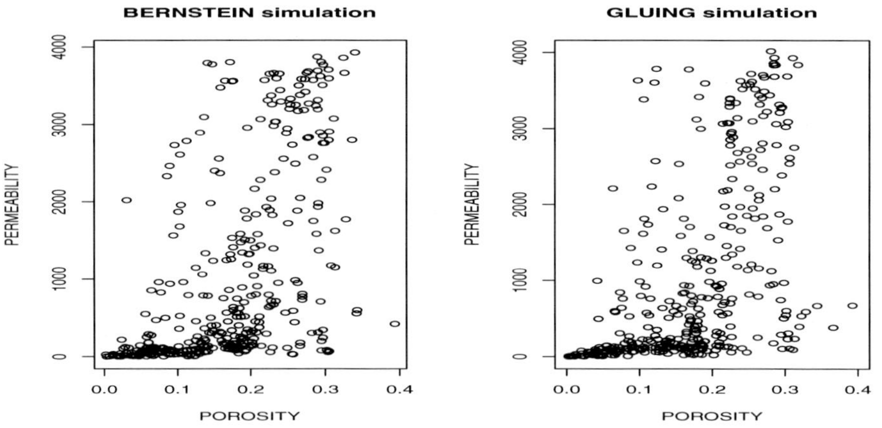


## 13.4 Final Remarks

For the data under consideration (see Figs. 13.1 and 13.2), eventhough there was not strong evidence against the symmetry of the underlying copula [6], it was not possible to find a single known parametric family of copulas that could avoid being

**Fig. 13.8** Median, first and third quartile regression curves. *Left:* Bernstein copula and Bernstein marginals. *Right:* Gluing copula (Clayton + Independence + asymmetric Clayton + asymmetric Ali-Mikhail-Haq) and Bernstein marginals, with *vertical lines* indicating *porosity gluing points* 0.168, 0.219, and 0.2786.

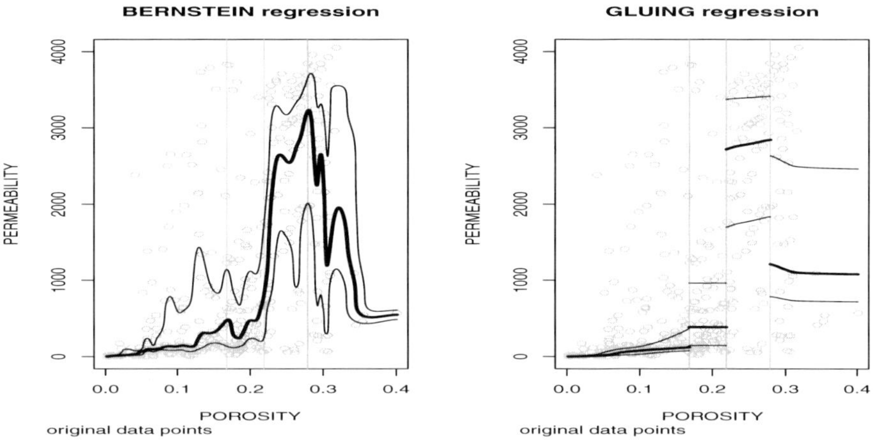


rejected by a goodness-of-fit-test (see Table 13.5), and an analogous situation for the marginal distributions. One way of tackling such situation is by a totally nonparametric approach, using Bernstein copula [15, 16] and Bernstein marginals [12], but paying the price of a noisy regression (see Fig. 13.8, Left) and being unable to model *upper and lower tail dependence,* as an immediate consequence of its definition.

**Table 13.5** Goodness-of-fit $p$-values for several families of copulas for the whole sample, calculated with R package "copula" [18].

| Copula | GoF $p$-value |
|---|---|
| Normal | 0.00535 |
| Plackett | 0.00535 |
| Frank | 0.00055 |
| Clayton | 0.00015 |
| Husler-Reiss | 0.00015 |
| t-Copula | $< 0.00005$ |
| Galambos | $< 0.00005$ |
| Gumbel | $< 0.00005$ |

An alternative found was a *gluing semiparametric approach:* decomposing the total sample into subsamples whose dependence structures (copulas) were simpler to model by symmetric and parametric families of copulas, maintaining a nonparametric estimation of the marginals. The result was specially satisfactory in terms of regression, since the discontinuous curve therein obtained essentially matches what it was expected by petrophysical experts [11], see Fig. 13.8 (Right): a moderate and close to linear increase of permeability for porosity values under a *percolation threshold* (porosity around 0.2), a stable level of permeability around the percolation threshold, and an explosive increase in permeability after such threshold. In addition, under this approach it was possible to model a significant lower tail dependence (subsample 1: Clayton copula) of $\lambda_U = 0.641$ (on a $[0,1]$ scale).

## References


1. Billingsley, P.: Probability and Measure, 3rd edn. Wiley, New York, NY (1995)
2. Deheuvels, P.: La fonction de dépendance empirique et ses propriétés. Un test non paramétrique d'indépendance. Acad. Roy. Belg. Bull. Cl. Sci. **65**(5), 274–292 (1979)
3. Díaz-Viera, M., Casar-González, R.: Stochastic simulation of complex dependency patterns of petrophysical properties using t-copulas. Proc. IAMG'05 GIS Spat. Anal. **2**, 749–755 (2005)
4. Dierckx, P.: Curve and Surface Fitting with Splines. Oxford University Press, New York, NY (1995)
5. Dorai-Raj, S.: DierckxSpline: R companion to "Curve and Surface Fitting with Splines". R package version 1.0–9 (2008)
6. Erdely, A., González-Barrios, J.M.: A nonparametric symmetry test for absolutely continuous bivariate copulas. IIMAS UNAM Preimpreso No. **151**, 1–26 (2009)
7. Fermanian, J-D., Radulovíc, D., Wegcamp, M.: Weak convergence of empirical copula processes. Bernoulli **10**, 547–860 (2004)
8. Genest, C., Quessy, J.-F, Rémillard, B.: Local efficiency of a Cramér–von Mises test of independence. J. Multivar. Anal. **97**, 274–294 (2006)
9. Genest, C., Rémillard, B., Beaudoin, D.: Goodness-of-fit tests for copulas: a review and a power study. Insur. Math. Econ. **44**, 199–213 (2009)
10. Hoeffding, W.: Scale-invariant correlation theory. In: Fisher, N.I., Sen, P.K. (eds.) The Collected Works of Wassily Hoeffding, pp. 57–107. Springer, New York, NY (1940)
11. Kazatchenko, E., Markov, M., Mousatov, A., Parra, J.: Carbonate microstructure determination by inversion of acoustic and electrical data: application to a South Florida Aquifer. J. Appl. Geophys. **59**, 1–15 (2006)
12. Muñoz-Pérez, J., Fernández-Palacín, A.: Estimating the quantile function by Bernstein polynomials. Comput. Stat. Data Anal. **5**, 391–397 (1987)
13. Nelsen, R.B.: An Introduction to Copulas, 2nd edn. Springer, New York, NY (2006)
14. Rüschendorf, L.: Asymptotic distributions of multivariate rank order statistics. Ann. Statist. **4**, 912–923 (1976)
15. Sancetta, A., Satchell, S.: The Bernstein copula and its applications to modeling and approximations of multivariate distributions. Econom. Theory **20**, 535–562 (2004)
16. Sancetta, A.: Nonparametric estimation of distributions with given marginals via Bernstein-Kantorovic polynomials: $L1$ and pointwise convergence theory. J. Multivar. Anal. **98**, 1376–1390 (2007)
17. Siburg, K.F., Stoimenov, P.A.: Gluing copulas. Commun. Stat. Theory Methods **37**, 3124–3134 (2008)
18. Yan, J., Kojadinovic, I.: R package 'copula'. Version 0.8–12 (2009)

# Chapter 14
# Testing Under the Extended Koziol-Green Model

Auguste Gaddah and Roel Braekers

**Abstract** In this chapter, we consider a non-parametric testing procedure for an extension of the Koziol-Green model under two types of informative censoring. For the first type of informative censoring, we allow the censoring time to depend on the lifetime through an Archimedean copula function. For the second type, we generalize the relationship between the marginal distributions of the censoring time and lifetime by means of another copula function on the observed time and censoring indicator. In addition, we describe a bootstrap procedure to approximate the null distribution of the test statistics and illustrate it on a practical data set on survival with malignant melanoma.

## 14.1 Introduction

In many scientific endeavors, researchers are interested in non-negative continuous response variables, which in most cases are expressed as the time until a certain event. For example, in a clinical study, this is the time until the recurrence of a cancer tumor while in industrial settings it is the time until the break down of a machine. In social studies, this time may be the duration of unemployment. Due to practical constraints, the time of interest (lifetime) is usually subjected to random right censoring and is not fully observed. That is, for each response variable of interest $Y_i$ $(i = 1, 2, \ldots, n)$, there exist another non-negative continuous variable $C_i$

Auguste Gaddah
Interuniversity Institute for Biostatistics and Statistical Bioinformatics, Universiteit Hasselt, Diepenbeek, Belgium
e-mail: auguste.gaddah@uhasselt.be

Roel Braekers
Interuniversity Institute for Biostatistics and Statistical Bioinformatics, Universiteit Hasselt, Diepenbeek, Belgium
e-mail: roel.braekers@uhasselt.be

P. Jaworski et al. (eds.), *Copula Theory and Its Applications*, Lecture Notes in Statistics 198,
DOI 10.1007/978-3-642-12465-5_14, © Springer-Verlag Berlin Heidelberg 2010

called censoring variable such that the observable data is the pair $(Z_i, \delta_i)$ where $Z_i = \min(Y_i, C_i)$ and $\delta_i = \mathbf{1}\{Y_i \leq C_i\}$.

Let $Y_i$, $C_i$, $Z_i$ and $\delta_i$ be independent copies of $Y, C, Z$ and $\delta$ respectively and denote $F(t) = P(Y \leq t)$, $G(t) = P(C \leq t)$, $H(t) = P(Z \leq t)$ the respective distribution functions of $Y, C$ and $Z$. In order to make inference about the lifetime, it is imperative to make a non-verifiable assumption about the relationship between $Y$ and $C$ [10]. It is common in time to event analysis to assume independence between these random variables. Under this assumption, the Kaplan-Meier product limit estimator is the standard estimator for $F$ [4]. In some applications, this assumption is not realistic. For instance, in a cancer study where the event of interest is the recurrence of a cancer tumor and the censoring event is death, or in industrial testing, an equipment may be taken away (censored) because it shows signs of future failure. This problem is solved by [9], where an Archimedean copula function is used to describe the join survival distribution of $Y$ and $C$. That is

$$S(t_1, t_2) = P(Y > t_1, C > t_2) = \varphi^{[-1]}\left(\varphi\left(\bar{F}(t_1)\right) + \varphi\left(\bar{G}(t_2)\right)\right) \tag{14.1}$$

with $\bar{F}(t) = 1 - F(t)$ and $\bar{G}(t) = 1 - G(t)$ being the survival functions of $Y$ and $C$ respectively, and $\varphi : [0,1] \to [0,\infty]$ is an Archimedean copula generator function with pseudo inverse $\varphi^{[-1]}$. See [7] for further details about $\varphi$.

In some settings however, the censoring time is additionally informative to the survival time through its distribution function. Koziol and Green tackled this latter problem by considering a sub-model of the general Kaplan-Meier estimator under the assumption that $Y$ and $C$ are independent [6]. In this sub-model, the authors assumed that the lifetime distribution and the censoring distribution satisfy

$$1 - G(t) = (1 - F(t))^{\beta} \quad , \quad \forall t \geq 0 \tag{14.2}$$

for some $\beta > 0$; and it can be shown that characterization (14.2) is satisfied if and only if the observable variables $Z$ and $\delta$ are independent. The validity of the Koziol-Green sub-model was investigated by [2] who developed a testing procedure and showed that there are many data sets for which relation (14.2) is not satisfied. To circumvent this limitation of the Koziol-Green sub-model, relation (14.2) is recently replaced by

$$\bar{G}(t) = \mu\left(\bar{F}(t)\right) \quad , \quad t \geq 0 \tag{14.3}$$

where $\mu(w)$ is a non-decreasing function of $w \in [0,1]$, $\mu(0) = 0$ and $\mu(1) = 1$. This latter assumption, generalizes (14.2) and allows for possible relationship between $Z$ and $\delta$. Moreso, the function $\mu(\cdot)$ can be found from some known copula function $\mathscr{C}$ satisfying

$$H^u(t) = P(Z \leq t, \delta = 1) = \mathscr{C}(\gamma, H(t)) \tag{14.4}$$

where $\gamma = P(\delta = 1)$ is the expected proportion of uncensored observations. Note that Eq. (14.4) is a direct consequence of Sklar's theorem presented in [7]. It describes the join distribution of $Z$ and $\delta$ and is different from (14.1). Obviously, it is desirable to extend the classical Koziol-Green sub-model to accommodate the information contained in (14.1) and (14.4). To do so, we follow [10] and deduce from

(14.1),

$$\frac{dH^u(t)}{dt} = -\frac{\partial}{\partial t_1} S(t_1,t_2)\bigg|_{t_1=t_2=t} = \frac{\varphi'(\bar{F}(t))}{\varphi'(S(t,t))}\frac{dF(t)}{dt} = \frac{\varphi'(\bar{F}(t))}{\varphi'(\bar{H}(t))}\frac{dF(t)}{dt}$$

with $\varphi'(u) = \frac{d}{du}\varphi(u)$ and $S(t,t) = \varphi^{-1}\left(\varphi(\bar{F}(t) + \varphi(\bar{G}(t)))\right) = 1 - H(t) = \bar{H}(t)$.

Reorganizing this equation, gives

$$\varphi'\left(\bar{F}(t)\right)\frac{dF(t)}{dt} = \varphi'\left(\bar{H}(t)\right)\frac{dH^u(t)}{dt}$$

By integrating on both sides and noting that $\varphi(\bar{F}(0)) = \varphi(1) = 0$, we obtain

$$\bar{F}(t) = \varphi^{[-1]}\left(-\int_0^t \varphi'(\bar{H}(s))dH^u(s)\right). \tag{14.5}$$

From the join distribution of $Z$ and $\delta$ given in (14.4), we find

$$dH^u(s) = \mathscr{C}_{01}\left(\gamma, H(s)\right)dH(s)$$

where $\mathscr{C}_{ij}(u,v) = \frac{\partial^{i+j}}{\partial u^i \partial v^j}\mathscr{C}(u,v)$ denotes the $i$th and $j$th partial derivatives of $\mathscr{C}(\cdot,\cdot)$ with respect to its first and second arguments respectively (i.e. $\mathscr{C}_{01}(u,v) = \frac{\partial}{\partial v}\mathscr{C}(u,v)$).

Introducing the preceding relation into (14.5) in conjunction with a variable transformation, we obtain

$$\bar{F}(t) = \varphi^{[-1]}\left(-\int_0^{H(t)} \varphi'(1-w)\mathscr{C}_{01}\left(\gamma, w\right)dw\right) \tag{14.6}$$

Next, we replace $\gamma$ and $H(t)$ in the above expression by their empirical counterparts defined respectively as $\gamma_n = \frac{1}{n}\sum_{i=1}^n \mathbf{1}\{\delta_i = 1\}$ and $H_n(t) = \frac{1}{n}\sum_{i=1}^n \mathbf{1}\{Z_i \le t\}$. Consequently, we find an estimator for the survival function $\bar{F}$ as

$$\bar{F}_n(t) = \varphi^{[-1]}\left(-\int_0^{H_n(t)} \varphi'(1-w)\mathscr{C}_{01}\left(\gamma_n, w\right)dw\right) \tag{14.7}$$

This estimator was recently proposed by [3] who obtained some asymptotic and numerical results pertaining to $\bar{F}_n(t)$. In that paper, the authors deduced from Sklar's theorem [7] that the copula function $\mathscr{C}$ in (14.4) is not unique, since the censoring indicator $\delta$ is a discrete random variable. However, they proceeded and showed that the non-uniqueness of $\mathscr{C}$ does not change the estimator $\bar{F}_n(t)$ in (14.7) for copulas with the same vertical $\gamma_n$-section as studied in [5]. The purpose of the current paper is therefore to determine when model (14.7) is valid in a real data set. This is equivalent to testing for the vertical $\gamma$-section of some copula function $\mathscr{C}$ such that (14.4) is satisfied. Hereto, we take as null hypothesis,

$$H_0 : H^u(t) - \mathscr{C}(\gamma, H(t)) = 0 \tag{14.8}$$

versus the general alternative

$$H_a : H^u(t) - \mathscr{C}(\gamma, H(t)) \neq 0$$

In the next section, we give some asymptotic results that underly the test statistics to be introduced in Sect. 14.3. Afterwards, we illustrate the testing procedure on a real data set in Sect. 14.4.

## 14.2 Asymptotic Results

Now we give two theorems that will serve as the bases for our testing procedure. In the first theorem, we give an asymptotic representation of the basic empirical quantity $H_n^u(t) - \mathscr{C}(\gamma_n, H_n(t))$ as the sum of $n$ independent and identically distributed random variables with a remainder term which is $O\left(n^{-1}\log(n)\right)$ almost surely, where $H_n^u(t) = \frac{1}{n}\sum_{i=1}^n \mathbf{1}\{Z_i \le t, \delta_i = 1\}$. In the second theorem, we establish the weak convergence of the basic empirical quantity to a zero mean Gaussian process. Before we state the theorems, we first give the following definition and assumption that are vital to the establishment of the theorems.

A.1 Let $\mathscr{C}_{ij}(u,v) = \frac{\partial^{i+j}}{\partial u^i \partial v^j}\mathscr{C}(u,v)$ denote the $i$th and $j$th partial derivatives of some copula function $\mathscr{C}(\cdot,\cdot)$ with respect to its first and second arguments respectively. We assume that $\mathscr{C}_{20}(u,v)$, $\mathscr{C}_{02}(u,v)$ and $\mathscr{C}_{11}(u,v)$ exist and are continuous for all $(u,v) \in ]0,1[^2$.

**Theorem 14.2.1.** *Under the null hypothesis $H_0$, assume A.1 is satisfied. Then, $\forall t \ge 0$*

$$H_n^u(t) - \mathscr{C}(\gamma_n, H_n(t)) = \frac{1}{n}\sum_{i=1}^n k(t; Z_i, \delta_i) + r_n(t)$$

*where*

$$k(t;Z_i,\delta_i) = \mathbf{1}\{Z_i \le t, \delta_i = 1\} - H^u(t) - (\mathbf{1}\{Z_i \le t\} - H(t))\mathscr{C}_{01}(\gamma, H(t)) \\ - (\mathbf{1}\{\delta_i = 1\} - \gamma)\mathscr{C}_{10}(\gamma, H(t))$$

*and*

$$\sup_{t\in[0,+\infty]} |r_n(t)| = O\left(n^{-1}\log(n)\right) \quad a.s.$$

*Proof.* Under the null hypothesis $H_0$, we can write

$$\begin{aligned} H_n^u(t) - \mathscr{C}(\gamma_n, H_n(t)) &= [H_n^u(t) - \mathscr{C}(\gamma_n, H_n(t))] - [H^u(t) - \mathscr{C}(\gamma, H(t))] \\ &= [H_n^u(t) - H^u(t)] - [\mathscr{C}(\gamma_n, H_n(t)) - \mathscr{C}(\gamma, H(t))] \end{aligned}$$

Using a Taylor expansion on the 2nd term in the preceding expression, we get

$$H_n^u(t) - \mathscr{C}(\gamma_n, H_n(t)) = \frac{1}{n}\sum_{i=1}^{n} k(t; Z_i, \delta_i) + r_n(t) \tag{14.9}$$

where

$$\begin{aligned} k(t; Z_i, \delta_i) = &\mathbf{1}\{Z_i \le t, \delta_i = 1\} - H^u(t) - [\mathbf{1}\{Z_i \le t\} - H(t)]\mathscr{C}_{01}(\gamma, H(t)) \\ &- [\mathbf{1}\{\delta_i = 1\} - \gamma]\mathscr{C}_{10}(\gamma, H(t)) \end{aligned}$$

and

$$\begin{aligned} r_n(t) = &\frac{1}{2}[\gamma_n - \gamma]^2 \mathscr{C}_{20}(\gamma^*, H^*(t)) + \frac{1}{2}[H_n(t) - H(t)]^2 \mathscr{C}_{02}(\gamma^*, H^*(t)) \\ &+ [\gamma_n - \gamma][H_n(t) - H(t)]\mathscr{C}_{11}(\gamma^*, H^*(t)) \end{aligned} \tag{14.10}$$

with $\gamma^*$ lying between $\gamma_n$ and $\gamma$; and $H^*(t)$ between $H_n(t)$ and $H(t)$. We now determine the rate of convergence of $r_n(t)$. We start by applying Bernstein's inequality and obtain for all $\varepsilon > 0$

$$P(|\gamma_n - \gamma| > \varepsilon) \le P\left(\left|\sum_{i=1}^{n} (\mathbf{1}\{\delta_i = 1\} - \gamma)\right| > n\varepsilon\right) \le 2\exp\left(-\frac{n\varepsilon^2}{2(\gamma + \varepsilon)}\right)$$

Taking $\varepsilon = \varepsilon_n = Kn^{-1/2}\log(n)^{1/2}$ for some $K > 0$, yields $\sum_{i=1}^{n} \exp\left(-\frac{n\varepsilon^2}{2(\gamma+\varepsilon)}\right) < \infty$. And, using the Borrel-Cantelli lemma, we obtain

$$|\gamma_n - \gamma| = O\left(n^{-1/2}\log(n)^{1/2}\right) \quad \text{a.s.} \tag{14.11}$$

Analogously, we use Dvoretzky, Kiefer and Wolfwitz theorem instead of Bernstein's inequality and obtain, for all $\varepsilon > 0$

$$\sup_{t \in [0, +\infty]} |H_n(t) - H(t)| = O\left(n^{-1/2}\log(n)^{1/2}\right) \quad \text{a.s.} \tag{14.12}$$

From (14.10), it follows that

$$\begin{aligned} \sup_{t\in[0,+\infty]} |r_n(t)| \le &|\gamma_n - \gamma|^2 \sup_{t\in[0,+\infty]} |\mathscr{C}_{20}(\gamma^*, H^*(t))| \\ &+ \sup_{t\in[0,+\infty]} |H_n(t) - H(t)|^2 \sup_{t\in[0,+\infty]} |\mathscr{C}_{02}(\gamma^*, H^*(t))| \\ &+ |\gamma_n - \gamma| \sup_{t\in[0,+\infty]} |H_n(t) - H(t)| \sup_{t\in[0,+\infty]} \mathscr{C}_{11}(\gamma^*, H^*(t)) \end{aligned}$$

Since $\gamma_n \to \gamma$ a.s and $H_n(t) \to H(t)$ a.s as $n \to \infty$, we know that $\gamma^* \to \gamma$ a.s and $H^*(t) \to H(t)$ a.s as $n \to \infty$. Therefore, under Assumption A.1, we use (14.11) and (14.12) to obtain

$$\sup_{t\in[0,+\infty]} |r_n(t)| = O\left(n^{-1}\log(n)\right) \quad \text{a.s}$$

□

**Theorem 14.2.2.** *Under the null hypothesis $H_0$, assume A.1 is satisfied. Then, as $n \to \infty$*

$$n^{1/2}\left(H_n^u(\cdot) - \mathscr{C}(\gamma_n, H_n(\cdot))\right) \to \psi(\cdot) \qquad in \qquad \ell^\infty[0,+\infty]$$

*where $\psi(\cdot)$ is a zero mean Gaussian process with variance-covariance function given by*

$$\begin{aligned}\Gamma_{st} &= [H^u(s\wedge t) - H^u(s)H^u(t)] + [H(s\wedge t) - H(s)H(t)]\mathscr{C}_{01}(\gamma, H(s))\mathscr{C}_{01}(\gamma, H(t)) \\ &+ \gamma[1-\gamma]\mathscr{C}_{10}(\gamma, H(s))\mathscr{C}_{10}(\gamma, H(t)) + [H^u(s) - \gamma H(s)]\mathscr{C}_{01}(\gamma, H(s))\mathscr{C}_{10}(\gamma, H(t)) \\ &+ [H^u(t) - \gamma H(t)]\mathscr{C}_{01}(\gamma, H(t))\mathscr{C}_{10}(\gamma, H(s)) - [H^u(s\wedge t) - H^u(s)H(t)]\mathscr{C}_{01}(\gamma, H(t)) \\ &- [H^u(s\wedge t) - H^u(t)H(s)]\mathscr{C}_{01}(\gamma, H(s)) - [H^u(s) - \gamma H(s)]\mathscr{C}_{10}(\gamma, H(t)) \\ &- [H^u(t) - \gamma H(t)]\mathscr{C}_{10}(\gamma, H(s))\end{aligned} \tag{14.13}$$

*for all $s \geq 0$ and all $t \geq 0$.*

*Proof.* Under the null hypothesis $H_0$, we show the weak convergence of the empirical quantity $n^{1/2}\left(H_n^u(\cdot) - \mathscr{C}(\gamma_n, H_n(\cdot))\right)$ to a zero mean Gaussian process with variance-covariance function $\Gamma_{st}$, for all $(s,t) \in [0,+\infty]$. To do so, we work in two steps. We first establish the finite dimensional distributions of the process under consideration and then append it with tightness in $\ell^\infty[0,+\infty]$. To start, we use the main term in representation (14.9) and denote $W_n(t) = \frac{1}{n}\sum_{i=1}^n k(t; Z_i, \delta_i)$. For some integer $q > 0$, we take distinct time points $0 = t_1 < t_2 < \cdots < t_q$. Then, by the multivariate central limit theorem, $(W_n(t_1), W_n(t_2), ..., W_n(t_q))$ converges to an asymptotic normal distribution with mean vector

$$E(W_n(t)) = E(k(t; Z, \delta)) = 0$$

and variance-covariance matrix equals

$$\Gamma_{t_j t_k} = \text{Cov}(k(t_j; Z, \delta)k(t_k; Z, \delta)) = E(k(t_j; Z, \delta)k(t_k; Z, \delta))$$

Computing the expectation in the preceding display, we obtain (14.13), with $s = t_j$ and $t = t_k$.

To show tightness, we first note that

$$\begin{aligned}\sup_{t\in[0,+\infty]} |k(t; Z, \delta)| \leq &\sup_{t\in[0,+\infty]} |\mathbf{1}\{Z \leq t, \delta = 1\} - H^u(t)| \\ &+ |\mathbf{1}\{\delta = 1\} - \gamma| \sup_{t\in[0,+\infty]} \mathscr{C}_{10}(\gamma, H(t)) \\ &+ \sup_{t\in[0,+\infty]} |\mathbf{1}\{Z \leq t\} - H(t)| \sup_{t\in[0,+\infty]} \mathscr{C}_{01}(\gamma, H(t)) \\ \leq &\, 3\end{aligned}$$

Secondly, we let $\mathscr{F} = \{k(t; Z, \delta) : t \geq 0\}$. Then, $\mathscr{F}$ consists of uniformly bounded function over $[0,+\infty]$. As such, their bracketing number is $N_{[]}(\alpha, \mathscr{F}, L_2(P))) = O\left(\exp(K\alpha^{-1})\right)$ for $\alpha < 6$ and some $K > 0$. For $\alpha > 6$, we take $N_{[]}(\alpha, \mathscr{F}, L_2(P))) =$

1. Furthermore, we note that proving tightness of the process is equivalent to showing that the class of functions $\mathscr{F}$ is Donsker. As a result, we apply Theorem 19.5 in [11] to obtain

$$\int_0^1 \sqrt{\log N_{[]}(\alpha,\mathscr{F},L_2(P))}d(3\alpha) = 3\int_0^1 \sqrt{\log N_{[]}(\alpha,\mathscr{F},L_2(P))}d\alpha$$
$$\leq 3\int_0^1 \sqrt{\frac{K}{\alpha}}d\alpha < \infty$$

This shows that the process under consideration is tight in $\ell^\infty[0,+\infty]$. Combining this with the convergence of the finite dimensional distributions completes the proof. □

## 14.3 Test Statistics

In this section, we give the test statistics based on the basic empirical process

$$\psi_n(\cdot) = n^{1/2}(H_n^u(\cdot) - \mathscr{C}(\gamma_n, H_n(\cdot)))$$

Specifically, we consider a Kolmogorov-Smirnov and Cramer-von Mises type statistics defined respectively by

$$T_{KS} = \sup_{t\in[0,+\infty]} |\psi_n(t)| \qquad \text{and} \qquad T_{CM} = \int_0^{+\infty} \psi_n(t)^2 d\mathscr{C}(\gamma_n, H_n(t))$$

As a consequence of Theorem 2, we now give a corollary that will serve as the basis for finding critical values of the test statistics.

**Corollary 14.3.1.** *Under the null hypothesis, assume A.1 is satisfied. Then,*

$$T_{KS} \to \sup_{t\in[0,+\infty]} |\psi(t)|$$
$$T_{CM} \to \int_0^{+\infty} \psi(t)^2 d\mathscr{C}(\gamma, H(t))$$

We do not give the proof of the Corollary, but it can easily be obtained by an application of Helly-Bray theorem (p. 117 in [8]) and Theorem 2.2.4 in [7].

For practical application of the test statistics, we use the following formulation. Let $Z_{(1)}, Z_{(2)}, \ldots, Z_{(n)}$ denote the order statistics of the $Z$ sample and $\delta_{(1)}, \delta_{(2)}, \ldots, \delta_{(n)}$ denote the induced $\delta$ sample. Further, let $r(= 1,2,\ldots,n)$ be the rank of $Z_{(r)}$ and denote the number of uncensored observations not greater than $Z_{(r)}$ by

$$N_r = \{1 \leq j \leq r : \delta_{(j)} = 1\}$$

Then, the test statistics can be expressed as

$$T_{KS} = n^{1/2} \max_{1\le r\le n} \left| \frac{N_r}{n} - \mathscr{C}\left(\gamma_n, \frac{r}{n}\right) \right|$$
$$T_{CM} = n \sum_{r=1}^{n} \left( \frac{N_r}{n} - \mathscr{C}\left(\gamma_n, \frac{r}{n}\right) \right)^2 \left( \mathscr{C}\left(\gamma_n, \frac{r}{n}\right) - \mathscr{C}\left(\gamma_n, \frac{r-1}{n}\right) \right) \tag{14.14}$$

At this point, it is clear that a valid test of the null hypothesis $H_0$ should be based on the null distribution of the test statistics. Due to its complicated variance-covariance function (14.13), it is not feasible to readily find critical values for the test. As a result, we instead describe a bootstrap procedure to approximate the null distribution of the critical values. The procedure consists of the following steps:

1. Given the observed data, we estimate $\gamma$ and $H(t)$ by

$$\gamma_n = \frac{1}{n} \sum_{i=1}^{n} \mathbf{1}\{\delta_i = 1\} \qquad \text{and} \qquad H_n(t) = \frac{1}{n} \sum_{i=1}^{n} \mathbf{1}\{Z_i \le t\}$$

2. For each $i$ $(i = 1, 2, \ldots, n)$,

   a. we generate two independent uniform (0,1) samples $u_i$ and $s_i$
   b. given the copula function $\mathscr{C}$ under the null hypothesis $H_0$, we set $v_i = (\mathscr{C}_{10})^{-1}(s_i)$, where $(\mathscr{C}_{10})^{-1}$ is the inverse of $\mathscr{C}_{10}$.
   c. we define the bootstrap pair $(Z_i^*, \delta_i^*)$ by

$$Z_i^* = \inf\{t : H_n(t) \ge v_i\} \qquad \text{and} \qquad \delta_i^* = \mathbf{1}\{u_i > 1 - \gamma_n\}$$

3. Given the bootstrap sequence $\{Z_n^*, \delta_n^*\}$ in the preceding step, we analogously compute the bootstrap counterparts $T_{KS}^*$ and $T_{CM}^*$ of $T_{KS}$ and $T_{CM}$ as given in (14.14).
4. We repeat steps 2 and 3 for a fixed bootstrap size $B$ and obtain the approximate $p$-values for the test statistics $T_{KS}$ and $T_{CM}$ respectively by

$$\frac{1}{B} \sum_{b=1}^{B} \mathbf{1}\left\{T_{KS_b}^* > T_{KS}\right\} \qquad \text{and} \qquad \frac{1}{B} \sum_{b=1}^{B} \mathbf{1}\left\{T_{CM_b}^* > T_{CM}\right\} \tag{14.15}$$

It is important to note that the validity of the bootstrap approximation can be assured by showing that the original empirical process $\psi_n(\cdot)$ and its bootstrap counterpart converge to the same limiting Gaussian process. We do not show this, but it analogous to the proof of Theorem 2.

## 14.4 Data Example: Survival with Malignant Melanoma

In this section, we illustrate the testing procedure on a real data set. The data concern 205 patients with malignant melanoma (cancer of skin). Of these patients, 57 (28%) died of malignant melanoma (event), 14 (7%) died of other causes and 134 (65%) were alive at the end of the study. See [1] for more details about this data set.

For the purpose of this section, we treat those observations corresponding to deaths due to other causes and those corresponding to the 134 survivors as censored observations. Furthermore, we believe that the occurrence of the censoring is an indirect manifestation of the surgical operation. Therefore, we suspect that the censoring time is informative to the survival time through its distribution function. As such, we can apply the extended estimator on the data set, provided an appropriate copula function can be found for the join distribution of the observed time and censoring indicator. Before applying the testing procedure, we perform a preliminary search of a potential copula function $\mathscr{C}$ by graphically investigating whether

$$H_n^u(t) = \mathscr{C}(\gamma_n, H_n(t))$$

nearly holds for all $t \geq 0$, where $H_n^u(t), H_n(t)$ and $\gamma_n$ are as previously defined. More specifically, we compare the vertical $\gamma_n$-section of the Fréchet-Hoeffding lower bound (W), Fréchet-Hoeffding upper bound (M), Clayton, Product, Plackett and Frank copulas to the empirical quantity $H_n^u(H_n^{-1}(p))$, where $H_n^{-1}(p) = \inf\{t : H_n(t) > p\}$ is the quantile function of $H_n(t)$. These copula functions are respectively given as $\mathscr{C}(u,v) = \max(u+v-1,0)$, $\mathscr{C}(u,v) = \min(u,v)$, $\mathscr{C}(u,v) = \left[\max\left(u^{-2}+v^{-2}-1,0\right)\right]^{1/2}$, $\mathscr{C}(u,v) = uv$, $\mathscr{C}(u,v) = \frac{1+4(u+v)-\sqrt{[1+4(u+v)]^2-80uv}}{8}$, and $\mathscr{C}(u,v) = -\frac{1}{4}\log\left(1+\frac{(e^{-4u}-1)(e^{-4v}-1)}{(e^{-4}-1)}\right)$.

Among these copula functions, we see from Fig. 14.1 that the Clayton copula gives the best approximation to the empirical quantity and suggests itself as a potential candidate for this data set. Using (14.14) together with the bootstrap procedure described in Sect. 14.3, we further test each of the copula functions above under the null hypothesis $H_0$. With a bootstrap size $B = 10{,}000$, we report the $p$-values of both test statistics in Table 14.1. Except for the Clayton copula, the table shows that the hypothesized copula functions are rejected at 5% significant level.

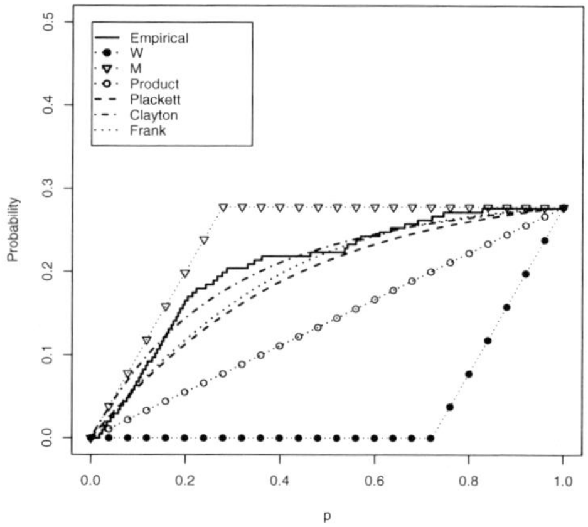


**Fig. 14.1** Graphical test of the copula function on the observed time and censoring indicator.

This supports the observation made in Fig 14.1 and concludes the Clayton copula (given above) as appropriate to describe the join distribution of the observed survival time and censoring indicator in this data set. With the chosen copula function, we further compare (not shown) the extended Koziol-Green estimates with that of the Kaplan-Meier and the classical Koziol-Green. We observe that unlike the classical Koziol-Green estimates, the extended Koziol-Green and Kaplan-Meier estimates are close to each other. This corroborate the appropriateness of the extended Koziol-Green model on this data set. Moreso, the extended Koziol-Green estimator uses the extra information contained in the join distribution of the observed time and censoring indicator, which results in estimates with smaller variances as demonstrated in [3].

**Table 14.1** Bootstrap test of the copula function on the observed time and censoring indicator.

| | Clayton | Product | Plackett | Frank |
|---|---|---|---|---|
| $T_{KS}$ | 0.3774 | 1.7682 | 0.7762 | 0.6611 |
| $p$-value | 0.2512 | 0.0000 | 0.0011 | 0.0069 |
| $T_{CM}$ | 0.0136 | 0.3355 | 0.0455 | 0.0295 |
| $p$-value | 0.0872 | 0.0000 | 0.0040 | 0.0175 |

**Acknowledgements** The authors gratefully acknowledge the financial support from the IAP research Network P6/03 of the Belgian Government (Belgian Science Policy).

## References

1. Andersen, K.P., Borgan, O. Gill, R.D.: Statistical Models Based on Counting Processes. Springer, New York, NY (1993)
2. Csörgó, S.: Testing for the Proportional Hazards Model of Random Censorship, Proceedings of the 4th Prague Symposium on Asymptotic Statistics, Prague (1988)
3. Gaddah, A., Braekers, R.: An extension of the Koziol-Green model under dependent censoring (submitted)
4. Kaplan, E.L., Meier, P.: Non-parametric estimation from incomplete observations. J. Am. Stat. Assoc. **53**, 457–481 (1958)
5. Klement, P.E., Kolesárová, A., Mesiar, R., Sempi, C.: Copulas constructed from horizontal sections. Commun. Stat. Theory Methods **36**, 2901–2911 (2007)
6. Koziol, J.A., Green, S.B.: A Cramér-von Mises statistic for randomly censored data. Biometrika **63**, 465–474 (1976)
7. Nelsen, R.B.: An Introduction to Copulas. Springer, New York, NY (2006)
8. Rao, C.R.: Linear Statistical Inference and Its Applications. Wiley, New York, NY (1973)
9. Rivest, L., Wells, M.T.: A martingale approach to the copula-graphic estimator for the survival function under dependent censoring. J. Multivar. Anal. **79**, 138–155 (2001)
10. Tsiatis, A.: A nonidentifiability aspect of the problem of competing risks, Proc. Natl. Acad. Sci. USA **72**, 20–22 (1975)
11. van der Vaart, A.W.: Asymptotic Statistics. Cambridge University Press, Cambridge (1998)

# Chapter 15
# Parameter Estimation and Application of the Multivariate Skew $t$-Copula

Tõnu Kollo and Gaida Pettere

**Abstract** Copula theory has got a rapid development in recent years. Most used copulas are symmetric: Archimedean are symmetric by construction while other continuous multivariate copulas are usually constructed from elliptical distributions and therefore are symmetric. From skewed copulas we can refer only to a copula introduced in [5], which the authors called skew $t$-copula. The construction of it differs from our approach. We introduce a multivariate $t$-copula which is based on the skew $t$-distribution introduced in [1]. Parameters of the copula have been estimated by method of moments and a simulation rule is given. The behaviour of estimates of the shape parameter of the skewed $t$-distribution is illustrated by simulation. The skew $t$-copula is used for modelling real data.

## 15.1 Introduction

Intensive development of the copula theory has taken place in recent years. A comprehensive presentation of the theory on introductory level can be found in [9]. Applications in different areas have stimulated research in constructing copulas, especially financial applications (see [3], for instance). Often data have skewed univariate marginal distributions of different type. For construction of a multivariate model with certain dependence structure and different marginals copula theory has been the only tool at hand so far. Most copulas in use are symmetric: Archimedian copulas are symmetric by construction while elliptical copulas (Gaussian copula, $t$-copula) are based on multivariate symmetric distributions. To join skewed marginals into a

Tõnu Kollo
Institute of Mathematical Statistics, University of Tartu, Tartu, Estonia
e-mail: tonu.kollo@ut.ee

Gaida Pettere
Department of Engineering Mathematics, Riga Technical University, Riga, Latvia
e-mail: gaida@latnet.lv

P. Jaworski et al. (eds.), *Copula Theory and Its Applications*, Lecture Notes in Statistics 198,
DOI 10.1007/978-3-642-12465-5_15, 

multivariate distribution it seems more natural to use a skew multivariate distribution. Since the multivariate skew-normal distribution was introduced in [2], there has been a rapid development in construction of multivariate skew distribution families. The theory as well as applications of multivariate skew distributions up to 2004 are presented in the collective monograph [7] including multivariate skew $t$-distribution. Different variants of the multivariate skew $t$-distribution can be found in [8, Chap. 5]. Skew-normal distribution has lighter than normal tails and therefore is not attractive for financial applications where data usually have heavy tail area. In [5] a non-symmetric multivariate copula has been examined which the authors called skew $t$-copula. They consider multivariate normal mean-variance mixture distributions and in a special case refer to it as the skew $t$-copula. This copula is constructed from a multivariate skew distribution which has the covariance matrix when the number of degrees of freedom $\nu > 4$. We are going to use multivariate skew $t$-distribution introduced in [1]. In this case we have a simpler expression for the density than in [5] and the covariance matrix exists when $\nu > 2$. This gives us a possibility for modelling distributions with heavier tail area than the one in [5].

## 15.2 Preliminary Notions and Notation

In this section we are going to present notation and formulae which we use for construction of the copula. Following [1], a random $p$-vector $\mathbf{Y} = (Y_1, ..., Y_p)^T$ is skew normally distributed with parameters $\Sigma$, $\mu$ and $\alpha$, when its density function is in the form:

$$f_{\mathbf{Y}}(\mathbf{y}) = 2\phi(\mathbf{y}-\mu;\ \Sigma)\Phi(\alpha^T\mathbf{W}^{-1}(\mathbf{y}-\mu)) \tag{15.1}$$

where $\phi(\mathbf{y}-\mu;\ \Sigma)$ is the density function of the normal distribution $N_p(\mu,\Sigma)$, $\Phi(\cdot)$ is the distribution function of $N(0,1)$ with $\mu \in R^p$, $\alpha \in R^p$, positive definite $p \times p$-matrix $\Sigma$, and $\mathbf{W}$ is the $p \times p$ diagonal matrix $\mathbf{W} = (\delta_{ij}\sqrt{\sigma_{ij}})$, $i,\ j = 1, ..., p$, where $\delta_{ij}$ is the Kronecker delta. Notation $\mathbf{Y} \sim SN_p(\mu,\Sigma,\alpha)$ is used for $\mathbf{Y}$ with the density (15.1).
A vector $\mathbf{Y} \sim SN_p(\mu,\Sigma,\alpha)$ has the following representation

$$\mathbf{Y} = \begin{cases} \mathbf{Z}, & \alpha^T\mathbf{W}^{-1}(\mathbf{Z}-\mu) > Z_0 \\ -\mathbf{Z}, & \text{otherwise} \end{cases}, \tag{15.2}$$

where $\mathbf{Z} \sim N_p(\mu,\Sigma)$ and $Z_0 \sim N(0,1)$.
There exist many different modifications and extensions of the standard multivariate $t$-distribution. An overview of these distributions is given in [8, Chap. 5]. We shall use notation $t_{p,\nu}$ when we refer to $p$-variate $t$-distribution with $\nu$ degrees of freedom. Similar notation is used for the density and distribution functions.
Following [1], a $p$-dimensional random vector $\mathbf{X} = (X_1, ..., X_p)^T$ is said to have $p$-variate $t$-distribution with $\nu$ degrees of freedom, mean vector $\mu$ and covariance matrix $\Sigma$, if its density function is given by:

$$t_{p,\nu}(\mathbf{x},\mu,\Sigma) = \frac{\Gamma\left(\frac{\nu+p}{2}\right)}{(\pi\nu)^{\frac{p}{2}}\Gamma\left(\frac{\nu}{2}\right)|\Sigma|^{\frac{1}{2}}}\left[1+\frac{(\mathbf{x}-\mu)^T\Sigma^{-1}(\mathbf{x}-\mu)}{\nu}\right]^{-\frac{\nu+p}{2}} . \tag{15.3}$$

Next we give the definition of the $p$-dimensional skew $t_{p,\nu}$-distribution following [1].

**Definition 15.2.1.** A random $p$-vector $\mathbf{X} = (X_1,...,X_p)^T$ has $p$-variate skew $t_{p,\nu}$-distribution with parameters $\mu$, $\alpha$ and $\Sigma$, if its density function is of the form

$$g_{p,\nu}(\mathbf{x};\mu,\Sigma,\alpha) = 2\cdot t_{p,\nu}(\mathbf{x};\mu,\Sigma)\cdot T_{1,\nu+p}\left[\alpha^T\mathbf{W}^{-1}(\mathbf{x}-\mu)\left(\frac{\nu+p}{Q+\nu}\right)^{\frac{1}{2}}\right] , \tag{15.4}$$

where $Q$ denotes the quadratic form

$$Q = (\mathbf{x}-\mu)^T\Sigma^{-1}(\mathbf{x}-\mu) ,$$

$T_{1,\nu+p}(\cdot)$ is the distribution function of the central univariate $t$-distribution with $\nu + p$ degrees of freedom and $\mathbf{W}$ is defined as in (15.1).

The parameter $\alpha$ is called the shape parameter and it regulates both, shape and location and $\mu$ is considered as the location or shift parameter.
From [1] and [10] we get the two first moments of the $p$-variate skew $t_{p,\nu}$-distribution:

$$E\mathbf{X} = \mu + \mathbf{W}\xi, \tag{15.5}$$

$$D\mathbf{X} = \frac{\nu}{\nu-2}\Sigma - \mathbf{W}\xi\xi^T\mathbf{W} \tag{15.6}$$

where

$$\xi = \left[\frac{\nu}{\pi(1+\alpha^T\mathbf{R}\alpha)}\right]^{\frac{1}{2}}\cdot\frac{\Gamma\left(\frac{\nu-1}{2}\right)}{\Gamma\left(\frac{\nu}{2}\right)}\cdot\mathbf{R}\alpha \tag{15.7}$$

and $\mathbf{R}$ is the correlation matrix:

$$\mathbf{R} = \mathbf{W}^{-1}\Sigma\mathbf{W}^{-1} \tag{15.8}$$

with $\mathbf{W}$ as defined in text after the formula (15.1).
The distribution given in (15.4) is easy to simulate. A random $p$-vector $\mathbf{X}$ with the density function (15.4) has the following representation

$$\mathbf{X} = \mu + V^{-\frac{1}{2}}\mathbf{Y} ,$$

where $\mathbf{Y} \sim SN_p(\Sigma,\alpha)$ and $\nu V \sim \chi^2_\nu$ is independent of $\mathbf{Y}$.

## 15.3 Construction of a Skew $t$-Copula

We are going to construct a skewed copula which is based on the multivariate skew $t_{p,\nu}$-distribution with the density function defined in (15.4). Let $X_1,...,X_p$ be continuous random variables with the strictly monotone distribution functions $F_i(x_i):\ R^1 \to I=[0,1]$, and the density functions $f_i(x_i):R^1 \to R^1$, $i=1,...,p$, respectively. Let their joint distribution function be $F_{\mathbf{X}}(x_1,...,x_p)$ and the density function $f_{\mathbf{X}}(x_1,..,x_p)$. From Sklar's theorem [9, p. 41] the distribution function $F_{\mathbf{X}}(x_1,...,x_p)$ can be presented through a copula $C(u_1,...,u_p):[0,1]^p \to [0,1]$, $\mathbf{u}=(u_1,...,u_p)^T \in [0,1]^p$:

$$F_{\mathbf{X}}(x_1,...,x_p)=C(F_1(x_1),...,F_p(x_p))\ . \tag{15.9}$$

The copula density $c(u_1,...,u_p)$ is obtained from the copula $C(u_1,...,u_p)$ by differentiation:

$$c(u_1,...,u_p)=\frac{\partial^p C(u_1,...,u_p)}{\partial u_1 ....\partial u_p}\ .$$

Taking into account (15.9) we can present the density $f_{\mathbf{X}}(x_1,...,x_p)$ through the copula density

$$f_{\mathbf{X}}(x_1,...,x_p)=c(F_1(x_1),...,F_p(x_p))f_1(x_1)\cdot ... \cdot f_p(x_p)\ .$$

From here the copula density $c(\mathbf{u}):I^p \to R$ can be expressed through the densities of $\mathbf{X}$ and $X_i$, $i=1,...,p$:

$$c(\mathbf{u})=\frac{f_{\mathbf{X}}(F_1^{-1}(u_1),...,F_p^{-1}(u_p))}{f_1(F_1^{-1}(u_1))\cdot ... \cdot f_p(F_p^{-1}(u_p))} \tag{15.10}$$

where $F_1(\cdot),...,F_p(\cdot):R^1 \to I$ are the univariate marginal distribution functions and $f_1(\cdot),...,f_p(\cdot):R^1 \to R^1$ the corresponding marginal densities.
In the following definition we summarize (15.9) and (15.10) to define a multivariate skew $t$-copula.

**Definition 15.3.1.** A copula is called skew $t_{p,\nu}$-copula, if

$$\begin{aligned} &C_{p,\nu}(u_1,...,u_p:\mu,\Sigma,\alpha) \\ &=G_{p,\nu}(G_{1,\nu}^{-1}(u_1;\mu_1,\sigma_{11},\alpha_1),...,G_{1,\nu}^{-1}(u_p;\mu_p,\sigma_{pp},\alpha_p);\mu,\Sigma,\alpha) \end{aligned} \tag{15.11}$$

where $G_{1,\nu}^{-1}(u_i;\mu_i,\sigma_{ii},\alpha_i):R^1 \to I$, $i \in \{1,...,p\}$ denotes the inverse of the univariate $t_{1,\nu}$-distribution function and $G_{p,\nu}$ is the distribution function of the $p$-variate skew $t_{p,\nu}$-distribution with the density (15.4).

The corresponding copula density is

$$c_{p,\nu}(\mathbf{u};\mu,\Sigma,\alpha) = \frac{g_{p,\nu}\left[\{G_{1,\nu}^{-1}(u_1;\mu_1,\sigma_{11},\alpha_1),\dots,G_{1,\nu}^{-1}(u_p;\mu_p,\sigma_{pp},\alpha_p)\};\mu,\Sigma,\alpha\right]}{\prod_{i=1}^{p} g_{1,\nu}\left[G_{1,\nu}^{-1}(u_i;\mu_i,\sigma_{ii},\alpha_i);\mu_i,\sigma_{ii},\alpha_i\right]} \tag{15.12}$$

where the density function $g_{p,\nu}(\mathbf{u};\mu,\Sigma,\alpha) : R^p \to R$ is defined by (15.4) and functions $G_{1,\nu}^{-1}(u_i;\mu_i,\sigma_{ii},\alpha_i)$ are as in Definition 15.3.1. As in numerator of the density expression in (15.12) stands density of a non-symmetric multivariate skew $t$-distribution, the resulting copula density also represents a non-symmetric skewed multivariate distribution. Several properties of the skew $t$-copula are still under investigation. For example, what is the maximum value of the Mardia's multivariate skewness characteristic of the copula; how tail dependence can be expressed between bivariate marginals, etc.

By integration we get from (15.12) an expression of the skew $t$-copula given in the next proposition.

**Proposition 15.3.1.** *Multivariate skew $t_{p,\nu}$-copula can be presented of the form:*

$$C_{t,\nu}(\mathbf{u};\mu,\Sigma,\alpha) = 2\int_{-\infty}^{G_{1,\nu}^{-1}(u_1;\mu_1,\sigma_1,\alpha_1)} \dots \int_{-\infty}^{G_{1,\nu}^{-1}(u_p;\mu_p,\sigma_p,\alpha_p)} t_{p,\nu}(\mathbf{x};\mu,\Sigma) \times T_{\nu+p}\left[\alpha^T\mathbf{W}^{-1}(\mathbf{x}-\mu)\left(\frac{\nu+p}{Q+\nu}\right)^{\frac{1}{2}}\right] d\mathbf{x} \tag{15.13}$$

*where $Q = (\mathbf{x}-\mu)^T\Sigma^{-1}(\mathbf{x}-\mu)$ and $\mathbf{W}$ are defined in Sect. 15.2.*

## 15.4 Parameter Estimation

Estimation methods for copula models are studied by Choros et al. [4] with special attention to maximum likelihood and pseudolikelihood methods. In the case of multivariate copulas the big number of parameters courses additional problems, therefore we have applied the moments' method.

Let us consider a special case of the two-parameter distribution and take the shift parameter $\mu = 0$. So we start from the density (15.3) where $\mu = 0$. Then the first two moments from formulas (15.5)–(15.8) will be of the form

$$E\mathbf{X} = \mathbf{W}\xi \tag{15.14}$$

$$D\mathbf{X} = \frac{\nu}{\nu-2}\Sigma - \mathbf{W}\xi\xi^T\mathbf{W} \tag{15.15}$$

where $\xi$ is given by (15.7) and (15.8).

We are going to use the moments' method. Let $\overline{\mathbf{X}}$ and $\mathbf{S_X}$ denote the sample mean and the sample covariance matrix, respectively. Estimates of $\Sigma$ and $\alpha$ are given in the next proposition.

**Proposition 15.4.1.** *Estimates of the parameters* $\Sigma$ *and* $\alpha$ *of the skew* $t_{p,\nu}$*-distribution with* $\nu > 2$ *are of the form:*

$$\widehat{\Sigma} = \frac{\nu-2}{\nu}(\mathbf{S_X} + \overline{\mathbf{X}}\,\overline{\mathbf{X}}^T); \tag{15.16}$$

$$\widehat{\alpha} = \frac{b(\nu)\beta}{\sqrt{b^2(\nu) - \overline{\mathbf{X}}^T\widehat{\Sigma}^{-1}\overline{\mathbf{X}}}}, \tag{15.17}$$

*where*

$$\beta = \frac{1}{b(\nu)}\widehat{\mathbf{W}}\widehat{\Sigma}^{-1}\overline{\mathbf{X}} \tag{15.18}$$

*and* $\widehat{\mathbf{W}} = (\delta_{ij}\sqrt{\widehat{\sigma}_{ij}})$, $i,\ j = 1,\ldots,p$, *where* $\delta_{ij}$ *is the Kronecker delta and*

$$b(\nu) = \left[\frac{\nu}{\pi}\right]^{\frac{1}{2}} \cdot \frac{\Gamma\left(\frac{\nu-1}{2}\right)}{\Gamma\left(\frac{\nu}{2}\right)}. \tag{15.19}$$

*Proof.* From (15.14) and (15.15) we get straightforwardly estimate of $\Sigma$ given in (15.16).
An expression of $\widehat{\alpha}$ can be obtained from (15.7):

$$\widehat{\alpha} = \frac{(1+\widehat{\alpha}^T\widehat{\mathbf{R}}\widehat{\alpha})^{\frac{1}{2}}}{b(\nu)}\widehat{\mathbf{W}}\widehat{\Sigma}^{-1}\overline{\mathbf{X}} \tag{15.20}$$

where $b(\nu)$ is defined in (15.19) and $\widehat{\mathbf{R}} = \widehat{\mathbf{W}}^{-1}\widehat{\Sigma}^{-1}\widehat{\mathbf{W}}^{-1}$.
Then from formula (15.20) using notation (15.18) follows

$$\widehat{\alpha} = (1+\widehat{\alpha}^T\widehat{\mathbf{R}}\widehat{\alpha})^{\frac{1}{2}}\beta$$

or

$$\frac{\widehat{\alpha}_2}{\widehat{\alpha}_1} = \frac{\beta_2}{\beta_1}, \ldots, \frac{\widehat{\alpha}_p}{\widehat{\alpha}_1} = \frac{\beta_p}{\beta_1}. \tag{15.21}$$

When we insert the last equalities into the equation for the first coordinate in formula (15.20), we obtain

$$\widehat{\alpha_1} = \beta_1(1 + (\widehat{\alpha}_1, \frac{\beta_2}{\beta_1}\widehat{\alpha}_1, \ldots, \frac{\beta_p}{\beta_1}\widehat{\alpha}_1)\widehat{\mathbf{R}}(\widehat{\alpha}_1, \frac{\beta_2}{\beta_1}\widehat{\alpha}_1, \ldots, \frac{\beta_p}{\beta_1}\widehat{\alpha}_1)^T)^{\frac{1}{2}}.$$

From here

$$\widehat{\alpha}_1^2 = \beta_1^2 + \widehat{\alpha}_1^2\beta^T\widehat{\mathbf{R}}\beta$$

and

$$\widehat{\alpha}_1 = \frac{\beta_1}{\sqrt{1-\beta^T\widehat{\mathbf{R}}\beta}}.$$

Any $\widehat{\alpha}_i$ can be expressed in the similar way using formula (15.21):

$$\widehat{\alpha}_i = \frac{\beta_i}{\sqrt{1-\beta^T\widehat{\mathbf{R}}\beta}} \,.$$

Applying (15.18) we get statement of the Proposition.

**Corollary 15.4.1.** *The estimates* (15.17) *can be found if*

$$\overline{\mathbf{X}}^T\widehat{\Sigma}^{-1}\overline{\mathbf{X}} < \frac{\nu}{\pi}\cdot\left[\frac{\Gamma\left(\frac{\nu-1}{2}\right)}{\Gamma\left(\frac{\nu}{2}\right)}\right]^2 \,.$$

*Proof.* From (15.17) it follows that the following inequality must hold:

$$b^2(\nu) - \overline{\mathbf{X}}^T\widehat{\Sigma}^{-1}\overline{\mathbf{X}} > 0 \,.$$

From here the Corollary follows straightforwardly taking into account expression of $b(\nu)$ in (15.19).

## 15.5 Simulation

In this section we are going first to examine behaviour of the estimate (15.17) of the shape parameter $\alpha$ in the bivariate case with $\nu = 3$. Let the parameters of the $t_{2,3}$-distribution be

$$\alpha^T = (0.36, 0.54) \quad \text{and} \quad \Sigma = \begin{pmatrix} 9.67 & 10.17 \\ 10.17 & 19.33 \end{pmatrix} \,.$$

Then

$$E\mathbf{X}^T = (2,3) \quad \text{and} \quad D\mathbf{X} = \begin{pmatrix} 25.0 & 24.5 \\ 24.5 & 49.0 \end{pmatrix} \,.$$

We simulate from the skew $t_{2,3}$ distribution samples of size 200. From every sample we calculate estimates $\widehat{\alpha}_1$ and $\widehat{\alpha}_2$. The number of replications is 300. Basic characteristics of the empirical distributions of $\widehat{\alpha}_1$ and $\widehat{\alpha}_2$ are presented in Table 15.1.

As one can see from Table 15.1, the estimates are slightly biased and they fluctuate rather symmetrically around their sample means.

Despite the expression of the skew $t_{p,\nu}$-copula density and distribution function are complicated, the distribution is easy to simulate. The simulation is based on the simulation rule for the skew $t_{p,\nu}$-distribution.

(1) Find the Cholesky decomposition $\mathbf{A}$ of $\Sigma$, ($\mathbf{A}\mathbf{A}^T = \Sigma$).
(2) Simulate $p$ independent values from N(0,1) and form $p$-vector $\mathbf{z}$.
(3) Set vector $\mathbf{x} = \mathbf{A}\mathbf{z}$.
(4) Simulate value $z_0$ from N(0,1).
(5) Get realization of the skew normal vector $\mathbf{y}$ putting

$$\mathbf{y} = \begin{cases} \mathbf{x} & \text{if } \alpha^T \mathbf{W}^{-1}\mathbf{x} > z_0 \\ -\mathbf{x} & \text{if } \alpha^T \mathbf{W}^{-1}\mathbf{x} \leq z_0 \end{cases} .$$

(6) Simulate $h : \chi^2_\nu$.
(7) Find vector $\mathbf{t} = \frac{\mathbf{y}}{\sqrt{h/\nu}}$.
(8) Set vector $\mathbf{u}$ so that every coordinate $u_i = G_{1,\nu}(t_i; 0, \sigma_{ii}, \alpha_i)$, $i \in [1, ..., p]$.
(9) Set vector $\mathbf{x} = (F_1^{-1}(u_1), ..., F_p^{-1}(u_p))$ where $F_i(\cdot)$ is the marginal distribution function of the initial random variable $X_i$.
(10) Repeat steps 2–9 $n$ times.

**Table 15.1** Descriptive statistics of estimates of the shape parameter $\alpha$.

| | $\alpha_1$ | $\alpha_2$ |
|---|---|---|
| **Average** | **0.41** | **0.62** |
| **Median** | **0.41** | **0.60** |
| Standard deviation | 0.28 | 0.29 |
| Skewness | 0.19 | 0.34 |
| Kurtosis | 0.42 | 0.46 |
| 25% quartile | 0.23 | 0.43 |
| 75% quartile | 0.56 | 0.80 |
| **Theoretical value** | **0.36** | **0.54** |

## 15.6 Application

We have used skew $t_{3,\nu}$-copula to model data with skewed marginals. We had anthropometrical data with three variables under consideration: height, waist and bust circuit measures from 177 girls at age from 7 to 10. The data had been collected from girls in different schools in Riga for optimal planning of production of clothes. Descriptive statistics of the variables are given in Table 15.2.

Linear correlation between height and waist circuit is 0.4737, between height and bust circuit is 0.6198 and between waist and bust circuit is 0.861. Marginal distributions were approximated by Gamma and lognormal distributions. The best fit for the height and for waist circuit was obtained by Gamma distribution, for bust circuit by lognormal distribution. The goodness-of-fit was measured by the Kolmogorov test (the 5% critical value equals 0.1065). Results of testing are shown in Table 15.3.

The shape parameter $\alpha$ and the scale parameter $\Sigma$ for skew $t_{3,\nu}$-copula were estimated from data using formulas (15.16), (15.17), (15.18) and (15.19). The number of degrees of freedom $\nu$ was taken 3 and 4 to be able to use the multivariate

**Table 15.2** Descriptive statistics of marginals.

| | Height | Waist circuit | Bust circuit |
|---|---|---|---|
| Mean | 133.21 | 57.74 | 67.66 |
| Median | 132.5 | 56.5 | 66.5 |
| Mode | 129 | 54 | 62 |
| Standard deviation | 7.69 | 6.27 | 6.05 |
| Kurtosis | −0.65 | 12.58 | 0.07 |
| Skewness | 0.37 | 2.42 | 0.76 |
| Range | 35.5 | 52.5 | 30 |
| Minimum | 116.5 | 47.5 | 56.0 |
| Maximum | 152.0 | 100.0 | 86.0 |
| Sample size | 177 | 177 | 177 |

**Table 15.3** Testing results for the marginal distributions.

| Measure | Used distribution | Parameters | |
|---|---|---|---|
| Height | Gamma | $\alpha$ | 300 |
| | | $\beta$ | 0.436 |
| | | Test value | 0.0579 |
| Waist girth | Gamma | $\alpha$ | 84.69 |
| | | $\beta$ | 0.68 |
| | | Test value | 0.0724 |
| Bust girth | Lognormal | $\mu$ | 4.2 |
| | | $\sigma$ | 0.089 |
| | | Test value | 0.0616 |

$t$-distribution with maximally heavy tail area. Closeness of the copula models to the data was examined in a simulation experiment. Additionally the best normal copula was found and compared with the $t_{3,\nu}$-copula models. The results are shown in Table 15.4.

The simulation rule from Sect. 15.5 was used for simulation from skew $t_{3,\nu}$-copula. For simulation from normal copula the well-known algorithm was used (see [3], for example). We have used the Genest and Rivest construction [6] to compare sample data and simulated data in each coordinate plane. The module of maximum distance between obtained univariate cumulative distribution functions was used to decide which copula fits the best to given data.

As it is possible to see from Table 15.4, both used $t_{3,\nu}$-copula models give a good fit in XY and XZ planes while in YZ plane the average maximum distance between distribution functions is the smallest in the case of normal copula. The same time 98% confidence intervals are always wider in the normal copula case.

To summarize we can say that the skew $t$-copula gives us a tool which can be effectively used in applications when analyzing and modelling skewed data.

**Table 15.4** Descriptive statistics for fitting measures of different copulas
(NC – Normal copula, STC1 – Skew $t_{3,\nu}$-copula $\nu = 3$, STC2 – Skew $t_{3,\nu}$-copula $\nu = 4$).

| | XY coordinate plane | | | XZ coordinate plane | | | YZ coordinate plane | | |
|---|---|---|---|---|---|---|---|---|---|
| Copula | NC | STC1 | STC2 | NC | STC1 | STC2 | NC | STC1 | STC2 |
| Mean | 0.0807 | 0.0736 | 0.0742 | 0.0758 | 0.0690 | 0.0693 | 0.0831 | 0.0846 | 0.0853 |
| Mode | 0.0565 | 0.0565 | 0.0621 | 0.0791 | 0.0734 | 0.0678 | 0.0791 | 0.0847 | 0.0847 |
| Median | 0.0734 | 0.0734 | 0.0734 | 0.0734 | 0.0678 | 0.0678 | 0.0847 | 0.0847 | 0.0847 |
| Stand. dev. | 0.0211 | 0.0203 | 0.0199 | 0.0191 | 0.0150 | 0.0146 | 0.0109 | 0.0121 | 0.0117 |
| 1% percent | 0.0452 | 0.0395 | 0.0395 | 0.0395 | 0.0452 | 0.0452 | 0.0565 | 0.0621 | 0.0621 |
| 99% percent | 0.1356 | 0.1243 | 0.1186 | 0.1130 | 0.1017 | 0.1017 | 0.1186 | 0.1186 | 0.1186 |
| 98% confid. interval | 0.0904 | 0.0848 | 0.0791 | 0.0735 | 0.0565 | 0.0565 | 0.0621 | 0.0565 | 0.0565 |

**Acknowledgements** The authors are thankful to a Referee for valuable comments. Tõnu Kollo is grateful to the Estonian Research Foundation for support through the grant GMTMS7435.

# References

1. Azzalini, A., Capitanio, A.: Distributions generated by perturbation of symmetry with emphasis on a multivariate skew t-distribution. J. R. Stat. Soc. Ser. B Stat. Methodol. **65**, 367–389 (2003)
2. Azzalini, A., Dalla Valle, A.: The multivariate skew normal distribution. Biometrika **83**, 715–726 (1996)
3. Cherubini, U., Luciano, E., Vecchiato, W.: Copula Methods in Finance. Wiley, New York, NY (2004)
4. Choroś, B., Ibragimov, R., Permiakova, E.: Copula estimation. In: Jaworski, P., Durante, F., Härdle, W., Rychlik, T. (eds.) Copula Theory and Its Applications, Proceedings of the Workshop, Warsaw, 25–26 Sept 2009, Springer, Dordrecht (2010)
5. Demarta, S., McNeil, A.J.: The *t* copula and related copulas. Int. Stat. Rev. **73**, 111–129 (2005)
6. Genest, C., Rivest, L.: Statistical inference procedures for bivariate Archimedean copulas. J. Am. Stat. Assoc. **88**, 1034–1043 (1993)
7. Genton, M.G. (ed.): Skew-Elliptical Distributions and Their Applications. A Journey Beyond Normality. Chapman & Hall/CRC, Boca Raton, FL (2004)
8. Kotz, S., Nadarajah, S.: Multivariate t Distributions and Their Applications. Cambridge University Press, Cambridge (2004)
9. Nelsen, R.B.: An Introduction to Copulas. Springer, New York, NY (1999)
10. Thompson, K.R., Shen, Y.: Coastal flooding and the multivariate skew-t distribution. In: Genton, M.G. (ed.) Skew-Elliptical Distributions and Their Applications. A Journey Beyond Normality, pp. 243–258. Chapman & Hall/CRC, Boca Raton, FL (2004)

# Chapter 16
# On Analytical Similarities of Archimedean and Exchangeable Marshall-Olkin Copulas

Jan-Frederik Mai and Matthias Scherer

**Abstract** While Archimedean copulas are parameterized by real-valued functions, exchangeable Marshall-Olkin copulas are defined via sequences of real numbers. From a probabilistic perspective, the models behind both families have a different motivation. Consequently, their statistical properties are also different. In this regard, their striking analytical similarities are even more surprising. Considering sequences as discretized functions, most statements about Archimedean copulas and their corresponding generator functions translate into equivalent statements about exchangeable Marshall-Olkin copulas and their parameterizing sequences. This chapter reviews classical and recent results on both families of copulas with a focus on completely monotone functions and sequences.

## 16.1 Introduction

A $d$-dimensional copula is called *Archimedean* if it is of the form

$$C(u_1,\ldots,u_d) = \psi\big(\psi^{-1}(u_1)+\ldots+\psi^{-1}(u_d)\big), \quad u_1,\ldots,u_d \in [0,1]. \tag{16.1}$$

In this case, the function $\psi : [0,\infty) \to [0,1]$ is called the *generator* of $C$. By definition, such a copula is symmetric, meaning that $C$ is invariant under permutations of its arguments. This parametric approach leaves sufficient freedom for various dependence structures, and, at the same time, makes this class of copulas analytically tractable. It is therefore no surprise that Archimedean copulas are widely used in ap-

Jan-Frederik Mai
HVB-Institute for Mathematical Finance, Technische Universität München, Garching, Germany
e-mail: mai@tum.de

Matthias Scherer
HVB-Institute for Mathematical Finance, Technische Universität München, Garching, Germany
e-mail: scherer@tum.de

P. Jaworski et al. (eds.), *Copula Theory and Its Applications*, Lecture Notes in Statistics 198,
DOI 10.1007/978-3-642-12465-5_16, 

plications. On a theoretical level, as well as considering applications, Archimedean copulas are studied in [13–15, 24–26].

We call a copula a $d$-dimensional *Marshall-Olkin copula* if it admits the form

$$C(u_1,\dots,u_d) = \prod_{\emptyset \neq I \subset \{1,\dots,d\}} \min_{z\in I}\Big\{u_z^{\frac{\eta_I}{\sum_{J:z\in J}\eta_J}}\Big\}, \quad u_1,\dots,u_d \in [0,1], \tag{16.2}$$

for a set of $2^d-1$ parameters $\eta_I > 0$, $\emptyset \neq I \subset \{1,\dots,d\}$. The copula (16.2) is the survival copula of the classical Marshall-Olkin distribution as introduced in [22]. Hence, [16] calls this a survival Marshall-Olkin copula. We omit the term "survival" for the sake of convenience. Note that the form (16.2) can alternatively be derived from [14, Eq. (6.36), p. 192]. In general, copulas of the form (16.2) are not symmetric. To maintain comparability with Archimedean copulas, only the subclass of *exchangeable Marshall-Olkin copulas* is studied in the present survey. This subclass is obtained from (16.2) by imposing the following technical requirement on the parameters $\eta_I$, $\emptyset \neq I \subset \{1,\dots,d\}$:

$$|I| = |J| \;\Rightarrow\; \eta_I = \eta_J, \tag{16.3}$$

where the cardinality of a finite set $I$ is denoted by $|I|$. Condition (16.3) reduces the number of required parameters from $2^d-1$ to $d$. These parameters are denoted by $\lambda_1,\dots,\lambda_d > 0$ and are defined via $\{\lambda_k\} := \{\eta_I \,|\, |I| = k\}$, i.e. all (original) parameters corresponding to subsets of size $k = 1,\dots,d$ are identical. In this case, the parametric form (16.2) can significantly be simplified to

$$C(u_1,\dots,u_d) = \prod_{k=1}^{d} u_{(k)}^{\frac{\sum_{i=0}^{d-k}\binom{d-k}{i}\lambda_{i+1}}{\sum_{i=0}^{d-1}\binom{d-1}{i}\lambda_{i+1}}}, \quad u_1,\dots,u_d \in [0,1],$$

where $u_{(1)} \leq \ldots \leq u_{(d)}$ denotes the ordered list of the numbers $u_1,\dots,u_d \in [0,1]$. Defining $a_{k-1} := \sum_{i=0}^{d-k}\binom{d-k}{i}\lambda_{i+1} / \sum_{i=0}^{d-1}\binom{d-1}{i}\lambda_{i+1}$, for $k = 1,\dots,d$, $C$ is more compactly given by

$$C(u_1,\dots,u_d) = \prod_{k=1}^{d} u_{(k)}^{a_{k-1}}, \quad u_1,\dots,u_d \in [0,1]. \tag{16.4}$$

Copulas of the form (16.4) are subsequently called *eMO-copulas*. The acronym "Marshall-Olkin" refers to the authors of the seminal paper [22], in which a multivariate distribution having this survival copula as dependence structure is defined and motivated. Further references studying this kind of distributions are [6, 8, 14, 16, 17, 19, 20].

Besides the fact that both families are symmetric, copulas of the forms (16.1) and (16.4) do not share immediate similarities. In fact, considering their probabilistic motivations, they appear to be very different. The construction of an Archimedean copula involves an $l_1$-norm symmetric distribution, i.e. a uniformly distributed random vector on the unit simplex, which is scaled by an independent random radius, see [24]. This probabilistic model implies that the singular component of an Archimedean copula can never contain the unit cube's diagonal, since the uniform distribution on the simplex is absolutely continuous and an atom of the random radius can only induce a singularity on a set with constant $l_1$-norm, excluding the diagonal. In contrast, *eMO*-copulas originate from a frailty-model in which components fail when hit by random shocks, see [22]. This implies that all (non-degenerate) *eMO*-copulas exhibit a singular component containing the diagonal of the unit cube, corresponding to a shock which kills all components simultaneously. Furthermore, this observation implies that the sole copula which is both Archimedean and of class *eMO* is the independence copula. Moreover, *eMO*-copulas are always extreme-value copulas,[1] whereas the only Archimedean copula of this kind is the Gumbel copula, see [9].

However, the results of [19, 20] reveal some appealing analytical similarities between both families, which are highlighted in the present article. The common ground are statements on completely monotone functions and sequences, defined in Sect. 16.2. Probabilistic models and the corresponding sampling strategies are discussed in Sect. 16.3.

## 16.2 Complete Monotonicity and $d$-Monotonicity

### *16.2.1 Definitions and Examples*

In the following, $\{a_k\}_{k\in\mathbb{N}_0}$ denotes a sequence of numbers in $[0,1]$ and $\psi:[0,\infty)\to[0,1]$ is a real-valued function. The function $\psi$ is called *completely monotone (c.m.)* if it is continuous on $[0,\infty)$, has derivatives $\psi^{(k)}$ of all orders $k\in\mathbb{N}_0$ on $(0,\infty)$, and

$$(-1)^k\,\psi^{(k)}(x)\geq 0,\quad \forall\,k\in\mathbb{N}_0,\quad \forall x>0.$$

In particular, c.m. functions are decreasing and convex. A similar concept is available for sequences, the difference being that the derivative, i.e. the limit of the difference quotient, is replaced by its discrete analogue. For $\{a_k\}_{k\in\mathbb{N}_0}$ the *difference operator* $\Delta$ is defined as $\Delta a_k := a_{k+1}-a_k$ for all $k\in\mathbb{N}_0$. Similarly to computing higher derivatives, this linear operator can also be applied iteratively. For example, write $\Delta^2 a_k$ for the expression $\Delta(\Delta a_k)=\Delta(a_{k+1}-a_k)=\Delta a_{k+1}-\Delta a_k = a_{k+2}-2a_{k+1}+a_k$. More general, write $\Delta^j a_k$ when $\Delta$ is applied $j$ times to $a_k$. The expression $\Delta^j a_k$ involves the $j+1$ numbers $a_k,\ldots,a_{k+j}$. Moreover, it is convenient to introduce the notation $\Delta^0 a_k := a_k$. In analogy with c.m. functions one defines the sequence $\{a_k\}_{k\in\mathbb{N}_0}$ to be *completely monotone (c.m.)* if

$$(-1)^j\,\Delta^j a_k\geq 0,\quad \forall\,k\in\mathbb{N}_0,\quad \forall\,j\in\mathbb{N}_0.$$

[1] A copula $C$ is called *extreme-value copula* if $C(u_1^t,\ldots,u_d^t)=C(u_1,\ldots,u_d)^t$ for all $t\geq 0$.

*Example 16.2.1 (C.m. sequences and functions).* Standard examples for c.m. functions are $\psi(x) = \exp(-\theta x)$ and $\psi(x) = (1+x)^{-\theta}$ for some $\theta > 0$. C.m. sequences are obtained, e.g., by $a_k := \psi(k)$, $k \in \mathbb{N}_0$, whenever $\psi$ is a c.m. function, see also [18] for a thorough discussion of such constructions. For instance, the sequence $\{(1+k)^{-\theta}\}_{k\in\mathbb{N}_0}$ is c.m. for every $\theta > 0$.

### 16.2.2 Probabilistic Interpretations

The aforementioned notion of complete monotonicity is purely analytic. Interestingly, since the 1920s a probabilistic interpretation is available for c.m. functions as well as for c.m. sequences. C.m. functions arise as Laplace transforms of non-negative random variables, which is known as Bernstein's Theorem [3]. More precisely,

$$\begin{aligned} &\text{a function } \psi : [0,\infty) \to [0,1] \text{ is c.m. with } \psi(0) = 1 \\ \Leftrightarrow\ &\exists \text{ a random variable } W : \Omega \to [0,\infty) \text{ s.t. } \psi(x) = \mathbb{E}\big[e^{-xW}\big], x \geq 0. \end{aligned}$$

The probabilistic interpretation of c.m. sequences is due to Hausdorff, see [11, 12]. He shows that

$$\begin{aligned} &\text{a sequence } \{a_k\}_{k\in\mathbb{N}_0} \text{ is c.m. and } a_0 = 1 \\ \Leftrightarrow\ &\exists \text{ a random variable } \tau : \Omega \to [0,1] \text{ s.t. } a_k = \mathbb{E}\big[\tau^k\big], k \in \mathbb{N}_0. \end{aligned} \tag{16.5}$$

Moreover, the random variable $\tau$ is uniquely determined by its moments. This classical result is also known as *little moment problem.*

An alternative characterization of c.m. sequences can be inferred as a corollary of a recent result of [10]. The latter reference links killed Lévy subordinators and c.m. sequences. Recall that a Lévy subordinator is a real-valued non-decreasing stochastic process which starts at zero, is stochastically continuous, and has stationary and independent increments. A *killed Lévy subordinator* $\Lambda = \{\Lambda_t\}_{t\geq 0}$ is a $[0,\infty]$-valued stochastic process which coincides with a Lévy subordinator until time $E$ and equals infinity thereafter. The random time $E$ is either $\infty$, in which case $\Lambda$ is an ordinary real-valued Lévy subordinator, or $E$ is an exponential random variable, independent of $\Lambda$. For further background on this subject, the reader is referred to the textbooks [1, 4, 5, 27]. A killed Lévy subordinator is uniquely characterized by its Laplace transforms, which have the form

$$\mathbb{E}\big[e^{-x\Lambda_t}\big] = e^{-t\,\Phi(x)}, \quad \forall x,t \geq 0,$$

for a function $\Phi : [0,\infty) \to [0,\infty)$ which has c.m. derivative $\Phi^{(1)}$ and satisfies $\Phi(0) = 0$, see [7, p. 450]. The function $\Phi$ is called the *Laplace exponent* of $\Lambda$ and is a so-called *Bernstein function*, see [1, p. 52]. In particular, it is increasing and concave. Gnedin and Pitman [10] show that a sequence $\{c_k\}_{k\in\mathbb{N}_0}$ with $c_0 = 0$

is completely alternating[2] (c.a.) if and only if there exists a killed Lévy subordinator $\Lambda$ with Laplace exponent $\Phi$ such that $c_k = \Phi(k)$ for all $k \in \mathbb{N}_0$. Moreover, $\Lambda$ is uniquely determined by this sequence. Basic computations show that c.m. sequences $\{a_k\}_{k\in\mathbb{N}_0}$ are therefore in one-to-one correspondence with Laplace exponents $\Phi$ of killed Lévy subordinators, respectively with Bernstein functions, via

$$a_k = \Phi(k+1) - \Phi(k), \quad \forall k \in \mathbb{N}_0. \tag{16.6}$$

This alternative view on c.m. sequences is useful for the construction of *eMO*-copulas in Sect. 16.3 below.

Finally, completely monotone functions and sequences play a fundamental role in the study of Archimedean and *eMO*-copulas, as the following two statements show. Regarding functions, it is shown in [15] that

$$\begin{aligned}&\text{Equation (16.1) defines a copula for all } d \geq 2\\ \Leftrightarrow\ &\text{the function } \psi \text{ is c.m. with } \psi(0) = 1 \text{ and } \lim_{x\to\infty} \psi(x) = 0.\end{aligned} \tag{16.7}$$

Regarding sequences, the *eMO*-counterpart of statement (16.7) is provided by [19, Corollary 3.4]:

$$\begin{aligned}&\text{Equation (16.4) defines a copula for all } d \geq 2\\ \Leftrightarrow\ &\text{the sequence } \{a_k\}_{k\in\mathbb{N}_0} \text{ is c.m. with } a_0 = 1.\end{aligned}$$

### 16.2.3 d-Monotonicity

A weaker notion, compared to complete monotonicity, is the closely related concept of $d$-monotonicity. For $d \geq 2$, a function $\psi : [0,\infty) \to [0,1]$ is called *$d$-monotone* if it is differentiable on $(0,\infty)$ up to the order $d-2$, if the derivatives satisfy

$$(-1)^j\, \psi^{(j)}(x) \geq 0, \quad \forall x > 0,\ j = 0,\ldots,d-2,$$

if further $(-1)^{d-2}\, \psi^{(d-2)}$ is non-increasing and convex on $(0,\infty)$, and if $\psi$ is continuous at zero, see [24, Definition 2.3].

Translating this concept to sequences, loosely speaking, one defines a finite sequence $(a_0,\ldots,a_{d-1})' \in \mathbb{R}^d$ of length $d$ to be $d$-monotone if $(-1)^j \Delta^j a_k \geq 0$ whenever this expression is well-defined. Note that the expression $(-1)^j \Delta^j a_k = \sum_{i=0}^{j} \binom{j}{i} (-1)^i a_{k+i}$ involves the numbers $a_k,\ldots,a_{k+j}$, which are not defined when $j+k > d-1$. More precisely, $(a_0,\ldots,a_{d-1})' \in \mathbb{R}^d$ is called *$d$-monotone* if it satisfies

$$(-1)^j \Delta^j a_k \geq 0, \quad k = 0,1,\ldots,d-1,\ j = 0,1,\ldots,d-k-1.$$

[2] $\{c_k\}_{k\in\mathbb{N}_0}$ is called *completely alternating* if $(-1)^j \Delta^j c_k \leq 0$ for all $k \in \mathbb{N}_0$, and for all $j \in \mathbb{N}$.

Obviously, a function is c.m. if and only if it is $d$-monotone for all $d \geq 2$. Moreover, the subsequence of the first $d$ elements of a c.m. sequence is $d$-monotone, and a sequence $\{a_k\}_{k \in \mathbb{N}_0}$ is c.m. if and only if the finite sequences $(a_0, \ldots, a_{d-1})'$ are $d$-monotone for all $d \geq 2$. In this regard, the concept of $d$-monotonicity is more general than that of complete monotonicity. A function that is $d$-monotone but not c.m. is said to be *proper d-monotone*. Similarly, a finite sequence $(a_0, \ldots, a_{d-1})'$ is called *proper d-monotone* if there is no $(d+1)$-monotone sequence $(b_0, \ldots, b_d)'$ satisfying $(a_0, \ldots, a_{d-1})' = (b_0, \ldots, b_{d-1})'$.

*Example 16.2.2 (Proper d-monotone sequences and functions).* A $d$-monotone function which is not c.m., i.e. which is not the Laplace transform of some positive random variable, is given by $\psi(x) = \max\{1-x, 0\}$. It is readily verified that $\psi$ is 2-monotone but not $d$-monotone for $d \geq 3$. In [24] it is carried out that $\psi$ is $d$-monotone with $\psi(0) = 1$ and $\lim_{x \to \infty} \psi(x) = 0$ if and only if there exists a positive random variable $R$ such that $\psi(x) = \mathbb{E}\big[\max\{1 - x/R, 0\}^{d-1}\big], x \geq 0$. This probabilistic interpretation of $d$-monotone functions is due to [29].

The sequence $(1, 1/2, 1/10)'$ is proper 3-monotone. Assume that $(b_0, \ldots, b_3)' = (1, 1/2, 1/10, b_3)'$ is 4-monotone. Then, it follows that

$$0 \leq \nabla^0 b_3 = b_3 = 1 - 3\frac{1}{2} + 3\frac{1}{10} - \nabla^3 b_0 \leq -\frac{1}{5},$$

which is a contradiction. Hence, $(1, 1/2, 1/10)'$ is a proper 3-monotone sequence. Unfortunately, a similar probabilistic interpretation as for $d$-monotone functions is not known, yet.

General characterizations of Archimedean and *eMO*-copulas in a fixed dimension $d \geq 2$ rely on the concept of $d$-monotonicity. On the one hand, [24, Theorem 3.1] shows that

$$\begin{aligned} &\text{Equation (16.1) defines a copula in dimension } d \geq 2 \\ &\Leftrightarrow \text{ the function } \psi \text{ is } d\text{-monotone with } \psi(0) = 1 \text{ and } \lim_{x \to \infty} \psi(x) = 0. \end{aligned} \tag{16.8}$$

On the other hand, [19, Theorem 3.1] states that

$$\begin{aligned} &\text{Equation (16.4) defines a copula in dimension } d \geq 2 \\ &\Leftrightarrow \text{ the sequence } (a_0, \ldots, a_{d-1})' \text{ is } d\text{-monotone with } a_0 = 1. \end{aligned} \tag{16.9}$$

Take notice of the fact that the, to some degree technical and not intuitive, definitions of $d$-monotonicity can be justified by statements (16.8) and (16.9). These show that the notion of $d$-monotonicity is of relevance in probability theory.

## 16.3 Probabilistic Models and Sampling

So far, both families of copulas have only been introduced analytically in (16.1) and (16.4). While this might be sufficient for the derivation of theoretical statements, a profound investigation and motivation of the dependence structure behind some copula should always include a probabilistic model. Besides, such a model has the advantage that the copula in concern can be sampled along the probabilistic construction.

For Archimedean and $eMO$-copulas it is appropriate to distinguish between c.m. and proper $d$-monotone functions, respectively sequences. As noted above, a proper $d$-monotone function (resp. sequence) is not $(d+1)$-monotone. Hence, it can only provide a copula, and therefore a corresponding random vector, in dimensions less than or equal to $d$. On the contrary, a c.m. function (resp. sequence) induces an infinite sequence of random variables, each $d$-dimensional subvector of which has the corresponding Archimedean (resp. $eMO$-) copula. In this case, according to De Finetti's theorem, see e.g. [2], the underlying probabilistic model relies on the technique of conditional independence. This means that it is possible to identify a common stochastic factor, conditioned on which the whole sequence is i.i.d.. This explains the separate treatment of the c.m. case in Sect. 16.3.1, whereas models for a fixed dimension $d$ are presented in Sect. 16.3.2.

### *16.3.1 The Completely Monotone Case*

On the one hand, the probabilistic model for Achimedean copulas with c.m. generator is due to [23]. Let $(\Omega, \mathscr{F}, \mathbb{P})$ be a probability space on which a positive random variable $W$ with Laplace transform $\psi$ and, independently, i.i.d. exponential random variables $E_1, E_2, \ldots$ with mean $\mathbb{E}[E_1] = 1$ are defined. An infinite sequence of random variables $\{\tau_k\}_{k\in\mathbb{N}}$ is given by

$$\tau_k := \frac{E_k}{W}, \quad k \in \mathbb{N}.$$

For each $d \geq 2$ the survival copula of the random vector $(\tau_1, \ldots, \tau_d)'$ is then given by Eq. (16.1). Moreover, the survival function $\bar{F}_k$ of $\tau_k$ equals $\psi$ for each $k$. In mathematical terms, with $C$ given by (16.1), it holds that

$$\mathbb{P}\big(\tau_1 > t_1, \ldots, \tau_d > t_d\big) = C\big(\psi(t_1), \ldots, \psi(t_d)\big), \quad t_1, \ldots, t_d \geq 0.$$

Notice that conditional on $W$, the random variables $\tau_k$ are i.i.d. exponential with mean $1/W$.

On the other hand, a model for copulas of the form (16.4) with a given c.m. sequence $\{a_k\}_{k\in\mathbb{N}_0}$ of parameters is developed in [21]. On a probability space $(\Omega, \mathscr{F}, \mathbb{P})$ let $E_1, E_2, \ldots$ be i.i.d. exponentially distributed with mean $\mathbb{E}[E_1] = 1$. It follows from the aforementioned correspondence (16.6) that there exists a (unique) killed Lévy subordinator $\Lambda = \{\Lambda_t\}_{t\geq 0}$ with Laplace exponent $\Phi$ satisfying $\Phi(k+1) - \Phi(k) = a_k$, $k \in \mathbb{N}_0$. Assuming that $\Lambda$ is defined on the given probability space independently of $E_1, E_2, \ldots$, one defines the sequence $\{\tau_k\}_{k\in\mathbb{N}}$ of random variables via

$$\tau_k := \inf\{t > 0 : \Lambda_t \geq E_k\}, \quad k \in \mathbb{N}.$$

Since the first-passage time of a killed Lévy subordinator across an independent exponential threshold level is again exponentially distributed, each $\tau_k$ is exponential with parameter $\Phi(1) = a_0 = 1$. Conditioned on the path of $\Lambda$, each $\tau_k$ has the survival function $t \mapsto \exp(-\Lambda_t)$, $t \geq 0$, since $\Lambda$ is non-decreasing. Furthermore, for $d \geq 2$ and $C$ given by (16.4) with $\{a_k\}_{k\in\mathbb{N}_0} = \{\Phi(k+1) - \Phi(k)\}_{k\in\mathbb{N}_0}$, it follows that

$$\mathbb{P}\big(\tau_1 > t_1, \ldots, \tau_d > t_d\big) = C\big(e^{-t_1}, \ldots, e^{-t_d}\big), \quad t_1, \ldots, t_d \geq 0.$$

This means that $C$ is the survival copula of $(\tau_1, \ldots, \tau_d)'$.

In both probabilistic models one can identify a $\sigma$-algebra $\mathcal{G} \subset \mathcal{F}$ such that conditioned on $\mathcal{G}$ the random variables $\{\tau_k\}_{k\in\mathbb{N}}$ are i.i.d.. In the Archimedean case $\mathcal{G} = \sigma(W)$, i.e. the common stochastic factor is the random variable $W$. In the *eMO* case $\mathcal{G} = \sigma(\Lambda_t : t \geq 0)$, i.e. the common stochastic factor is the stochastic process $\Lambda$. Considering applications, this conditional independence approach allows for an efficient pricing approach for portfolio credit derivatives in both cases. More precisely, by integrating out the common factor one can approximate the loss distribution of a large homogeneous credit portfolio, see [21, 28].

## *16.3.2 The Proper d-Monotone Case*

The following probabilistic models reveal that the aforementioned similarities between the copula families in concern are quite surprising. Both models are very different in nature, which in particular results in quite dissimilar distributional properties, as already indicated in the introduction. Figure 16.1 illustrates the distributional properties via three-dimensional scatter-plots.

Following [24], to construct a $d$-dimensional random vector with the Archimedean survival copula (16.1) when the generator $\psi$ is proper $d$-monotone, one may proceed as follows. On a probability space $(\Omega, \mathcal{F}, \mathbb{P})$ let $E_1, \ldots, E_d$ be i.i.d. exponential random variables with mean $\mathbb{E}[E_1] = 1$. By [29], there exists a positive random variable $R > 0$ such that $\psi(x) = \mathbb{E}[\max\{1 - x/R, 0\}^{d-1}]$, $x \geq 0$. Assuming that $R$ is defined on $(\Omega, \mathcal{F}, \mathbb{P})$, independently of $E_1, \ldots, E_d$, it holds that the random vector $(\tau_1, \ldots, \tau_d)'$ given by

$$\tau_k := R\,\frac{E_k}{E_1 + \ldots + E_d}, \quad k = 1, \ldots, d,$$

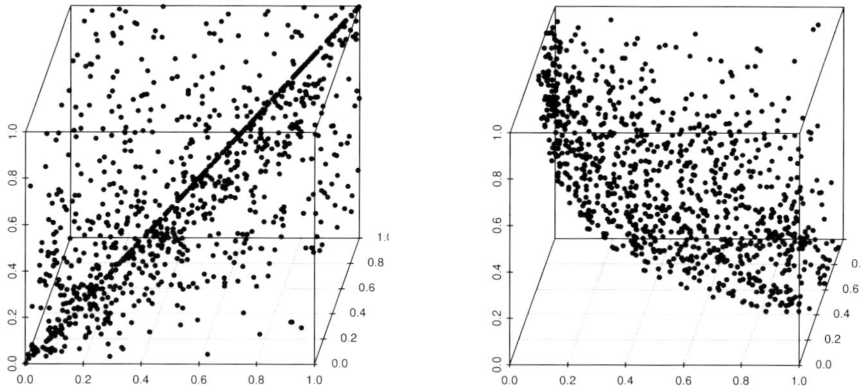

**Fig. 16.1** *Left*: Three-dimensional scatter-plot of 1,000 samples from an *eMO*-copula of the form (16.4) with parameters $(a_0,a_1,a_2) = (1,2/5,1/10)$. The singularity on the diagonal is easy to observe. *Right*: Three-dimensional scatter-plot of 1,000 samples from an Archimedean copula of the form (16.1) with generator $\psi(x) = \big(1-x^{\theta}-2x\,\theta/(\theta-1)\,(1-x^{\theta-1})+x^2\,\theta/(\theta-2)\,(1-x^{\theta-2})\big)\,\mathbf{1}_{\{x<1\}}$ for $\theta = 3.5$. This generator corresponds to the radius $R := U^{1/\theta}$ with $U \sim U[0,1]$.

has survival copula (16.1) and each $\tau_k$ has survival function $\psi$. Notice that the random vector constructed in this way equals a uniform distribution on the $d$-dimensional unit simplex, which is scaled by an independent radial variable $R > 0$. Conditioned on the factor $R$, the induced survival copula equals the Archimedean copula with generator $\psi(x) = \max\{1-x,0\}^{d-1}$, being a pointwise lower bound for all Archimedean copulas, see [24, Proposition 4.6]. In this regard, instead of conditional independence (as in the case of a c.m. generator) one could speak of "conditional countermonotonicity". Opposed to the constructions in the c.m. case, the present construction might not be extended to dimension $d+1$ (with proper $d$-monotone $\psi$).

Given a proper $d$-monotone sequence $(a_0,\dots,a_{d-1})'$ with $a_0 = 1$, in order to construct a random vector with *eMO*-survival copula (16.4) it is not possible to apply the aforementioned construction via Lévy subordinators. Still, the original approach of [22] provides a probabilistic model. On a probability space $(\Omega,\mathscr{F},\mathbb{P})$ let $\bigcup_{j=1}^{d}\{E_{i_1,\dots,i_j}\,|\,1 \le i_1 < \dots < i_j \le d\}$ be a collection of $2^d-1$ independent exponential random variables. For each $j = 1,\dots,d$, the parameter $\lambda_j$ of all exponential random variables in the set $\{E_{i_1,\dots,i_j}\,|\,1 \le i_1 < \dots < i_j \le d\}$ is supposed to be given by $\lambda_j := (-1)^{j-1}\,\Delta^{j-1}a_{d-j} \ge 0$. Notice that $a_0 = 1$ and $d$-monotonicity of $(a_0,\dots,a_{d-1})'$ guarantee that at least one $\lambda_j$ is strictly positive. If some $\lambda_j$ is zero, this means that the corresponding random variable $E_{i_1,\dots,i_j} \equiv \infty$ is almost surely infinite (and, hence, not exponentially distributed in this special case).[3] It then follows that the random vector $(\tau_1,\dots,\tau_d)'$ with

[3] One can check that at least one $\lambda_j$ is strictly positive, ensuring that all $\tau_k$ are well-defined.

$$\tau_k := \min \bigcup_{j=1}^{d} \left\{ E_{i_1,\ldots,i_j} \middle| k \in \{i_1,\ldots,i_j\} \right\}, \quad k = 1,\ldots,d,$$

has survival copula (16.4) and each $\tau_k$ is exponentially distributed with parameter $\sum_{j=1}^{d} \binom{d-1}{j-1} \lambda_j$. Intuitively, this construction of [22] is motivated by a system of initially functional components, which are destroyed by exogenous shocks arriving at the times $E_{i_1,\ldots,i_j}$ and killing the components $i_1,\ldots,i_j$. Finally, it is important to note that this original construction for *eMO*-copulas relies on $2^d - 1$ independent exponential random variables and becomes inefficient for $d \gg 2$ when used for sampling.

## References

1. Applebaum, D.: Lévy Processes and Stochastic Calculus. Cambridge University Press, Cambridge (2004)
2. Aldous, D.J.: Exchangeability and Related Topics. École d'Été de Probabilités de Saint-Flour XIII-1983. Lecture Notes in Mathematics, vol. 1117, pp. 1–198. Springer, Berlin (1985)
3. Bernstein, S.: Sur les fonctions absolument monotones. Acta Mathematica **52**(1), 1–66 (1929)
4. Bertoin, J.: Lévy Processes. Cambridge University Press, Cambridge (1996)
5. Bertoin, J.: Subordinators: Examples and Applications. École d'Été de Probabilités de Saint-Flour XXVII-1997. Lecture Notes in Mathematics, vol. 1717, pp. 1–91. Springer, Berlin (1999)
6. Embrechts, P., Lindskog, F., McNeil, A.J.: Modelling dependence with copulas and applications to risk management. In: Rachev, S. (ed.) Handbook of Heavy Tailed Distributions in Finance, pp. 329–384. Elsevier/North-Holland, Amsterdam (2003)
7. Feller, W.: An Introduction to Probability Theory and Its Applications, vol. II, 2nd edn. Wiley, New York, NY (1966).
8. Galambos, J., Kotz, S.: Characterizations of Probability Distributions. Lecture Notes in Mathematics, vol. 675. Springer, Heidelberg (1978)
9. Genest, C., Rivest, L.-P.: A characterization of Gumbel's family of extreme value distributions. Stat. Probab. Lett. **8**(3), 207–211 (1989)
10. Gnedin, A., Pitman, J.: Moments of convex distribution functions and completely alternating sequences. Probab. Stat. Essays in Honor of David A. Freedman **2**, 30–41 (2008)
11. Hausdorff, F.: Summationsmethoden und Momentfolgen I. Mathematische Zeitschrift **9**(3–4), 74–109 (1921)
12. Hausdorff, F.: Momentenproblem für ein endliches Intervall. Mathematische Zeitschrift **16**, 220–248 (1923)
13. Hofert, M.: Sampling Archimedean copulas. Comput. Stat. Data Anal. **52**(12), 5163–5174 (2008)
14. Joe, H.: Multivariate Models and Dependence Concepts. Chapman and Hall/CRC, London (1997)
15. Kimberling, C.H.: A probabilistic interpretation of complete monotonicity. Aequationes Mathematicae **10**, 152–164 (1974)
16. Li, H.: Tail dependence comparison of survival Marshall-Olkin copulas. Methodol. Comput. Appl. Probab. **10**(1), 39–54 (2008)
17. Li, H.: Orthant tail dependence of multivariate extreme value distributions. J. Multivar. Anal. **100**(1), 243–256 (2009)
18. Lorch, L., Newman, D.J.: On the composition of completely monotonic functions and completely monotonic sequences and related questions. J. Lond. Math. Soc. **s2–28**(1), 31–45 (1983)

19. Mai, J.-F., Scherer, M.: Lévy-frailty copulas. J. Multivar. Anal. **100**(7), 1567–1585 (2009)
20. Mai, J.-F., Scherer, M.: Reparameterizing Marshall-Olkin copulas with applications to sampling. J. Stat. Comput. Simul. In press (2010)
21. Mai, J.-F., Scherer, M.: A tractable multivariate default model based on a stochastic timechange. Int. J. Theor. Appl. Finance **12**(2), 227–249 (2009)
22. Marshall, A.W., Olkin, I.: A multivariate exponential distribution. J. Am. Stat. Assoc. **62**(317), 30–44 (1967)
23. Marshall, A.W., Olkin, I.: Families of multivariate distributions. J. Am. Stat. Assoc. **83**(403), 834–841 (1988)
24. McNeil, A.J., Nešlehová, J.: Multivariate Archimedean copulas, *d*-monotone functions and $l_1$-norm symmetric distributions. Ann. Stat. **37**(5B), 3059–3097 (2009)
25. Müller, A., Scarsini, M.: Archimedean copulae and positive dependence. J. Multivar. Anal. **93**(2), 434–445 (2005)
26. Nelsen, R.B.: An introduction to copulas. Springer, New York, NY (1999)
27. Sato, K.-I.: Lévy Processes and Infinitely Divisible Distributions. Cambridge University Press, Cambridge (1999)
28. Schönbucher, P.J.: Taken to the limit: simple and not-so-simple loan loss distributions. Working paper, retrievable from http://www.gloriamundi.org/picsresources/pjs.pdf (2002)
29. Williamson, R.E.: Multiply monotone functions and their Laplace transforms. Duke Math. J. **23**(2), 189–207 (1956)

# Chapter 17
# Relationships Between Archimedean Copulas and Morgenstern Utility Functions

Jaap Spreeuw

**Abstract** The (additive) generator of an Archimedean copula is a strictly decreasing and convex function, while Morgenstern utility functions (applying to risk aversion decision makers) are nondecreasing and concave. In this presentation, relationships between generators and utility functions are established. For some well known Archimedean copula families, links between the generator and the corresponding utility function are demonstrated. Some new copula families are derived from classes of utility functions which appeared in the literature, and their properties are discussed. It is shown how dependence properties of an Archimedean copula translate into properties of the utility function from which they are constructed.

## 17.1 Introduction

Archimedean copulas are constructed using a one-dimensional function, the generator, which is nonincreasing and convex. Von Neumann-Morgenstern utility functions, on the other hand, are nondecreasing (decision makers prefer more to less) and concave (decision makers are risk averse). Therefore, an affine transformation of a utility function, with sign changed, could act as a generator for an Archimedean copula, subject to some additional conditions. Applying this methodology can lead to copula families that are either new or well known.

This chapter examines relationships between (generators of) Archimedean copulas and Von Neumann-Morgenstern utility functions. In particular, it will be shown how properties of a utility function translate into the type of dependence induced by the Archimedean copula generated from it. For the sake of brevity, we will confine ourselves to notions of positive dependence only.

---

Jaap Spreeuw
Cass Business School, City University London, London, UK
e-mail: `J.Spreeuw@city.ac.uk`

P. Jaworski et al. (eds.), *Copula Theory and Its Applications*, Lecture Notes in Statistics 198,
DOI 10.1007/978-3-642-12465-5_17, 

Section 17.2 gives a brief definition of generators of Archimedean copulas, while Sect. 17.3 elaborates on the aforementioned method of obtaining generators from utility functions.

Avérous and Dortet-Bernadet [2] derive relationships between dependence properties of Archimedean copulas and aging properties of their generator. In Sect. 17.4, links between copula and utility function are exhibited. Section 17.5 considers several utility functions which appeared in the literature as examples. Particular attention will be devoted to utility functions with Decreasing Absolute Risk Aversion (DARA), a class that is widely applied in economics and decision theory. Conclusions are presented in Sect. 17.6.

## 17.2 Archimedean Copulas

We define $C(\cdot,\cdot)$ to be a two dimensional copula. An Archimedean copula can be specified as:

$$C_{\varphi}(v_1,v_2)=\varphi\left(\varphi^{[-1]}(v_1)+\varphi^{[-1]}(v_2)\right); \qquad 0\leq v_1,v_2\leq 1, \tag{17.1}$$

with $\varphi$ nonincreasing and convex, $\varphi(0)=1$ and $\varphi(s)=0$ for $s\geq s^*$ for some nonnegative $s^*$. The generator is strict if $\lim_{s\to\infty}\varphi(s)=0$ (so $s^*=\infty$), and non-strict if $s^*$ is finite. The function $\varphi^{[-1]}$ is defined as the generalized inverse of $\varphi$:

$$\varphi^{[-1]}(s)=\begin{cases}\varphi^{-1}(s) & \text{for } 0<s\leq 1\\ s^* & \text{for } \quad s=0\end{cases}.$$

*Remark 17.2.1.* In certain literature about Archimedean copulas, the generator is defined in terms of $\varphi^{[-1]}$ rather than $\varphi$ ("inverse operator inside, rather than outside, the brackets"). However, we prefer the notation above, as it leads to somewhat simpler expressions. So for instance, the function $s\mapsto\exp[-s]$ is used as generator for the independence copula, rather than $s\mapsto-\log[s]$.

*Remark 17.2.2.* The generator, as specified in this chapter, is invariant to multiplication of the argument by a positive constant. For $\varepsilon>0$, $\varphi(s)$ and $\varphi(\varepsilon s)$ lead to the same copula. This often leads to simplifications.

## 17.3 Utility Functions

A utility function $\psi: I\to\mathbb{R}$, with $I$ being a subset of $\mathbb{R}$, is of a von Neumann-Morgenstern type if it is nondecreasing and concave. In this paper we will assume $\psi'(s)>0$ for $s\in I$. Hence, the function $-\psi:\mathbb{R}\to\mathbb{R}$ is strictly decreasing and convex. This does not mean that $-\psi$ could serve as a generator of an Archimedean copula, since in general the combination of additional requirements $\psi(0) = -1$ and $\lim_{s\to\infty} \psi(s) \geq 0$ is not satisfied.

However, generators of Archimedean copulas can be constructed from affine transformations of utility functions. An important measure for risk perception in utility theory is the degree of absolute risk aversion, defined by Pratt [9] as

$$r_{\psi}(s) = -\frac{\psi''(s)}{\psi'(s)} \geq 0, \quad s \in \mathbb{R}. \tag{17.2}$$

(The subscript $\psi$ in $r_{\psi}$indicates that the degree of absolute risk aversion is related to the utility function). We define $u(s) = \alpha + \beta\psi(s)$, with $\alpha$ and $\beta$ real, $\beta > 0$. It is easy to verify that $r_u(s) = r_{\psi}(s)$, so risk perception is invariant up to an affine transformation. Then for $s \geq 0$, $\max[-u(s), 0]$ could serve as a generator of an Archimedean copula, provided that: (a) $u(0) = -1$; (b) $\lim_{s\longrightarrow\infty} u(s) \geq 0$. Applying the first condition gives $\alpha = -1 - \beta\psi(0)$. The corresponding generator will then be

$$\varphi(s) = \max[1 + \beta(\psi(0) - \psi(s)), 0], \quad s \geq 0. \tag{17.3}$$

A necessary condition for strictness of the generator is that $\lim_{s\to\infty} \psi(s) = \psi(\infty) < \infty$. Then satisfaction of the second condition requires $\beta \geq (\lim_{s\to\infty} \psi(s) - \psi(0))^{-1}$, and a strict generator is obtained for $\beta = (\lim_{s\to\infty} \psi(s) - \psi(0))^{-1}$, reducing (17.3) to

$$\varphi(s) = \frac{\psi(\infty) - \psi(s)}{\psi(\infty) - \psi(0)}, \quad s \geq 0. \tag{17.4}$$

In all other cases, and also in all cases with $\lim_{s\to\infty} \psi(s) = \infty$, the generator is not strict. The inverse of the generator is

$$\varphi^{[-1]}(s) = \psi^{-1}\left(\psi(0) + \frac{1-s}{\beta}\right).$$

*Remark 17.3.1.* Obviously, we can only derive generators in this way for $\psi(0)$ well defined and finite. This condition is e.g. not met for the widely applied utility functions $\psi(s) = \log s$, and $\psi(s) = -s^{1-\gamma}$ with $\gamma > 1$.

Three observations can be made regarding the role of the parameter $\beta$:

1. As defined in Nelsen [8], the upper tail dependence coefficient, denoted by $\lambda_u$, is

$$\lambda_u = 2 - \lim_{s\downarrow 0} \frac{1 - \varphi(2s)}{1 - \varphi(s)} = 2 - \lim_{s\downarrow 0} \frac{\psi(2s) - \psi(0)}{\psi(s) - \psi(0)},$$

   implying that $\lambda_u$ does not depend on $\beta$.
2. According to Nelsen [8], a sufficient condition for the copula generated by (17.3) to be negatively ordered in terms of $\beta$ (in the sense that $C_{\beta_1} \leq C_{\beta_2}$ for $\beta_1 > \beta_2$, where $C_{\beta_i}$ indicates the copula with parameter $\beta_i$, $i \in \{1, 2\}$) is that $\left(\varphi_{\beta_1}^{[-1]}\right)'(s) \Big/ \left(\varphi_{\beta 2}^{[-1]}\right)'(s)$ is nondecreasing for $s \in (0, 1)$ (here $\varphi_{\beta_i}^{[-1]}(s)$ indicates the inverse generator with parameter $\beta_i$, $i \in \{1, 2\}$). We have that

$$\frac{\partial \left(\varphi_{\beta_1}^{[-1]}\right)'(s) \Big/ \left(\varphi_{\beta 2}^{[-1]}\right)'(s)}{\partial s}$$
$$= \left(\psi^{-1}\right)'\left(\psi(0)+\frac{1-s}{\beta_1}\right)\left(\psi^{-1}\right)'\left(\psi(0)+\frac{1-s}{\beta_2}\right)$$
$$\left(\frac{r^{-1}\left(\psi(0)+\frac{1-s}{\beta_2}\right)}{\beta_2}-\frac{r^{-1}\left(\psi(0)+\frac{1-s}{\beta_1}\right)}{\beta_1}\right), \tag{17.5}$$

defining

$$r^{-1}(x)=\frac{\left(\psi^{-1}\right)''(x)}{\left(\psi^{-1}\right)'(x)}, \quad x \in (\psi(0), \psi(\infty)).$$

Note that $r^{-1}(x) \geq 0$, since $\left(\psi^{-1}\right)'(x) \geq 0$ and $\left(\psi^{-1}\right)''(x) \geq 0$ for all $x \in (\psi(0), \psi(\infty))$ (the inverse of a nondecreasing and concave function is nondecreasing and convex). Hence, (17.5) is nonnegative for $r^{-1}\left(\psi(0)+\frac{1-s}{\beta}\right)\Big/\beta$ decreasing in $\beta$. It will transpire that all the copulas obtained from the generators derived as above in this paper are negatively ordered in $\beta$.

3. Assuming that the assumption of $r^{-1}\left(\psi(0)+\frac{1-s}{\beta}\right)\Big/\beta$ decreasing in $\beta$ (as in Observation 2 above) holds, Theorem 4.4.7 of Nelsen [8] can be applied to find out if this family includes $W$ (Fréchet-Höffding's lower bound) as a limiting member for $\beta \to \infty$. Applying this Theorem 4.4.7 leads to

$$\lim_{\beta\to\infty}\frac{\varphi^{[-1]}(s)}{\left(\varphi^{[-1]}\right)'(t)}=\lim_{\beta\to\infty}\frac{\psi^{-1}\left(\psi(0)+\frac{1-s}{\beta}\right)}{\left(\psi^{-1}\right)'\left(\psi(0)+\frac{1-t}{\beta}\right)\beta^{-1}}.$$

Using de l'Hopital's rule gives

$$\lim_{\beta\to\infty}\frac{\varphi^{[-1]}(s)}{\left(\varphi^{[-1]}\right)'(t)}$$
$$=\lim_{\beta\to\infty}\frac{\left(\psi^{-1}\right)'\left(\psi(0)+\frac{1-s}{\beta}\right)(s-1)\beta^{-2}}{\left(\psi^{-1}\right)'\left(\psi(0)+\frac{1-t}{\beta}\right)\beta^{-2}+\left(\psi^{-1}\right)''\left(\psi(0)+\frac{1-t}{\beta}\right)(1-t)\beta^{-3}}$$
$$=(s-1)\lim_{\beta\to\infty}\frac{\left(\psi^{-1}\right)'\left(\psi(0)+\frac{1-s}{\beta}\right)}{\left(\psi^{-1}\right)'\left(\psi(0)+\frac{1-t}{\beta}\right)}\frac{1}{1+r^{-1}\left(\psi(0)+\frac{1-t}{\beta}\right)(1-t)\beta^{-1}}.$$

For $\left(\psi^{-1}\right)'(\psi(0)) \neq 0$, this limit equals $s-1$, which means that $W$ is then obtained as a limiting member for $\beta \to \infty$.

## 17.4 Relationships Between Properties of Utility Functions and Properties of Generators

Using concepts from reliability theory, Avérous and Dortet-Bernadet [2] derive several relationships between type of dependence of a copula, and aging properties of the generator, which is in fact a survival function. Given the expressions (17.3) and (17.4), these aging characteristics translate into properties of the corresponding utility function. In this section, links between type of dependence of copulas and behavior of corresponding utility functions will be investigated.

For the sake of brevity, we will restrict ourselves to concepts of positive dependence. All notions of positive dependence that appeared in the literature, including the weakest one of Positive Quadrant Dependence (PQD) as defined by Lehmann [6], require the generator to be strict.

For this reason we will focus on strict generators. It should be stated that most applications in the literature are based on copulas with a strict generator. A non strict generator implies that $C(u_1, u_2) = 0$ for some $u_1, u_2 > 0$. It can sometimes be hard to justify that two events have a nonzero chance of happening individually, but cannot happen jointly. Furthermore, applying the pseudomaximum likelihood method as in Genest et al. [4] requires the copula to be absolutely continuous, which is implied by a strict generator.

In the sequel, we consider two continuous random variables $X$ and $Y$, and either an Archimedean distribution copula $C_\varphi$ with generator $\varphi$ defined in (17.1) such that

$$\Pr[X \leq x, Y \leq y] = C_\varphi(\Pr[X \leq x], \Pr[Y \leq y]),$$

or an Archimedean survival copula defined as $\widehat{C}_{\widehat{\varphi}}$ with generator $\widehat{\varphi}$ such that

$$\Pr[X > x, Y > y] = \widehat{C}_{\widehat{\varphi}}(\Pr[X > x], \Pr[Y > y]).$$

The notation $\widehat{\psi}$ refers to the Morgenstern utility function, from which the generator $\widehat{\varphi}$ is constructed, just as in (17.4).

Apart from PQD, we will consider SI (Stochastically Increasing) (also from Lehmann [6]) and both LTD (Left Tail Decreasing) and RTI (Right Tail Increasing) (from Esary and Proschan [3]) as notions of dependence. The definitions are as below:

**Definition 17.4.1.** $(X,Y)$ is PQD $\Leftrightarrow \Pr[X \leq x, Y \leq y] \geq \Pr[X \leq x]\Pr[Y \leq y]$

**Definition 17.4.2.** $Y$ is LTD in $X \Leftrightarrow \Pr[Y \leq y | X \leq x]$ is nonincreasing in $x$ for all $y$.

**Definition 17.4.3.** $Y$ is RTI in $X \Leftrightarrow \Pr[Y > y | X > x]$ is nondecreasing in $x$ for all $y$.

**Definition 17.4.4.** $Y$ is SI in $X \Leftrightarrow \Pr[Y \leq y | X = x]$ is nonincreasing in $x$ for all $y$.

As pointed out in Avérous and Dortet-Bernadet [2], SI implies LTD and RTI, each of which in turn imply PQD. This can also be shown by using (conditional) hazard functions. Assuming $X$ and $Y$ to be continuous random variables, we define the unconditional hazard functions $\mu_Y$ in the usual way

$$\mu_X(x) = \frac{\frac{\partial}{\partial x}\Pr[X \leq x]}{\Pr[X > x]}; \quad \mu_Y(y) = \frac{\frac{\partial}{\partial y}\Pr[Y \leq y]}{\Pr[Y > y]}.$$

Furthermore, we define some conditional hazard functions. For instance, we define

$$\mu_Y(y\,|X = x) = \frac{\frac{\partial}{\partial y}\Pr[Y \leq y\,|X = x]}{\Pr[Y > y\,|X = x]}, \tag{17.6}$$

as the conditional hazard function of $Y$ at $y$ given $X = x$. In a similar way, we define the conditional hazard functions $X$ at $x$ given $Y = y$ as

$$\mu_X(x\,|Y = y) = \frac{\frac{\partial}{\partial x}\Pr[X \leq x\,|Y = y]}{\Pr[X > x\,|Y = y]}.$$

Likewise, we define the conditional hazard functions $\mu_Y(y\,|X \leq x)$ and $\mu_Y(y\,|X = x)$ and so on. This leads to the following propositions, proofs of which are straightforward. Note that $X$ and $Y$ can be interchanged.

**Proposition 17.4.1.** *$(X,Y)$ is PQD $\Leftrightarrow \mu_Y(y\,|X > x) \leq \mu_Y(y) \leq \mu_Y(y\,|X \leq x)$ for all $x$ and $y$.*

**Proposition 17.4.2.** *$Y$ is LTD in $X \Leftrightarrow \mu_Y(y\,|X \leq x) \geq \mu_Y(y\,|X = s)$ for all $x$ and $s$ with $x < s$.*

**Proposition 17.4.3.** *$Y$ is RTI in $X \Leftrightarrow \mu_Y(y\,|X > x) \leq \mu_Y(y\,|X = s)$ for all $x$ and $s$ with $s < x$.*

**Proposition 17.4.4.** *$Y$ is SI in $X \Leftrightarrow \mu_Y(y\,|X = x)$ is nonincreasing in $x$ for all $y$.*

The representation in terms of hazard functions also shows that SI is closely related to the notion of *long-term dependence* as defined in Hougaard [5].

**Definition 17.4.5.** Let $X$ and $Y$ be continuous random variables representing lifetimes. Then $X$ and $Y$ exhibit *long-term dependence* if $\mu_X(x\,|Y = y)$ is constant or decreasing as a function of $y \in [0,x]$ (or alternatively, if $\mu_Y(y\,|X = x)$ is constant or decreasing as a function of $x \in [0,y]$).

When comparing definitions, one sees that SI requires the conditional hazard function $\mu_Y(y\,|X = x)$ to be nonincreasing also for $x > y$. This condition is not required for long-term dependence.

*Remark 17.4.1.* Whether or not long-term dependence is a desirable feature in a model is a different question. As discussed in Spreeuw [11], long-term dependence seems to be a realistic assumption in many applications of reliability theory. For coupled lives, on the other hand, the presumption of long-term dependence seems dubious. The "broken heart syndrome", experienced in some empirical studies, indicates that bereaved lives whose partner died recently have a higher mortality than those who lost their partner years ago.

The following proposition shows the connection between the dependence properties of either the distribution copula $C_\varphi$ or the survival copula $\widehat{C}_{\widehat{\varphi}}$, and the risk perception properties of the utility functions $\psi$ and $\widehat{\psi}$, respectively.

**Proposition 17.4.5.**
*i) $C_\varphi$ or $\widehat{C}_{\widehat{\varphi}}$ is PQD*
$\Longleftrightarrow (\psi(\infty)-\psi(s))(\psi(\infty)-\psi(t)) \le (\psi(\infty)-\psi(s+t))(\psi(\infty)-\psi(0))$
*or* $(\widehat{\psi}(\infty)-\widehat{\psi}(s))(\widehat{\psi}(\infty)-\widehat{\psi}(t)) \le (\widehat{\psi}(\infty)-\widehat{\psi}(s+t))(\widehat{\psi}(\infty)-\widehat{\psi}(0))$*, respectively;*
*ii) $C_\varphi$ is LTD*$\Longleftrightarrow \log[\psi(\infty)-\psi(s)]$ *is convex in s;*
*iii) $\widehat{C}_{\widehat{\varphi}}$ is RTI*$\Longleftrightarrow \log[\widehat{\psi}(\infty)-\widehat{\psi}(s)]$ *is convex in s;*
*iv) $C_\varphi$ or $\widehat{C}_{\widehat{\varphi}}$ is SI* $\Longleftrightarrow r_\psi(s) = -\frac{\psi''(s)}{\psi'(s)}$ *is nonincreasing in s or* $r_{\widehat{\psi}}(s) = -\frac{\widehat{\psi}''(s)}{\widehat{\psi}'(s)}$ *is nonincreasing in s, respectively.*

*Proof.* i) and iv) Follows from the proofs of Proposition 1 in Avérous and Dortet-Bernadet [2], in connection with Eq. (17.4).

ii) Follows from the proof of Proposition 3 in Avérous and Dortet-Bernadet [2], in connection with Eq. (17.4).

iii) Observe that $\widehat{C}_{\widehat{\varphi}}$ has the RTI property if and only if

$$\frac{\widehat{\varphi}\left(\widehat{\varphi}^{[-1]}(s)+\widehat{\varphi}^{[-1]}(t)\right)}{s} \ge \frac{\widehat{\varphi}\left(\widehat{\varphi}^{[-1]}(s')+\widehat{\varphi}^{[-1]}(t)\right)}{s'} \qquad \forall 0<t<1;\ \forall 0<s<s'<1.$$

Following the proof of Proposition 3 in Avérous and Dortet-Bernadet [2], it follows that $-\log\widehat{\varphi}(s)$ is concave in $s$, and hence $\log[\widehat{\psi}(\infty)-\widehat{\psi}(s)]$ is convex in $s$.

*Remark 17.4.2.* For survival copulas (of the Archimedean type) Spreeuw [11] shows that long term dependence is equivalent to $r_{\widehat{\psi}}(s)$ nonincreasing in $s$.

There seems to be general consensus in the economic literature that the coefficient of absolute risk aversion should be decreasing (or at least nonincreasing) in terms of wealth. Arguments in favor of this property were already given in Arrow [1] and Pratt [9]. For this reason, most utility functions share the property of Decreasing Absolute Risk Aversion (DARA). This means that several utility functions can be used to construct copulas with the SI property, provided that $\psi(\infty)$ is finite.

Most examples of utility functions as given in the next section, do feature DARA. As we shall see, the generators constructed from some utility functions belong to well established copula families, but new generators do arise as well.

## 17.5 Examples

### 17.5.1 Classical Cases

#### 17.5.1.1 Linear Utility

Linear utility is equivalent to risk neutrality. The corresponding generator of an Archimedean copula is $\varphi(s) = \max[1-\beta s, 0]$ reducing to $\max[1-s, 0]$, (since, as stated above, a generator determines a copula, up to a constant positive factor) being the generator of the Fréchet-Höffding lower bound copula $C(v_1, v_2) = \max[v_1 + v_2 - 1, 0]$. Linear utility corresponds to $r(s) \equiv 0$.

#### 17.5.1.2 Constant Absolute Risk Aversion (CARA)

CARA functions correspond to $\psi(s) = -\exp[-\gamma s], \quad \gamma > 0$. They derive their name from $r(s) \equiv \gamma$, being independent of $s$. The corresponding generator of an Archimedean copula is therefore $\varphi(s) = \max[1-\beta(1-\exp[-s]), 0]$, requiring $\beta \geq 1$, generating the copula $C(v_1, v_2) = \max\left[\frac{1}{\beta}v_1 v_2 + \left(1-\frac{1}{\beta}\right)(v_1 + v_2 - 1), 0\right]$ which is Family 4.2.7 of Table 4.1 as in Nelsen [8]. The family is negatively ordered in $\beta$. The generator is strict only if $\beta = 1$, giving the independence copula.

#### 17.5.1.3 Constant Relative Risk Aversion (CRRA)

As stated in Pratt [9], there are three cases of utility functions with CRRA, i.e. $s\,r(s)$ is constant (and therefore DARA):

$$\psi(s) = \begin{cases} s^{1-\gamma} & \text{if } 0 < \gamma = s\,r(s) < 1 \\ \log s & \text{if } \quad s\,r(s) = 1 \\ -s^{-(\gamma-1)} & \text{if } \gamma = s\,r(s) > 1. \end{cases}.$$

CRRA utility functions are widely applied in the economic literature. But in spite of the DARA property, one cannot derive a generator of a copula that is SI (or features a weaker type of positive dependence). It is only in the first case that the utility function $\psi(s)$ is well defined for $s = 0$. This gives $\varphi(s) = \max\left[1-s^{1-\gamma}, 0\right]$, which is Family (4.2.2) of Table 4.1 as in Nelsen [8]. This generator is not strict, since $\varphi(1) = 0$.

### 17.5.2 The HARA Family

This family (Hyperbolic Absolute Risk Aversion), which contains several utility functions discussed above as special cases, has been introduced in Merton [7]. It is specified as:

$$\psi(s) = \frac{1-\gamma}{\gamma}\left(\frac{s}{1-\gamma}+\varepsilon\right)^{\gamma}; \quad \gamma \notin \{0,1\}; \frac{s}{1-\gamma}+\varepsilon > 0; \ \varepsilon = 1 \text{ if } \gamma = -\infty.$$

This utility function has risk aversion coefficient $r(s) = \left(\frac{s}{1-\gamma}+\varepsilon\right)^{-1}$. Given that the utility function must be well-defined for $s = 0$, $\varepsilon \geq 0$ is required, with strict inequality for $\gamma < 0$ or $\gamma > 1$. For $\varepsilon = 0$ (requiring $0 < \gamma < 1$) we get the CRRA case already seen before. Otherwise, the generator $\varphi(s) = \max\left[1+\beta\frac{1-\gamma}{\gamma}\left(1-\left(\frac{s}{1-\gamma}+1\right)^{\gamma}\right), 0\right]$ is obtained. The case $\gamma \to 1$ involves linear utility (and hence $W$), while $\gamma \to \pm\infty$ leads to CARA (and therefore Family 4.2.7 of Table 4.1 as in Nelsen [8]). Strictness of the generator requires $\gamma < 0$, implying DARA, leading to $\varphi(s) = (s+1)^{\gamma}$ being the generator of the Clayton copula, which is a standard example of an SI copula.

### 17.5.3 The Expo Power Utility

This family has been introduced in Saha [10]. A closely related family (which is essentially the same) is considered in Xie [12]. The expo-power utility function is as given below:

$$\psi(s) = \alpha - \exp[-\delta s^{\gamma}]; \quad \alpha > 1; 0 < \gamma \leq 1 \quad \text{and } \delta > 0.$$

We have $r(s) = \frac{1-\alpha+\alpha\delta s^{\gamma}}{s}$, which is decreasing in $s$. We obtain

$$\varphi(s) = \max[1-\beta(1-\exp[-s^{\gamma}]), 0].$$

The family is negatively ordered in both $\gamma$ and $\beta$. For $\beta \to \infty$, the generator reduces to the one for Family (4.2.2) of Table 4.1 as in Nelsen [8] (see above, for CRRA). The generator is strict for $\beta = 1$, leading to the Gumbel-Hougaard copula, which is also a standard SI case.

### 17.5.4 Other Examples of Decreasing Absolute Risk Aversion (DARA) as in Pratt [9]

Pratt developed a few more examples, that will be discussed below.

#### 17.5.4.1 Equation (37a) of Pratt

This concerns the case

$$\psi(s) = \arctan[s+\delta]; \quad \delta \geq 0.$$

Actually, Pratt imposed the restriction $\delta \geq 1$ to ensure that the utility function features DARA, but a valid generator is obtained for $0 \leq \delta < 1$ as well. We get as generator $\varphi(s) = \max[1+\beta(\arctan[\delta] - \arctan[s+\delta]), 0]$, $s \geq 0$, requiring $\beta \geq \frac{1}{\frac{1}{2}\pi - \arctan\delta}$. Like all other types considered in this note, this family is negatively ordered in $\beta$, with $W$ attained for $\beta \to \infty$. The generator is strict if $\beta = \frac{1}{\frac{1}{2}\pi - \arctan\delta}$, reducing the generator to $\varphi(s) = \frac{\frac{1}{2}\pi - \arctan[s+\delta]}{\frac{1}{2}\pi - \arctan\delta}$. Kendall's tau is in that case $\frac{4\delta(\pi - 2\arctan[\delta]) - 2}{(\pi - 2\arctan[\delta])^2} + 1$, which is increasing in $\delta$ with lower bound $1 - \frac{8}{\pi^2} \approx 0.18943$ (for $\delta = 0$) and upper bound $\frac{1}{3}$ (for $\delta \to \infty$) so the range of dependence is limited. This utility function features DARA (implying SI for the copula) for $\delta \geq 1$ (the restriction imposed by Pratt) while convexity of $\log[\psi(\infty) - \psi(s)]$ (implying LTD or RTI) is obtained for $\delta \geq 0.35735$.

#### 17.5.4.2 Equation (37b) of Pratt

This concerns the case

$$\psi(s) = \ln\left[1 - (s+\delta)^{-1}\right]; \quad \delta \geq 1.$$

We get

$$\varphi(s) = \max\left[1 + \beta \ln\left[\frac{(\delta - 1)(s+\delta)}{\delta(s+\delta-1)}\right], 0\right], \quad s \geq 0.$$

It is required that $\beta \geq \left(\ln\left[\frac{\delta}{\delta-1}\right]\right)^{-1}$. This family is also negatively ordered in $\beta$, with Fréchet-Höffding's lower bound reached for $\beta$ approaching infinity. The generator is strict if $\beta = \left(\ln\left[\frac{\delta}{\delta-1}\right]\right)^{-1}$ leading to $\varphi(s) = \frac{\ln[s+\delta] - \ln[s+\delta-1]}{\ln[\delta] - \ln[\delta-1]}$, $s \geq 0$. Then Kendall's tau is equal to $4\frac{2-(2\delta-1)\ln\left[\frac{\delta}{\delta-1}\right]}{\left(\ln\left[\frac{\delta}{\delta-1}\right]\right)^2} + 1$, varying in value between 1 (for $\delta \downarrow 0$) and $\frac{1}{3}$(for $\delta \to \infty$). The case $\delta \to \infty$ leads to $\varphi(s) = (s+1)^{-1}$ which is the generator of the copula $C(v_1, v_2) = \frac{v_1 v_2}{v_1 + v_2 - v_1 v_2}$.

#### 17.5.4.3 Equation (39) of Pratt

This concerns the case

$$\psi(s) = -c_1 e^{-\gamma s} - c_2 e^{-\delta s}; \quad \gamma, c_1, c_2, \delta > 0.$$

We get

$$\varphi(s) = \max\left[1 - \beta\left(c_1\left(1 - e^{-s}\right) - c_2 e^{-\delta s}\right)\right], \quad s \geq 0.$$

This is an example of a generator that has no analytical inverse, so the analysis that can be performed is limited. It is required that $\beta \geq (c_1 + c_2)^{-1}$. The strict generator gives

$$\varphi(s) = \frac{c_1}{c_1 + c_2} e^{-s} + \frac{c_2}{c_1 + c_2} e^{-\delta s},$$

being the Laplace transform of a two point frailty distribution with probabilities $\frac{c_1}{c_1+c_2}$ at point 1 and $\frac{c_2}{c_1+c_2}$ and point $\delta$. Since the utility function is DARA, and $\psi(\infty)$ is finite, the copula generated is SI. This is no surprise given that all shared frailty distributions constitute long term dependence, as pointed out in Spreeuw [11]. Writing $c = \frac{c_1}{c_1+c_2}$, we obtain $c(1-c)\left(\frac{d-1}{d+1}\right)^2$ as the expression for Kendall's tau, taking a minimum of zero for $c = 0$, $c = 1$ or $d = 1$. Comonotonicity is obtained for $c = 0.5$ and $d \to \infty$.

## 17.6 Conclusion

A flexible family of Archimedean copulas has been presented that can cover a large range of dependence, including countermonotonicity. In most examples, a strict generator is contained as a special case. Given the general consensus in economic theory that utility functions should feature Decreasing Absolute Risk Aversion, the connection between this property and the Stochastic Increasing notion of the corresponding copula is particularly useful.

Most examples concentrate on strict generators, as this is a requirement for any notion of positive dependence. In the future, we intend to study (17.3) in a more general sense, considering negative dependence as well.

## References

1. Arrow, K.J.: Essays in the Theory of Risk Bearing. Chicago, IL: Markham Publishing (1971)
2. Avérous, J., Dortet-Bernadet, J.-L.: Dependence for Archimedean copulas and aging properties of their generating functions. Sankhyā Indian J. Stat. **66**(4), 1–14 (2004)
3. Esary, J.D., Proschan, F.: Relationships among some concepts of bivariate dependence. Ann. Math. Stat. **43**(2), 651–655 (1972)
4. Genest, C., Ghoudi, K., Rivest, L.-P.: A semiparametric estimation procedure of dependence parameters in multivariate families of distributions. Biometrika **82**, 543–552 (1995)
5. Hougaard, P.: Analysis of Multivariate Survival Data. Springer, New York, NY (2000)
6. Lehmann, E.L.: Some concepts of dependence. Ann. Math. Stat. **37**, 1137–1153 (1966)
7. Merton, R.C.: Optimum consumption and portfolio rules in a continuous-time model. J. Econ. Theory **3**, 373–413 (1971)
8. Nelsen, R.B.: An Introduction to Copulas, 2nd edn. Springer, New York, NY (2006)
9. Pratt, J.W.: Risk aversion in the small and in the large. Econometrica **32**(1–2), 122–136 (1964)

10. Saha, A.: Expo-power utility: a flexible form for absolute and relative risk aversion. Am. J. Agric. Econ. **75**, 905–913 (1993)
11. Spreeuw, J.: Types of dependence and time-dependent association between two lifetimes in single parameter copula models. Scand. Actuar. J. **5**, 286–309 (2006)
12. Xie, D.: Power risk aversion utility functions. Ann. Econ. Finance **1**, 265–282 (2000)

# Index

absolutely monotone
function, 151, 154
ageing, 237–252
Archimax, *see* copula
Archimedean, *see* copula/fully-nested, *see* copula/Lévy, *see* copula/nested, *see* copula/partially-nested, *see* copula, *see* generator, *see* pair-copula, *see* semi-copula
Augé, *see* copula/family of/Cuadras-Augé

Bayesian, *see* estimation
Bernstein, *see* copula
Blomqvist's beta, 209, 219, 220
Brownian
bridge, 211
motion, 35, 50, 56
Burr, *see* distribution

Chen, *see* estimator/Chen's Beta
Clayton, *see* copula/family of/Clayton, *see* copula/Lévy-Clayton, *see* pair-copula
collateralized debt obligation (CDO), 88
completely monotone (c.m.)
function, 147, 151–153, 157, 159, 299, 301, 303
sequence, 299, 301, 303
convergence
almost sure, 268
poinwise, 10
uniform, 182
vague, 163
weak, 135, 211, 212, 217, 282, 284
wide, 163
copula
absolutely continuous, 18–20, 85, 171, 172, 187, 196, 198, 201, 202
Archimax, 128
Archimedean, 17, 18, 38, 94, 128, 132, 135, 147–153, 155, 157, 172–177, 217, 238, 239, 241, 242, 244–246, 248, 251, 252, 279, 280, 289, 299–301, 303–307, 311–321
Bernstein, 224, 267–271, 276, 277
comonotonic, 131, 169
consistent semimartingale, 49
construction of
distorsion method, 21
geometric methods, 22
gluing technique, 270
nested method, 20, 148–159
ordinal sum method, 21
pair-copula method, 20, 93–107, 148
pointwise composition, 22
shuffles, 22
with given diagonal section, 22, 170
with given lower tail, 167
diagonal, 170, 171
diffusion, 51, 60, 68
elliptical, 16, 17, 93, 289
empirical, 209, 211, 216, 219, 226, 227, 269, 274
exchangeable, 198
extreme value, 127–141, 177, 180
family of
Clayton, 18, 19, 80, 81, 84, 85, 87, 89, 124, 157, 175, 277, 278, 287, 288, 319
Cuadras-Augé, 169
Eyraud-Farlie-Gumbel-Morgenstern (EFGM), 19, 80, 85
Frank, 19, 80, 263, 277, 287
Galambos, 133, 277
Gumbel, 18, 84, 85, 89, 127, 132, 137, 157, 176, 277, 301, 319

P. Jaworski et al. (eds.), *Copula Theory and Its Applications*, Lecture Notes in Statistics 198,
DOI 10.1007/978-3-642-12465-5, 

Gumbel-Hougaard, 132, 319, *see* copula/family of/Gumbel
Hüsler-Reiss, 133, 277
Mardia-Takahasi-Clayton, 19, *see* copula/family of/Clayton
Marshall-Olkin, 48, 299–308
Plackett, 277, 287
Fréchet-Hoeffding bounds, 210, 228, 287
fully-nested Archimedean, 149, 152, 155, 156
Gaussian, 15, 17, 33, 56, 80, 84, 85, 88, 89, 93, 133, 134, 172, 222, 277, 289, 297
hierarchical Archimedean, *see* nested Archimedean
homogeneous, 169
independence, 16, 104, 131, 169, 213, 215, 224, 274, 287, 312
Lévy, 34–38, 49, 68
Lévy Archimedean, 37
Lévy-Clayton, 38
logistic, 132, *see* copula/family of/Gumbel
Mardia's skewness characteristic of, 293
Markov, 34, 54, 57–63, 68–71
max-stable, 129
negative logistic, 133
nested Archimedean, 86, 94, 147–149, 151–157, 159
normal, *see* copula/Gaussian
of componentwise maxima, 127
of extreme value distribution, 129
of order statistics, 195, 199
partially-nested Archimedean, 149, 150, 155
Poisson, 68
product, *see* copula/independence
semimartingale, 34, 44, 45, 47–51, 54, 69
skew, t-, 290, 292, 293, 295–297
static, 34
Student, *t*-, 172
Student, t-, 85, 93, 134, 268, 277, 289, 297
survival, 85, 89, 132, 174, 181, 230, 231, 239, 243–246, 248, 249, 251, 308, 315
symbolic, 68
symbolic Markov, 59, 67
symbolic product, 68
tail conditional, 179
threshold, 179
vine, 93–107, 172
Cox, *see* process
Cramer, *see* test statistics/Cramer-von Mises
credit default swaps (CDS), 88
Cuadras, *see* copula/family of/Cuadras-Augé

d-monotone
function, 18, 132, 151, 173, 303
sequence, 303
Deheuvels, *see* estimator
dependence
left tail decreasing (LTD), 239, 241, 242, 244, 245, 315–317, 320
lower tail dependent, 229
monotone regression dependence, 134, *see* stochastically increasing (SI)
positive K-dependent (PKD), 241, 242, 244, 245
positive quadrant dependence (PQD), 131, 241, 242, 244–246, 315–317
right tail increasing (RTI), 245, 246, 315–317, 320
stochastically increasing (SI), 241, 242, 244–246, 315–321, *see* monotone regression dependence
tail ... , *see* tail/dependence ...
tail independent, 229
upper tail dependent, 229
distribution
asymptotic, 140, 217
asymptotic normal, 284
Beta, 224
Burr, 18
comonotonic, 120
countermonotonic, 120
decreasing failure rate (DFR), 239, 241, 242, 246
decreasing failure rate in average (DFRA), 241, 242
elliptical, 16, 17, 100
empirical, 268, 295
exchangeable, 189
exponential, 11, 138, 141, 237, 240, 307
extreme value, 129, 141, 177, 229, 230
Eyraud-Farlie-Gumbel-Morgenstern, 19
function
empirical, 137, 211, 269
Gaussian, 134
Kendall, 135
marginal, 211
of order statistic, 192, 193, 201, 202
Gamma, 296
Gaussian, 16, 100, 133, 222, 290
Gumbel, 139
increasing failure rate (IFR), 241, 242, 247
increasing failure rate in average (IFRA), 240, 242
Kendall, 245, 246
lognormal, 296
Marshall-Olkin, 141, 300
meta-elliptical, 17
new better than used (NBU), 240, 242

new worse that used (NWU), 241, 242, 246
of order statistic, 194
of test statistics, 279, 286
Pareto, 18
skew, 290
skew, t-, 290–295, 297
skew-normal, 290
strongly unimodal (SU), 241, 242
Student, t, 16, 86, 134, 222, 290
survival, 280
symmetric, 289
vine, 97
Weibull, 137
domain of attraction, 129

elasticity, 173, 174, 184
equity derivatives, 257, 259, 261, 262
estimation
inference function for margins method, 79
method of moments, 289, 293
nonparametric Bayesian approach, 140
nonparametric bootstrap method, 217, 220
nonparametric method, 81, 138, 209, 216, 221
of parameters
Bayesian approach, 94, 103–106
maximum likelihood method, 78–80, 94, 101, 102, 106, 137
maximum pseudo-loglikelihood method, 137
stepwise approach, 94, 101, 102, 104–106
robust, 77
semiparametric method, 80, 84, 268, 277
estimator
CFG, 139, 140
Chen's Beta, 224
Deheuvels, 138
full sample, 86
HAC, 86
Hall-Tajvidi, 139
histogram, 223
inference functions for margins (IFM), 79, 80
k-nearest neighbour, 223
Kaplan-Meier, 280, 288
Koziol-Green, 288
maximum likelihood (MLE), 79, 80
of Blomqvist's beta, 220
of Hoeffding's phi square, 226
of Kendall's tau, 218
of relative entropy, 223
of scale parameter, 296
of shape parameter, 295, 296
of Spearman's rho, 216
of survival function, 281
of tail dependence coefficient, 229, 231
Pickands, 138–140
semiparametric, 80
exponent of regular variation, 181
extremal coefficient, 135
Eyraud, *see* copula/family of/Eyraud-Farlie-Gumbel-Morgenstern, *see* distribution/Eyraud-Farlie-Gumbel-Morgenstern

Farlie, *see* copula/family of/Eyraud-Farlie-Gumbel-Morgenstern, *see* distribution/Eyraud-Farlie-Gumbel-Morgenstern
Feller, *see* process
Fréchet, *see* copula/Fréchet-Hoeffding bounds
Frank, *see* copula/family of/Frank

Galambos, *see* copula/family of/Galambos
Gamma, *see* distribution
Gaussian, *see* copula, *see* distribution, *see* pair-copula, *see* process
generator
of Archimedean copula, 17–19, 132, 147–149, 152–154, 156–159, 173–177, 217, 238, 239, 251, 280, 299, 305–307, 311–321
of elliptical distribution, 16, 134
of Markov chain, 42, 51
of Markov process, 35, 57, 62
Gini's gamma, 209, 220, 221
goodness-of-fit, 276
Granger causality, 260
Gumbel, *see* copula/family of/Eyraud-Farlie-Gumbel-Morgenstern, *see* copula/family of/Gumbel, *see* distribution/Eyraud-Farlie-Gumbel-Morgenstern, *see* distribution, *see* pair-copula

Hüsler, *see* copula/family of/Hüsler-Reiss
Hall, *see* estimator/Hall-Tajvidi
hazard function, 316
Hoeffding, *see* copula/Fréchet-Hoeffding bounds
Hoeffding's phi square, 225, 271
Hougaard, *see* copula/family of/Gumbel-Hougaard

Kaplan, *see* estimator/Kaplan-Meier
kappa (measure of association), 227
Kendall, *see* distribution
Kendall's tau, 102, 103, 134, 209, 215, 217–220, 263, 271

Kolmogorov, *see* test statistics/Kolmogorov-Smirnov
Koziol-Green
  estimator, *see* estimator/Koziol-Green
  model, 279–288
Kullback-Leibler divergence, *see* relative entropy

Lévy, *see* copula, *see* measure, *see* process
leading part of expansion, 163, 164, 167, 170–172, 175–180, 182, 184

malignant melanoma, 279, 286
Mardia, *see* copula/family of/Mardia-Takahasi-Clayton, *see* copula/Mardia's skewness characteristic of
Markov
  ..., *see* copula/symbolic, *see* copula, *see* generator, *see* process
  chain, 51, 52, 70, 71
Marshall, *see* copula/family of/Marshall-Olkin, *see* distribution/Marshall-Olkin
martingale, 41, 46, 47, 257–261
  local, 41
  pricing approach, 257
  problem, 46–48, 50, 66
  property, 46, 259, 261
  semi-, 34, 39–45, 48, 49, 54
measure
  counting, 42, 46, 47, 51, 52
  jump, 48
  Lévy, 35, 36, 67, 68
  limit, 163, 166, 168
  of association, 209–232
  of concordance, 213–216, 218, 221, 271
  of functional dependence, 210
  Poisson, 49
  pricing, 71
  random, 40, 45–49, 52, 54
  relatively invariant, 168
  spectral, 129, 131, 137, 141, 177
  statistical, 71
medial correlation coefficient, *see* Blomqvist's beta
Meier, *see* estimator/Kaplan-Meier
Mises, von, *see* test statistics/Cramer-von Mises
Morgenstern, *see* copula/family of/Eyraud-Farlie-Gumbel-Morgenstern, *see* distribution/Eyraud-Farlie-Gumbel-Morgenstern

Olkin, *see* copula/family of/Marshall-Olkin, *see* distribution/Marshall-Olkin
option
  altiplano, 261, 262
  basket, 263
  multivariate digital, 261
  spread, 263
order statistics, 187–197, 199–202, 205, 206, 269, 285

pair-copula
  Archimedean, 99
  C-vine, 97
  canonical vine, 94, 96
  Clayton, 99, 104
  construction, 93, 94, 106, 107
  D-vine, 94, 95
  decomposition, 95, 96
  density, 95, 100
  Gaussian, 99, 104
  Gumbel, 99, 104
  regular vine, 94
  selection indicators, 104
  Students, t-, 99
  vine tree, 94, 96–99, 102, 103, 106
Pareto, *see* distribution
petrophysical modeling, 268–278
Pickands
  dependence function, 130, 133–135, 137, 138, 140, 141, 178
  estimator, *see* estimator/Pickands
Plackett, *see* copula/family of/Plackett
Poisson, *see* copula, *see* measure, *see* process
process
  Cox, 56
  diffusion, 50, 51, 59, 68
  empirical, 285, 286
  empirical copula, 81, 211
  Feller, 60, 64–68
  GARCH, 87
  Gaussian, 211, 212, 282, 284, 286
  indicator, 56
  Lévy, 35–39, 49, 64, 68, 259
  Lévy subordinator, 302
  Markov, 34, 54–61, 63, 64, 69, 71, 82–85, 258, 260
  martingale, 46, 259
  Poisson, 45, 49, 56, 68
  rating, 71, 73
  recovery, 69
  regime switching, 88
  semimartingale, 39–41
  trajectory, 39
  weak convergence, 212, 217
product of copulas, 82

rank association coefficient, *see* Gini's gamma
rank correlation coefficient, *see* Spearman's rho
rank order statistics, 211, 217
regression curve, 270
regular variation, 134, 140, 173, 181, 182
Reiss, *see* copula/family of/Hüsler-Reiss
relative entropy, 221–224
relatively invariant measure, 168
risk aversion
  coefficient, 317
  constant absolute (CARA), 318
  constant relative (CRRA), 318
  decreasing absolute (DARA), 317, 319, 321
  hyperbolic absolute (HARA), 319

Samaniego, *see* signature of system
scale invariance, 210
Schur, *see* survival model
section
  diagonal, 170, 171, 200, 270
  vertical, 281
semi-copula
  Archimedean, 239, 242, 248, 251
shape parameter, 291
shift parameter, 291
sigma (measure of association), 227
signature of system, 189–191, 193, 194, 196, 199, 205
  maximal, 191–193
  minimal, 191–193, 205
  Samaniego, 192, 194
Sklar's theorem, 3, 4, 12–14, 17, 37, 39, 78, 93, 95, 147, 161, 269, 280, 281
Smirnov, *see* test statistics/Kolmogorov-Smirnov
Spearman's rho, 105, 134, 209, 215–218, 220, 227, 231, 271
Student, *see* copula, *see* distribution, *see* pair-copula
survival model
  Schur-concave, 244, 247
  Schur-constant, 237, 239, 243–246, 249, 251, 252
  Schur-convex, 244
  time-transformed exponential (TTE), 244, 249, 251

system
  coherent, 187–194, 205
  k-out-of-n, 187–191, 193, 195, 197–200, 202–204
  mixed, 189–194, 196, 199, 205
  parallel, 187, 205
  series, 187, 205

tail
  conditional copula, 179
  dependence coefficient, 93, 99, 103, 135, 136, 162, 209, 229–231, 313
  dependence function (wrt extreme value), 129, 130, 135, 141, 177, 180, *see* tail/exponent measure function
  dependence function (wrt tail asymptotics), 162, 164, 166, 182, 230
  dependence measure, 230, 231
  expansion, 162, 163, 169–178, 181
  exponent measure function, 230, *see* tail/dependence function(wrt extreme value)
  independence, 229, 230
  integral, 36, 37, 68
  limit measure, 163, 168
  multivariate dependence, 209, 228
  of parent reliability function, 192
  spectral measure, 129, 137, 141, 177
  uniform expansion, 163, 164, 173, 174, 178
Tajvidi, *see* estimator/Hall-Tajvidi
Takahasi, *see* copula/family of/Mardia-Takahasi-Clayton
test statistics
  Cramer-von Mises, 273, 285
  Kolmogorov-Smirnov, 285
  t-, 86

utility function, 311–321

Value at Risk, 93, 103, 111, 114–120, 122–124, 182
VaR, *see* Value at Risk

weak convergence, 135, 211, 212, 217, 282, 284
Weibull, *see* distribution
wide convergence of measures, 163

## Lecture Notes in Statistics

For information about Volumes 1 to 144, go to http://www.springer.com/series/694

144: L. Mark Berliner, Douglas Nychka, and Timothy Hoar (Editors), Case Studies in Statistics and the Atmospheric Sciences. x, 208 pp., 2000.

145: James H. Matis and Thomas R. Kiffe, Stochastic Population Models. viii, 220 pp., 2000.

146: Wim Schoutens, Stochastic Processes and Orthogonal Polynomials. xiv, 163 pp., 2000.

147: Jürgen Franke, Wolfgang Härdle, and Gerhard Stahl, Measuring Risk in Complex Stochastic Systems. xvi, 272 pp., 2000.

148: S.E. Ahmed and Nancy Reid, Empirical Bayes and Likelihood Inference. x, 200 pp., 2000.

149: D. Bosq, Linear Processes in Function Spaces: Theory and Applications. xv, 296 pp., 2000.

150: Tadeusz Caliński and Sanpei Kageyama, Block Designs: A Randomization Approach, Volume I: Analysis. ix, 313 pp., 2000.

151: Håkan Andersson and Tom Britton, Stochastic Epidemic Models and Their Statistical Analysis. ix, 152 pp., 2000.

152: David Ríos Insua and Fabrizio Ruggeri, Robust Bayesian Analysis. xiii, 435 pp., 2000.

153: Parimal Mukhopadhyay, Topics in Survey Sampling. x, 303 pp., 2000.

154: Regina Kaiser and Agustín Maravall, Measuring Business Cycles in Economic Time Series. vi, 190 pp., 2000.

155: Leon Willenborg and Ton de Waal, Elements of Statistical Disclosure Control. xvii, 289 pp., 2000.

156: Gordon Willmot and X. Sheldon Lin, Lundberg Approximations for Compound Distributions with Insurance Applications. xi, 272 pp., 2000.

157: Anne Boomsma, Marijtje A.J. van Duijn, and Tom A.B. Snijders (Editors), Essays on Item Response Theory. xv, 448 pp., 2000.

158: Dominique Ladiray and Benoît Quenneville, Seasonal Adjustment with the X-11 Method. xxii, 220 pp., 2001.

159: Marc Moore (Editor), Spatial Statistics: Methodological Aspects and Some Applications. xvi, 282 pp., 2001.

160: Tomasz Rychlik, Projecting Statistical Functionals. viii, 184 pp., 2001.

161: Maarten Jansen, Noise Reduction by Wavelet Thresholding. xxii, 224 pp., 2001.

162: Constantine Gatsonis, Bradley Carlin, Alicia Carriquiry, Andrew Gelman, Robert E. Kass Isabella Verdinelli, and Mike West (Editors), Case Studies in Bayesian Statistics, Volume V. xiv, 448 pp., 2001.

163: Erkki P. Liski, Nripes K. Mandal, Kirti R. Shah, and Bikas K. Sinha, Topics in Optimal Design. xii, 164 pp., 2002.

164: Peter Goos, The Optimal Design of Blocked and Split-Plot Experiments. xiv, 244 pp., 2002.

165: Karl Mosler, Multivariate Dispersion, Central Regions and Depth: The Lift Zonoid Approach. xii, 280 pp., 2002.

166: Hira L. Koul, Weighted Empirical Processes in Dynamic Nonlinear Models, Second Edition. xiii, 425 pp., 2002.

167: Constantine Gatsonis, Alicia Carriquiry, Andrew Gelman, David Higdon, Robert E. Kass, Donna Pauler, and Isabella Verdinelli (Editors), Case Studies in Bayesian Statistics, Volume VI. xiv, 376 pp., 2002.

168: Susanne Rässler, Statistical Matching: A Frequentist Theory, Practical Applications and Alternative Bayesian Approaches. xviii, 238 pp., 2002.

169: Yu. I. Ingster and Irina A. Suslina, Nonparametric Goodness-of-Fit Testing Under Gaussian Models. xiv, 453 pp., 2003.

170: Tadeusz Caliński and Sanpei Kageyama, Block Designs: A Randomization Approach, Volume II: Design. xii, 351 pp., 2003.

171: D.D. Denison, M.H. Hansen, C.C. Holmes, B. Mallick, B. Yu (Editors), Nonlinear Estimation and Classification. x, 474 pp., 2002.

172: Sneh Gulati, William J. Padgett, Parametric and Nonparametric Inference from Record-Breaking Data. ix, 112 pp., 2002.

173: Jesper Møller (Editor), Spatial Statistics and Computational Methods. xi, 214 pp., 2002.

174: Yasuko Chikuse, Statistics on Special Manifolds. xi, 418 pp., 2002.

175: Jürgen Gross, Linear Regression. xiv, 394 pp., 2003.

176: Zehua Chen, Zhidong Bai, Bimal K. Sinha, Ranked Set Sampling: Theory and Applications. xii, 224 pp., 2003.

177: Caitlin Buck and Andrew Millard (Editors), Tools for Constructing Chronologies: Crossing Disciplinary Boundaries, xvi, 263 pp., 2004.

178: Gauri Sankar Datta and Rahul Mukerjee, Probability Matching Priors: Higher Order Asymptotics, x, 144 pp., 2004.

179: D.Y. Lin and P.J. Heagerty (Editors), Proceedings of the Second Seattle Symposium in Biostatistics: Analysis of Correlated Data, vii, 336 pp., 2004.

180: Yanhong Wu, Inference for Change-Point and Post-Change Means After a CUSUM Test, xiv, 176 pp., 2004.

181: Daniel Straumann, Estimation in Conditionally Heteroscedastic Time Series Models, x, 250 pp., 2004.

182: Lixing Zhu, Nonparametric Monte Carlo Tests and Their Applications, xi, 192 pp., 2005.

183: Michel Bilodeau, Fernand Meyer, and Michel Schmitt (Editors), Space, Structure and Randomness, xiv, 416 pp., 2005.

184: Viatcheslav B. Melas, Functional Approach to Optimal Experimental Design, vii., 352 pp., 2005.

185: Adrian Baddeley, Pablo Gregori, Jorge Mateu, Radu Stoica, and Dietrich Stoyan, (Editors), Case Studies in Spatial Point Process Modeling, xiii., 324 pp., 2005.

186: Estela Bee Dagum and Pierre A. Cholette, Benchmarking, Temporal Distribution, and Reconciliation Methods for Time Series, xiv., 410 pp., 2006.

187: Patrice Bertail, Paul Doukhan and Philippe Soulier, (Editors), Dependence in Probability and Statistics, viii., 504 pp., 2006.

188: Constance van Eeden, Restricted Parameter Space Estimation Problems, vi, 176 pp., 2006.

189: Bill Thompson, The Nature of Statistical Evidence, vi, 152 pp., 2007.

190: Jérôme Dedecker, Paul Doukhan, Gabriel Lang, José R. León, Sana Louhichi Clémentine Prieur, Weak Dependence: With Examples and Applications, xvi, 336 pp., 2007.

191: Vlad Stefan Barbu and Nikolaos Liminos, Semi-Markov Chains and Hidden Semi-Markov Models toward Applications, xii, 228 pp., 2007.

192: David B. Dunson, Random Effects and Latent Variable Model Selection, 2008.

193: Alexander Meister. Deconvolution Problems in Nonparametric Statistics, 2008.

194: Dario Basso, Fortunato Pesarin, Luigi Salmaso, Aldo Solari, Permutation Tests for Stochastic Ordering and ANOVA: Theory and Applications with R, 2009.

195: Alan Genz and Frank Bretz, Computation of Multivariate Normal and *t* Probabilities, viii, 126 pp., 2009.

196: Hrishikesh D. Vinod, Advances in Social Science Research Using R, xx, 207 pp., 2010.

197: M. González, I.M. del Puerto, T. Martinez, M. Molina, M. Mota, A. Ramos (Eds.), Workshop on Branching Processes and Their Applications, xix, 296 pp., 2010.

198: P. Jaworski, F. Durante, W. Härdle, T. Rychlik (Eds.), Copula Theory and Its Applications - Proceedings of the Workshop Held in Warsaw 25-26 September 2009, xviii, 327 pp., 2010.

199: Hannu Oja, Multivariate Nonparametric Methods with R, xii, 241 pp, 2010.